BAKER COLLEGE OF
CLINTON TWP. LIBRARY

Environmentally Conscious Manufacturing

Environmentally Conscious Manufacturing

Edited by
Myer Kutz

John Wiley & Sons, Inc.

This book is printed on acid-free paper. ∞

Copyright © 2007 by John Wiley & Sons, Inc. All rights reserved

Published by John Wiley & Sons, Inc., Hoboken, New Jersey
Published simultaneously in Canada

Wiley Bicentennial Logo: Richard J. Pacifico

No part of this publication may be reproduced, stored in a retrieval system, or transmitted in any form or by any means, electronic, mechanical, photocopying, recording, scanning, or otherwise, except as permitted under Section 107 or 108 of the 1976 United States Copyright Act, without either the prior written permission of the Publisher, or authorization through payment of the appropriate per-copy fee to the Copyright Clearance Center, 222 Rosewood Drive, Danvers, MA 01923, (978) 750-8400, fax (978) 646-8600, or on the Web at www.copyright.com. Requests to the Publisher for permission should be addressed to the Permissions Department, John Wiley & Sons, Inc., 111 River Street, Hoboken, NJ 07030, (201) 748-6011, fax (201) 748-6008, or online at www.wiley.com/go/permissions.

Limit of Liability/Disclaimer of Warranty: While the publisher and the author have used their best efforts in preparing this book, they make no representations or warranties with respect to the accuracy or completeness of the contents of this book and specifically disclaim any implied warranties of merchantability or fitness for a particular purpose. No warranty may be created or extended by sales representatives or written sales materials. The advice and strategies contained herein may not be suitable for your situation. You should consult with a professional where appropriate. Neither the publisher nor the author shall be liable for any loss of profit or any other commercial damages, including but not limited to special, incidental, consequential, or other damages.

For general information about our other products and services, please contact our Customer Care Department within the United States at (800) 762-2974, outside the United States at (317) 572-3993 or fax (317) 572-4002.

Wiley also publishes its books in a variety of electronic formats. Some content that appears in print may not be available in electronic books. For more information about Wiley products, visit our Web site at www.wiley.com.

Library of Congress Cataloging-in-Publication Data:

Environmentally conscious manufacturing / edited by Myer Kutz.
    p. cm.
 Includes bibliographical references and index.
 ISBN 978-0-471-72637-1
 1. Manufacturing processes—Environmental aspects. I. Kutz, Myer.
TS155.7.E5845 2007
670.28'6—dc22

2006029440

Printed in the United States of America

10 9 8 7 6 5 4 3 2 1

To my grandson Jayden and my grandnephew Carlos—may we engineers make this earth clean and healthy for you.

Contents

Contributors   ix
Preface   xi

1. Environmentally Benign Manufacturing   1
   William E. Biles

2. Design for the Environment   29
   Jack Jeswiet

3. Organization, Management, and Improvement of Manufacturing Systems   45
   Keith M. Gardiner

4. Manufacturing Systems Evaluation   79
   Walter W. Olson

5. Prevention of Metalworking Fluid Pollution: Environmentally Conscious Manufacturing at the Machine Tool   95
   Steven J. Skerlos

6. Metal Finishing and Electroplating   123
   Timothy C. Lindsey

7. Air Quality in Manufacturing   145
   John W. Sutherland, Donna J. Michalek, and Julio L. Rivera

8. Environmentally Conscious Electronic Manufacturing   179
   Richard Ciocci

9. Disassembly for End-of-life Electromechanical Products   211
   Hong C. Zhang, Liu Zhifeng, Gao Yang, and Chen Qing

10. Industrial Energy Efficiency   265
    Bhaskaran Gopalakrishnan, Deepak P. Gupta, Yogesh Mardikar, and Subodh Chaudhari

11. Industrial Environmental Compliance Regulations   295
    Thomas J. Blewett and Jack Annis

Index   325

Contributors

Jack Annis
University of Wisconsin—Stevens Point
Stevens Point, Wisconsin

William E. Biles
University of Louisville
Louisville, Kentucky

Thomas J. Blewett
University of Wisconsin Extension
Madison Wisconsin

Subodh Chaudhari
West Virginia University
Morgantown West Virginia

Richard Ciocci
Penn State University
Middletown, Pennsylvania

Keith M. Gardiner
Lehigh University
Bethlehem, Pennsylvania

Bhaskaran Gopalakrishnan
West Virginia University
Morgantown West Virginia

Deepak P. Gupta
West Virginia University
Morgantown West Virginia

Jack Jeswiet
Queen's University
Kingston, Ontario Canada

Tim Lindsey
Illinois Waste Management and Research Center
Champaign, Illinois

Yogesh Mardikar
West Virginia University
Morgantown West Virginia

Donna J. Michalek
Michigan Technological University
Houghton, Michigan

Walter W. Olson
University of Toledo
Toledo, Ohio

Chen Qing
Heifi University of Technology
Heifi, China

Julio L. Rivera
Michigan Technological University
Houghton, Michigan

Steven Skerlos
University of Michigan
Ann Arbor, Michigan

John W. Sutherland
Michigan Technological University
Houghton, Michigan

Gao Yang
Heifi University of Technology
Heifi, China

Hong C. Zhang
Texas Tech University
Lubbock, Texas

Liu Zhifeng
Heifi University of Technology
Heifi, China

Preface

Many readers will approach this series of books in environmentally conscious engineering with some degree of familiarity with, knowledge about, or even expertise in one or more of a range of environmental issues, such as climate change, pollution, and waste. Such capabilities may be useful for readers of this series, but they aren't strictly necessary. The purpose of this series is not to help engineering practitioners and managers deal with the *effects* of man-induced environmental change. Nor is it to argue about whether such effects degrade the environment only marginally or to such an extent that civilization as we know it is in peril, or that any effects are nothing more than a scientific-establishment-and-media-driven hoax and can be safely ignored. (Other authors, fiction and nonfiction, have already weighed in on these matters.) By contrast, this series of engineering books takes as a given that the overwhelming majority in the scientific community is correct, and that the future of civilization depends on minimizing environmental damage from industrial, as well as personal, activities. However, the series goes beyond advocating solutions that emphasize only curtailing or cutting back on these activities. Instead, its purpose is to exhort and enable engineering practitioners and managers to reduce environmental impacts—to engage, in other words, in *environmentally conscious engineering*, a catalog of practical technologies and techniques that can improve or modify just about anything engineers do, whether they are involved in designing something, making something, obtaining or manufacturing materials and chemicals with which to make something, generating power, or transporting people and freight.

Increasingly, engineering practitioners and managers need to know how to respond to challenges of integrating environmentally conscious technologies, techniques, strategies, and objectives into their daily work, and, thereby, find opportunities to lower costs and increase profits while managing to limit environmental impacts. Engineering practitioners and managers also increasingly face challenges in complying with changing environmental laws. So companies seeking a competitive advantage and better bottom lines are employing environmentally responsible design and production methods to meet the demands of their stakeholders, who now include not only owners and stockholders, but also customers, regulators, employees, and the larger, even worldwide community.

Engineering professionals need references that go far beyond traditional primers that cover only regulatory compliance. They need integrated approaches centered on innovative methods and trends in design and manufacturing that help them focus on using environmentally friendly processes and creating green products. They need resources that help them participate in strategies for designing environmentally responsible products and methods, resources that provide a foundation for understanding and implementing principles of environmentally conscious engineering.

To help engineering practitioners and managers meet these needs, I envisioned a flexibly connected series of edited handbooks, each devoted to a broad topic under the umbrella of environmentally conscious engineering, starting with three volumes that are closely linked—environmentally conscious mechanical design, environmentally conscious manufacturing, and environmentally conscious materials and chemicals processing.

The intended audience for the series is practicing engineers and upper-level students in a number of areas—mechanical, chemical, industrial, manufacturing, plant, and environmental—as well as engineering managers. This audience is broad and multidisciplinary. Some of the practitioners who make up this audience are concerned with design, some with manufacturing, and others with materials and chemicals processing, and these practitioners work in a wide variety of organizations, including institutions of higher learning, design, manufacturing, and consulting firms, as well as federal, state, and local government agencies. So what made sense in my mind was a series of relatively short handbooks, rather than a single, enormous handbook, even though the topics in each of the smaller volumes have linkages and some of the topics (*design for environment, DfE*, comes to mind) might be suitably contained in more than one freestanding volume. In this way, each volume is targeted at a particular segment of the broader audience. At the same time, a linked series is appropriate because every practitioner, researcher, and bureaucrat can't be an expert on every topic, especially in so broad and multidisciplinary a field, and may need to read an authoritative summary on a professional level of a subject that he or she is not intimately familiar with but may need to know about for a number of different reasons.

The Environmentally Conscious Engineering series is composed of practical references for engineers who are seeking to answer a question, solve a problem, reduce a cost, or improve a system or facility. These handbooks are not research monographs. The purpose is to show readers what options are available in a particular situation and which option they might choose to solve problems at hand. I want these handbooks to serve as a source of practical advice to readers. I would like them to be the first information resource a practicing engineer reaches for when faced with a new problem or opportunity—a place to turn to even before turning to other print sources, even any officially sanctioned ones, or to sites on the Internet. So the handbooks have to be more than references or collections of background readings. In each chapter, readers should feel that they are in the hands of an experienced consultant who is providing sensible advice that can lead to beneficial action and results.

I developed the second volume in the series, the *Handbook of Environmentally Conscious Manufacturing*, to provide readers with information about improving design and manufacturing processes and systems. This handbook covers in detail metalworking and metalworking fluids, metal finishing and electroplating processes, and particulate emissions and air quality. Key chapters discuss electronics manufacturing, as well as disassembly and recycling at the end of useful life for electromechanical products. The handbook closes with important chapters on industrial energy conservation and U.S. industrial environmental compliance regulations.

I asked the contributors, who are from North America (with one exception, a writing team from China), to provide short statements about the contents of their chapters and why the chapters are important. Here are their responses:

William E. Biles (University of Louisville in Louisville, Kentucky), who contributed the chapter on **Environmentally Benign Manufacturing**, writes, "This chapter stresses the importance of instituting environmentally benign manufacturing practices in several classes of manufacturing processes, including machining, metal casting, metal joining, metal forming, and plastics injection molding. Potentially offensive pollutants and byproducts are identified for each class of manufacturing process, and protective measures are discussed. It is shown that environmental impacts are sustained at each step in the supply chain, from supplier, to manufacturer, to distribution, and to the customer."

Jack Jeswiet (Queen's University in Kingston, Ontario, Canada), who contributed the chapter on **Design for Environment (DfE)**, writes, "Designing for the environment encompasses an awful lot of territory. It is something all designers should be required to include in their design process. Some of the elements of design must include consideration of energy, water use, incoming and outgoing effluents, materials used, CO_2 produced, product use and end of life. This is but a short list; there is much more. In the end, or some might say the beginning, the driver for new designs or redesign is consumer demand. For each screw, nut, and bolt used there is an environmental penalty. There will always be an environmental penalty, and total penalty must be kept to a minimum through designing for the environment."

Keith M. Gardiner (Lehigh University in Bethlehem, Pennsylvania), who contributed the chapter on **Organization, Management, and Improvement of Manufacturing Systems**, writes, "Chapter 3 opens with an industrial archaeological approach examining the whole context of the manufacturing system and its importance. Organizational and behavioral aspects of developing, operating, and continuously improving the systems are likened to practices in sports. In the latter, close attention is paid to the life cycle of the athletes with continual training, retraining, and ultimately a timely migration to another career stage. The role of the manufacturing system as an embedded and essential wealth generating and sustainable contributor to society is emphasized."

Walter W Olson (University of Toledo in Toledo, Ohio), who contributed the chapter on **Manufacturing Systems Evaluation**, writes," Manufacturing systems evaluation is an essential part of environmentally conscious manufacturing because it both establishes the baseline of where a manufacturer's plant is at today and what is needed to improve current systems and processes. Without an honest and frank appraisal of the current situation, a manufacturer is most likely to expend resources for work that does not need to be done while not fixing the real problems that may exist. Manufacturing systems evaluation is a well-defined procedure for performing such an appraisal."

Steven Skerlos (University of Michigan in Ann Arbor), who contributed the chapter on **Prevention of Metalworking Pollution: Environmentally Conscious Manufacturing at the Machine Tool**, writes, "The current utilization of metalworking fluids (MWFs) in manufacturing operations is harmful to the environment and to the health of workers, resulting in economic and societal pressures to redesign metalworking fluid systems utilizing ecological materials and green manufacturing principles. In this chapter, we review a set of strategies to minimize the life-cycle environmental and health impact of metalworking fluids."

Julio L. Rivera (Michigan Technological University in Houghton), who contributed the chapter on **Air Quality in Manufacturing** along with John W. Sutherland and Donna J. Michalek, writes, " Air quality in the work environment has long been identified as a challenge in the planning and management of industrial systems. Poor air quality in the workplace resulting from fumes, smoke, mist, and other airborne particulates have been linked to a variety of occupation-related diseases. Historically, these particulates have been controlled with filters and collection systems with varying degrees of success; these control techniques serve to increase capital and operating costs. The chapter introduces the basics of particulate and process emissions, and examines environmentally and economically responsible approaches to reduce/eliminate particulate emissions. Research gaps and emerging air-quality issues are identified that should be considered as we continue to strive to provide healthy workplaces."

Richard Ciocci (Pennsylvania State Harrisburg Campus in Middletown), who contributed the chapter on **Environmentally Conscious Electronics Manufacturing**, writes, "Encouraged by technological advancements, the international electronics industry has long been a leader in manufacturing innovation. Directed by changing legislation, such as the European Union's ban on lead and other materials in electronics, the industry continues its focus on manufacturing improvement. By including environmental responsibility within their product and process designs, electronics manufacturers are again demonstrating leadership in applying innovative methods."

Hong C. Zhang (Texas Tech University in Lubbock and Hefei University of Technology, Hefei, Anhui Province, China), who contributed the chapter on **Disassembly for End-of-Life Electro-Mechanical Products** along with Liu

Zhifeng, Gao Yang, and Chen Qing, writes, "Disassembly is the first step of recycle and recovery of end-of-life electro-mechanical products. This chapter systematically addresses the issue of the end-of-life electro-mechanical products disassembly and recycling. The chapter also addresses the issue how the disassembly work can feedback product designs by means of environmentally benign product design."

Tim Lindsey (University of Illinois in Champagne), who contributed the chapter on **Metal Finishing and Electroplating**, writes, "Metal-finishing processes are utilized by many industrial sectors to improve the appearance and/or performance of their products. Unlike other manufacturing operations, the vast majority of chemicals that metal finishers use end up as waste, some of which contain highly toxic or carcinogenic ingredients that are difficult to destroy or stabilize. Consequently, development and utilization of metal-finishing processes that are more environmentally friendly and safer is of utmost importance in the establishment of a sustainable industrial base."

Bhaskaran Gopalakrishnan (West Virginia University in Morgantown), who contributed the chapter on **Industrial Energy Efficiency** along with Deepak P. Gupta, Yogesh Mardikar, and Subodh Chaudhari, writes, "The chapter on industrial energy efficiency focuses on the importance of energy conservation in the industrial sector and the resulting benefits in terms of economics and the environment. Methods used for industrial energy conservation are described. The analysis of the industrial assessment database produced from the work of the U.S. Department of Energy–funded Industrial Assessment Centers (IAC) is presented. The chapter is important because energy supplies are finite and recent increases in energy costs are likely to impact the standard of living of citizens around the world, thus providing the rationale for developing industrial energy efficiency measures that are attractive for manufacturing facilities to implement on account of cost reduction, increased profitability, and reduced environmental impacts."

Thomas J. Blewett and Jack Annis (University of Wisconsin Extension in Madison), who contributed the chapter on **Industrial Environmental Compliance Regulations**, write, "Understanding environmental regulation is a difficult and complicated issue that holds a significant amount of potential liability for each engineer involved in product manufacture and services today. The regulatory chapter of this book describes how environmental regulation has evolved in the United States and provides the reader with a common sense approach to understanding this subject."

That ends the contributors' comments. I would like to express my heartfelt thanks to all of them for having taken the opportunity to work on this book. Their lives are terribly busy, and it is wonderful that they found the time to write thoughtful and complex chapters. I developed the handbook because I believed it could have a meaningful impact on the way many engineers approach their daily work, and I am gratified that the contributors thought enough of the idea that they were willing to participate in the project. Thanks also to my editor, Bob Argentieri, for his faith in the project from the outset. And a special note of thanks to my wife Arlene, whose constant support keeps me going.

Myer Kutz
Delmar, New York

CHAPTER 1

ENVIRONMENTALLY BENIGN MANUFACTURING

William E. Biles, Ph.D., P.E.
Department of Industrial Engineering, University of Louisville, Louisville, Kentucky

1	INTRODUCTION	1	4 MANUFACTURING PROCESSES	3
2	ENVIRONMENTALLY BENIGN MANUFACTURING	1	4.1 Machining Processes	4
			4.2 Metal Casting	5
			4.3 Metal-forming Processes	12
3	MANUFACTURING AND THE SUPPLY CHAIN	2	4.4 Metal Joining Processes	16
			4.5 Plastic Injection Molding	20
	3.1 Tier I and Tier II Suppliers	3	5 THE MANUFACTURED PRODUCT	26
	3.2 Transporters	3		

1 INTRODUCTION

How might mankind enjoy the fruits of an advanced civilization without endangering the viability of planet Earth for future generations? That is the fundamental challenge that we confront in the 21st century. In a time when the comforts and pleasures that can be derived from the products of modern technology are accessible for a significant portion of the world's population, how can we manufacture and deliver those products in an environmentally benign fashion?

2 ENVIRONMENTALLY BENIGN MANUFACTURING

The *environmentally benign manufacturing* movement addresses the dilemma of maintaining a progressive worldwide economy without continuing to damage our environment. How can companies—driven by the necessity for manufacturing the products sought by their customers in a cost-effective manner while maintaining market share and providing gains for their stockholders—also heed the growing clamor for a safe environment? This dilemma is fundamentally a trade-off between the needs of current generations and those of future generations. Will we seek creature comforts for ourselves without regard to the safety and well-being of our children and our children's children? Or will we reach a compromise that allows current generations to reap the benefits of our modern technological society while assuring the same benefits for future generations? The challenge for environmentally conscious manufacturers is to find ways to factor both economic and environmental considerations into their business plans.

The fundamental issue in environmentally benign manufacturing is to align business needs with environmental needs. That is, how do we manufacture market-competitive products without harming the air, water, or soil on planet Earth? How do we motivate companies to behave unilaterally to adopt environmentally benign manufacturing practices? Will nation-states unilaterally recognize the need to impose environmental standards on companies manufacturing products within their national boundaries? Recent experience informs us that progress is being made on each of these fronts, but that we have a long way to go to fully protect the environment from the offenses committed by the worldwide manufacturing community.

3 MANUFACTURING AND THE SUPPLY CHAIN

The issue of environmentally benign manufacturing is not isolated on the manufacturing function. Environmental issues abound from tier I and II suppliers to the manufacturing system all the way through the supply chain to the consumer. Figure 1 shows the position of the manufacturing function in the overall supply chain.

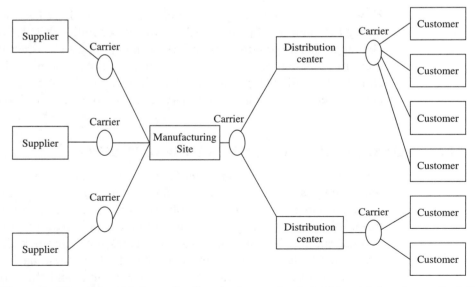

Figure 1 Material and information flow in the supply chain: Material flow is usually left-to-right, information flow right-to-left.

3.1 Tier I and Tier II Suppliers

Each tier I or II supplier has its own *manufacturing processes,* each with its own environmental impacts. It is incumbent upon the primary manufacturer to qualify its tier I and II suppliers not only in terms of quality, cost, and on-time delivery, but on their environmental performance as well. Suppliers must be made to understand that their very financial viability depends on their adopting sound environmental practices. Their role in the supply chain cannot be ignored. It is the responsibility of the primary manufacturer to ensure that its tier I and II suppliers adhere to environmental standards.

3.2 Transporters

The transportation function in the supply chain is also important in terms of its environmental impacts. *Transporters* are those entities that move materials and products from one point to another in the supply chain. Transporters are typically selected and retained according to their cost and reliability performance. Scant attention is paid to the issue of energy expenditure per unit delivery. In an *environmentally conscious manufacturing* approach, primary manufacturers must give closer attention to *energy expenditure per unit delivery* in selecting the mode of transportation from among highway, rail, air, water, and pipeline.

Cost and delivery-time considerations must be balanced against energy expenditure in choosing the transportation mode. For example, consider the case of a refrigerator manufactured in the United States, which is to be shipped to a distribution facility located 500 miles away. It is probably reasonable to immediately exclude pipeline (infeasible), water (not accessible), and air (too costly) transporters from consideration in this application. The trade-off between highway and rail—both of which are feasible, accessible, and within acceptable cost boundaries for the transport of refrigerators—should incorporate a comparison of the energy expenditure per unit (refrigerator) transported. Such a comparison would very likely come down in favor of rail transportation in terms of both cost and energy expenditure, and in favor of highway in terms of delivery time. At present, the delivery-time consideration dominates the transporter selection decision in favor of highway transportation. The entire transporter selection issue needs to be reexamined to consider environmental effects of the supply chain transportation function.

4 MANUFACTURING PROCESSES

The manufacturing process itself is perhaps the most important stage in the supply chain in terms of overall environmental impact. Here we shall consider five manufacturing processes that apply to metals and plastics: (1) machining processes, (2) metal casting, (3) metal forming, (4) metal joining, and (5) plastics injection molding.

4 Environmentally Benign Manufacturing

4.1 Machining Processes

Machining processes include such manufacturing operations as turning, milling, drilling, boring, thread cutting and forming, shaping, planning, slotting, sawing, shearing, and grinding.[1] Each of these processes involves the removal of metal from stock such as a cylindrical billet, cylindrical bar stock, or a cubical block. Metal-cutting economics seek to (1) minimize the cost of the metal cutting operation, (2) maximize tool life, or (3) maximize production rate. An environmentally benign manufacturing approach would add *minimizing environmental impact* to this list of economic objectives.

The achievement of these economic objectives in machining requires the use of *cutting fluids,* which act as coolants and/or lubricants in the machining process. The four major types of cutting fluids are: (1) soluble oil emulsions with water-to-oil ratios ranging from 20:1 to 80:1; (2) oils; (3) chemicals and synthetics; and (4) air. Cuttings fluids have six major roles in machining:

1. Removing the heat of friction
2. Minimizing part deformation due to heat
3. Reducing friction among chips, tool and work piece
4. Washing away chips
5. Reducing possible corrosion on both the work piece and machine
6. Preventing built-up edges on the product or part

The environmental impacts of machining processes are principally of two types: (1) the accumulation of metal chips; and (2) the release of cutting fluids into the environment. The best solution to the problem of chip accumulation is to recycle them by incorporating them as charge into the metal-casting operation. But recycling may involve transporting the chips to a distant site, thereby incurring the *transporter* impact. The best way to handle cutting fluids is to recycle them back to the machining operation, which requires that chips be separated from the machining effluent and that the cutting fluid be reconstituted to as close to its original state as possible. Each of these steps incurs an economic cost, which must be balanced against the cost of the environmental impact of simply placing the chips and used cutting fluid into a waste site.

Electrical discharge machining (EDM) removes electrically conductive material from the raw material stock by means of rapid, repetitive spark discharges from a pulsating D.C. power supply, with dielectric flowing between the workpiece and the tool (Figure 2). The cutting tool (electrode) is made of an electrically conductive material, usually carbon. The shaped tool is fed into the workpiece under servocontrol. A spark discharge then breaks down the dielectric fluid. The frequency and energy per spark are set and controlled with a D.C. power source. The servocontrol maintains a constant gap between the tool and the workpiece while advancing the electrode. The dielectric oil acts as a cutting fluid, cooling and flushing out the vaporized and condensed material while

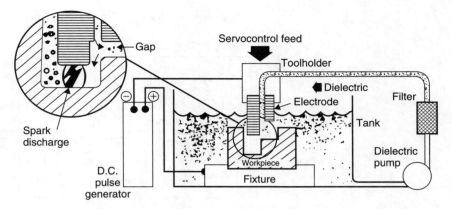

Figure 2 Electrical discharge machining.

reestablishing insulation in the gap. Material removal rate ranges from 16 to 245 cm^3/h. EDM is suitable for cutting materials regardless of their hardness or toughness. Round or irregularly shaped holes 0.002 inches (0.05 mm) in diameter can be produced with L/D ratio of 20:1. Narrow slots with widths as small as 0.002 to 0.010 inches (0.05–0.25 mm) can be cut by EDM.

4.2 Metal Casting

Metal-casting processes are divided according to the specific type of molding method, as follows: (1) sand casting; (2) die casting; (3) investment casting; (4) centrifugal casting; (5) plaster-mold casting; and (6) permanent casting. This section discusses the first three of these.[2]

Sand Casting

Sand casting is one of the most ancient forms of metalworking. The first sand casting of copper dates to about 6,000 years ago. Sand casting consists of pouring molten metal into shaped cavities formed in a sand mold, as shown in Figure 3. The sand used in fabricating the mold may be natural, synthetic, or artificially blended material.

Sand casting is a relatively simple process and consists of the following steps:

1. Mold preparation
2. Core preparation
3. Core setting
4. Metal preparation
5. Metal pouring
6. Part shakeout
7. Part cleaning

6 Environmentally Benign Manufacturing

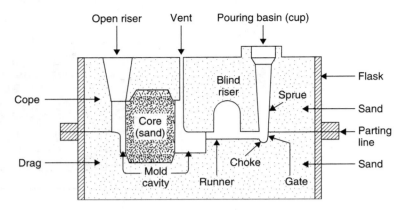

Figure 3 Sectional view of a sand-casting mold.

8. Sand reclamation
9. Sprue and gate reclamation

This section describes each step.

Mold Preparation. A mold is fabricated from foundry sand. It is created by pouring and compacting sand around a pattern. Once the sand is compacted the pattern is withdrawn, leaving a cavity in the shape of the part to be produced. The cavity holds the molten metal in the desired shape until it cools. Molding *sand* is a mixture of approximately 85 percent sand, from 4 to 10 percent clay, and from 2 to 5 percent water by mass. Small quantities of additives are used to prevent the metal from oxidizing as it cools. These additives are usually bituminous coal, anthracite, or ground coke.

Core Preparation. Cores are necessary for parts that are especially complicated or have internal cavities. Cores are created from sand and a binder—usually in the form of a resin—that cures through heat or gasification. The sand and binder are put in a mold called a *core box* that forms the desired shape. They are then removed from the core box and allowed to cure before placing them in the sand casting mold. The placement of the core is illustrated in the left half of Figure 3.

Core Setting. Once the mold and cores have been prepared, the cores are set in place inside the mold and the mold is closed.

Metal Preparation. Metal—usually iron, steel, or aluminum—is prepared by melting ingots or scrap with the additives or alloying materials needed to give the finished product its desired properties. Most sand casting is accomplished by melting and blending scrap material.

Metal Pouring. Metal is poured manually from a ladle or tilting furnace, or most commonly from an automatic pouring ladle, that is charged from holding furnaces.

Shakeout. Molds containing cooled parts are transferred by conveyor to a large rotary drum, where the sand molds are broken and the sand is separated from the newly molded parts.

Part Cleaning. Usable parts are separated from gates and risers, and damaged or incompletely formed parts are sorted out. Further cleaning may also be accomplished in the form of pressing, hand grinding, sandblasting, or tumbling the parts to remove the parting lines and rough edges as well as any burnt sand.

Sand Reclamation. All modern foundries reclaim molding sand for reuse. The sand is run through a process where lumps are broken up and any solids are removed by screening. New sand, clay, and water are added as needed to return the sand to a usable condition. Some sand that cannot be reclaimed is discarded. Most foundries have a sand laboratory whose responsibility is to monitor and manipulate the condition of the molding sand.

Sprue and Gate Reclamation. Any metal that is not a usable part is returned to the scrap area to be used in a future melt.

Environmental Concerns with Sand Casting

With respect to sand casting, the environmentally benign manufacturing function is concerned with minimizing the impact of the manufacturing steps just listed on the environment by changing or replacing processes that produce an environmentally offensive result or hazard. Consideration will be given here to each of the sand-casting subprocesses.

Molds and Cores. Molds and cores are made from sand. For every ton of castings produced, the process requires about 5.5 tons of sand. Problems occur when the sand and binders are exposed to the heat of the molten metal and sometimes during curing processes of mold preparation. This releases a wide variety of organic pollutants that are regulated by the *Clean Air Act* or the *Clean Water Act*. These pollutants come primarily from the chemical binders use to make cores stronger, or in some cases from binders added to the sand. When stronger molds are required, chemical binders are added to the mold sand. These binders include furanes, phenolic urethanes, and phenolic esters. The binder is chosen depending on the strength required for the metal being cast and the size of the mold. Other concerns are molding sand additives used to prevent the metal from oxidizing as it cools. These additives are usually bituminous coal, anthracite, or ground coke. Although these additives are a very small component by mass,

Table 1 Some Pollutants Associated with Binders Used in Mold Preparation

	Benzene	Methanol	Phenol	Toluene	Formaldehyde	MMDI
Furane	●	●	●	●		
Phenol urethane			●		●	●
Phenol ester			●		●	

Note: MMDI is an acronym for Monomeric Methylene Diphenyl Diisocyanate.

as they burn off on contact with the molten metal they create an assortment of hazardous air pollutants. Table 1 shows several pollutants associated with binders used in mold preparation.

Metal Preparation and Pouring. While the use of scrap metal can contribute to pollutants, the most significant contributor for these subprocesses are related to the heat input to melt the metal. Pollutants include large amounts of particulates and carbon monoxide, as well as smaller amounts of SO_2 and VOCs. Emissions are dependent on the type of furnace being used. Electric furnaces have a reduced environmental impact as compared to coke-fired furnaces of older foundries. Many foundries use pollution-control technology in the form of scrubbers to clean air before releasing it to the outside. These are used on all types of furnaces. Wet scrubbers are also used, but are less common and are used primarily on coke-fired cupola furnaces. These methods are effective at controlling air emissions, but they produce waste streams in the form of solid waste or contaminated water, which must be processed further. Table 2 shows pollutants generated by melting metal for several types of furnaces. Table 3 gives the energy requirements at the foundry for both fuel-fired and electric furnaces.

Table 2 Approximate On-site Emissions from Various Furnaces in lb./ton of Metal

	PM	CO	So_2	VOC
Fuel-fired Reverberatory Furnace	2.2	Unknown	N/A	Unknown
Induction Furnace	1	~0	~0	Unknown
Elec Arc Furnace	12.6	1–38	~0	0.06–0.30
Coke-fired Cupola	13.8	146	1.25+	Unknown

Note: Does not include emissions from electricity generation or fuel extraction.

Table 3 Energy Requirements at the Foundry in MBtu/ton Saleable Cast Material for Foundry Furnaces

Fuel Source	Furnace Type	MBtu/ton
Fuel-fired	Crucible	1.8–6.8
	Reverberatory	2.5–5.0
	Cupola (coke)	5.8
	Cupola (NG)	1.6
Electric	Induction	4.3–4.8
	Electric arc	4.3–5.2
	Reverberatory	5.2–7.9
	Cupola	1.1

Cleaning. Cleaning the product can involve the use of organic solvents, abrasives, pressurized water, or acids, often followed by protective coatings. Techniques used to remove sand and flashing include vibrating, wire-brushing, blast cleaning band saws, cutoff wheels, and grinders.

Removing Sprues, Runners, and Flashing. Although particle, HAP, and effluent pollutants are created in this stage, they are largely contained by filters and closed systems.

Sand Reclamation. Up to 90 percent of molding sand can be reused in a green sand foundry after filtration for fine dust and metal particles. Sand with chemical binders can be used only in small quantities, however. Sand that is not reused is sometimes used in road bases and asphalt concrete.[3] In the United States, from 7 to 8 million tons of mold sand (about 0.5 tons of sand/ton of cast metal) per year ends up in landfills. Spent sand makes up almost 70 percent of foundry solid wastes,[4] AIS, 1999.

From an environmental perspective, the foundry industry has improved remarkably in recent years. U.S. and off-shore foundries have been forced by both legislation and automakers to reduce their pollutants and waste streams—hence, the positive influence of manufacturers (automakers) on tier I suppliers (foundries). Foundries are relying more on electric and natural gas furnaces, thereby reducing the amount of input energy required and minimizing the amount of pollutants. Sand reclamation and use of spent sand for other purposes reduces the impact on landfills. The recent use of trimming presses helps to eliminate the need to grind parts to remove gates and sprues. One of the areas that could benefit from continued research is the development of benign binders for core and mold making processes. Redesigning parts to eliminate cores would also be helpful.

Another environmental concern for the sand casting process is the generation of waste from the machining of cast metal parts. Machining allowances are required in many cases because of unavoidable surface impurities, warpage, and surface variations. Average machining allowances are given in Table 4. Good practice dictates use of the minimum section thickness compatible with the design. The normal section recommended for various metals is shown in Table 5 (see Zohdi and Biles, 2006).

Die Casting

Die casting may be classified as a permanent-mold casting system. However, it differs from the process just described in that molten metal is forced into the mold or die under high pressure (1000–30,000 psi [6.89–206.8 MPa]). The metal solidifies rapidly (within a fraction of a second) because the die is water-cooled. Upon solidification, the die is opened. Ejector pins automatically eject the casting from the die. If the parts are small, several of them may be made at one time in what is termed a *multicavity die*.

There are two main types of machines used: the hot-chamber and the cold-chamber types.

Hot-Chamber Die Casting. In the hot-chamber machine, the metal is kept in a heated holding pot. As the plunger descends, the required amount of alloy is automatically forced into the die. As the piston retracts, the cylinder is again filled with the right amount of molten metal. Metals such as aluminum, magnesium,

Table 4 Machining Allowances for Sand Castings (in./ft.)

Metal	Casting Size	Finish Allowance
Cast irons	Up to 12 in.	3/32
	13–24 in.	1/8
	25–42 in.	3/16
	43–60 in.	1/4
	61–80 in.	5/16
	81–120 in.	3/8
Cast Steels	Up to 12 in.	1/8
	13–24 in.	3/16
	25–42 in.	5/16
	43–60 in.	3/8
	61–80 in.	7/16
	81–120 in.	1/2
Malleable irons	Up to 8 in.	1/16
	9–12 in.	3/32
	13–24 in.	1/8
	25–36 in.	3/16
Nonferrous metals	Up to 12 in.	1/16
	13–24 in.	1/8
	25–36 in.	5/32

Table 5 Minimum Sections for Sand Castings (in./ft.)

Metal	Section
Aluminum alloys	3/16
Copper alloys	3/32
Gray irons	1/8
Magnesium alloys	5/32
Malleable irons	1/8
Steels	1/4
White irons	1/8

and copper tend to alloy with the steel plunger and cannot be used in the hot chamber.

Cold-Chamber Die Casting. This process gets its name from the fact that the metal is ladled into the cold chamber for each shot. This procedure is necessary to keep the molten-metal contact time with the steel cylinder to a minimum. Iron pickup is prevented, as is freezing of the plunger in the cylinder.

Advantages and Limitations. Die-casting machines can produce large quantities of parts with close tolerances and smooth surfaces. The size is limited only by the capacity of the machine. Most die castings are limited to about 75 pounds (34 kg) of zinc; 65 pounds (30 kg) of aluminum; and 44 pounds (20 kg) of magnesium. Die casting can provide thinner sections than any other casting process. Wall thicknesses as thin as 0.015 inch (0.38 mm) can be achieved with aluminum in small items. However, a more common range on larger sizes will be 0.105 to 0.180 inch.

Some difficulty is experienced in getting sound castings in the larger capacities. Gases tend to be entrapped, which results in low strength and annoying leaks, causing an air pollution problem. One way to reduce metal sections without sacrificing strength is to add ribs and bosses into the product design. An approach to the porosity problem has been to operate the machine under vacuum.

The surface quality of the casting is dependent on that of the mold. Parts made from new or repolished dies may have a surface roughness of 24 μin. (0.61 μm). A high surface finish means that, in most cases, coatings such as chromeplating, anodizing, and painting may be applied directly. More recently, decorative texture finishes are obtained by photoetching. This technique has been used to simulate woodgrain finishes, as well as textile and leather finishes, and to obtain checkering and crosshatching patterns in the surface finish.

Investment Casting. Casting processes in which the pattern is used only once are variously referred to as *lost-wax* or *precision-casting* processes. These involve making a pattern of the desired form out of wax or plastic (usually polystyrene). The expendable pattern may be made by pressing the wax into a split mold or by using an injection-molding machine. The patterns may be gated together so that several parts can be made at once. A metal flask is placed around the assembled patterns, and a refractory mold slurry is poured in to support the patterns and form the cavities. A vibrating table equipped with a vacuum pump is used to eliminate all the air from the mold. Formerly, the standard procedure was to dip the patterns in the slurry several times until a coat was built up. This is called the *investment process*. After the mold material has set and dried, the pattern material is melted and allowed to run out of the mold.

The completed flasks are heated slowly to dry the mold and to melt out the wax, plastic, or whatever pattern material was used. When the molds have reached a temperature of 100°F (37.8°C), they are ready for pouring. Vacuum may be applied to the flasks to ensure complete filling of the mold cavities. When the metal has cooled, the investment material is removed by vibrating hammers or by tumbling. As with other castings, the gates and risers are cut off and ground down.

Ceramic Process. The ceramic process is somewhat similar to the investment casting in that a creamy, ceramic slurry is poured over a pattern. In this case, however, the pattern, made out of plastic, plaster, wood, metal, or rubber, is reusable. The slurry hardens on the pattern almost immediately and becomes a strong green ceramic of the consistency of vulcanized rubber. It is lifted off the pattern, while it is still in the rubberlike phase. The mold is ignited with a torch to burn off the volatile portion of the mix. It is then put in a furnace and baked at 1,800°F (982°C), resulting in a rigid refractory mold. The mold can be poured while still hot.

Full-mold Casting. Full-mold casting may be considered a cross between conventional sand casting and the investment technique of using lost wax. In this case, instead of a conventional pattern of wood, metals, or plaster, a polystyrene foam or Styrofoam is used. The pattern is left in the mold and is vaporized by the molten metal as it rises in the mold during pouring. Before molding, the pattern is usually coated with a zirconite wash in an alcohol vehicle. The wash produces a relatively tough skin separating the metal from the sand during pouring and cooling. Conventional boundry sand is used in backing up the mold.

4.3 Metal-forming Processes

Metal-forming processes use a remarkable property of metals—their ability to flow plastically in the solid state without concurrent deterioration of properties. Moreover, by simply moving the metal to the desired shape, there is little or no waste. Figure 4 shows some of the metal-forming processes. Metal-forming processes are classified into two categories: hot-working processes and cold-working processes.

4 Manufacturing Processes 13

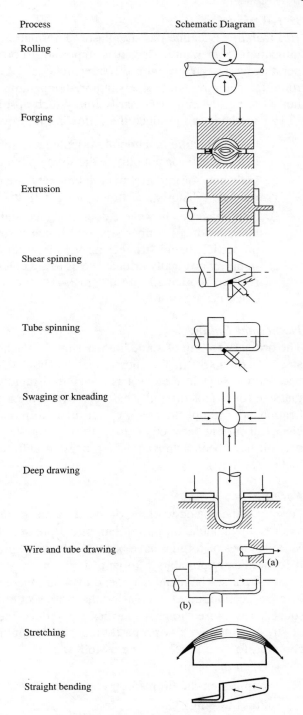

Figure 4 Several metal-forming processes.

Hot-Working

Hot working is defined as the plastic deformation of metals above their recrystallization temperature. Here it is important to note that the crystallization temperature varies greatly with different materials. Lead and tin are hot worked at room temperature, while steels require temperatures of 2,000°F (1,100°C). Thus, hot working does not necessarily imply high absolute temperatures.

Hot working can produce the following improvements in metal products:

1. Grain structure is randomly oriented and spherically shaped, which results in a net increase not only in the strength but also in ductility and toughness.
2. Inclusions or impurity material in metal are reoriented. The impurity material often distorts and flows along with the metal.
3. This material, however, does not recrystallize with the base metal and often produces a fiber structure. Such a structure clearly has directional properties, being stronger in one direction than in another. Moreover, an impurity originally oriented so as to aid crack movement through the metal is often reoriented into a "crack-arrestor" configuration perpendicular to crack propagation.

Isothermal Rolling

The ordinary rolling of some high-strength metals, such as titanium and stainless steels, particularly in thicknesses below about 0.15 inch (3.8 mm), is difficult because the heat in the sheet is transferred rapidly to the cold and much more massive rolls. This difficulty has been overcome by *isothermal rolling*. Localized heating is accomplished in the area of deformation by the passage of a large electrical current between the rolls, through the sheet. Reductions up to 90 percent per roll have been achieved. The process usually is restricted to widths below 2 inches (50 mm).

Forging

Forging is the plastic working of metal by means of localized compressive forces exerted by manual or power hammers, presses, or special forging machines. Various types of forging have been developed to provide great flexibility, making it economically possible to forge a single piece or to mass produce thousands of identical parts. The metal may be: drawn out, increasing its length and decreasing its cross section; upset, increasing the cross section and decreasing the length; or squeezed in closed impression dies to produce multidirectional flow. The state of stress in the work is primarily uniaxial or multiaxial compression. The most common forging processes are as follows:

- Open-die hammer
- Impression-die drop forging
- Press forging
- Upset forging

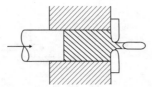

Figure 5 The metals extrusion process.

- Roll forging
- Swaging

Extrusion

In the extrusion process shown in Figure 5, metal is compressively forced to flow through a suitably shaped die to form a product with a reduced cross-section. Although extrusion may be performed either hot or cold, hot extrusion is employed for many metals to reduce the forces required, to eliminate cold-working effects, and to reduce directional properties. The stress state within the material is triaxial compression.

Lead, copper, aluminum, and magnesium, and alloys of these metals are commonly extruded, taking advantage of the relatively low yield strengths and extrusion temperatures. Steel is more difficult to extrude. Yield strengths are high and the metal has a tendency to weld to the walls of the die and confining chamber under the conditions of high temperature and pressures. With the development and use of phosphate-based and molten glass lubricants, however, substantial quantities of hot steel extrusions are now produced. These lubricants adhere to the billet and prevent metal-to-metal contact throughout the process.

Almost any cross-section shape can be extruded from the nonferrous metals. Hollow shapes can be extruded by several methods. For tubular products, the stationary or moving mandrel process is often employed. For more complex internal cavities, a spider mandrel or torpedo die is used. Obviously, the cost for hollow extrusions is considerably greater than for solid ones, but a wide variety of shapes can be produced that cannot be made by any other process.

Drawing

Drawing, shown in Figure 6, is a process for forming sheet metal between an edge-opposing punch and a die (draw ring) to produce a cup, cone, box, or shell-like part. The work metal is bent over and wrapped around the punch nose. At the same time, the outer portions of the blank move rapidly toward the center of the blank until they flow over the die radius as the blank is drawn into the die cavity by the punch. The radial movement of the metal increases the blank thickness as the metal moves toward the die radius; as the metal flows over the die radius, this thickness decreases because of the tension in the shell wall between the punch nose and the die radius and (in some instances) because of the clearance between the punch and the die.

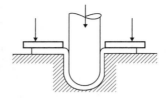

Figure 6 Deep drawing of a metal part.

4.4 Metal Joining Processes

The most common forms of metal joining are welding, soldering, and brazing. Each of these processes has the potential to be environmentally offensive, by generating noxious gases as part of the joining process or by producing metal wastes that must be disposed. Degarmo, Black, Kohser, and Klamecki provide an excellent discussion of these various joining processes (and indeed, any of the manufacturing processes discussed in the chapter).[5] Figure 7 gives the various classifications of welding processes employed in manufacturing.

Welding is the most common metal joining process. The principle classes of welding processes include: (1) gas-flame welding, which utilizes a high-temperature gas to melt selected surfaces of the mating parts; (2) arc-welding processes, which utilize an electric arc to produce molten material between

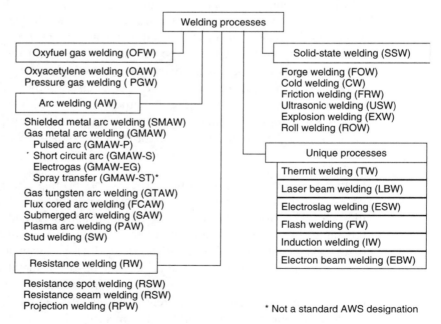

Figure 7 Classification of several common welding processes.

mating parts; and (3) resistance-welding processes, which utilize both heat and pressure to induce coalescence. Brazing and soldering are utilized when the mating surfaces cannot sustain the high temperatures required for welding mating parts. The ensuing sections give brief discussions of each of these joining processes and describe how environmental offenses can be avoided.

Welding Processes.
As just stated, three of the most common classes of welding processes used in manufacturing are oxyfuel gas welding, arc welding, and resistance welding. The coalescence between two metals requires sufficient proximity and activity between the atoms of the pieces being joined to cause the formation of common crystals.

Gas-flame Processes. Oxyfuel gas-welding processes utilize as their heat source the flame produced by the combustion of a fuel gas and oxygen. The combustion of *acetylene* (C_2H_2)—commonly known as the oxyacetylene torch—produces temperatures as high as 5850°F (3250°C). Three types of flames can be obtained by varying the oxygen/acetylene ratio: (1) If the ratio is between 1:1 and 1.15:1, all oxygen-acetylene reactions are carried to completion and a *neutral flame* is produced; (2) If the ratio is closer to 1.5:1, an *oxidizing flame* is produced, which is hotter than the neutral flame but similar in appearance; (3) Excess fuel produces a *carburizing flame*.

Almost all oxyfuel gas welding is of the *fusion* type, which means that the metals to be joined are simply melted at the interfacing surfaces and no pressure is required. This process is best suited to steels and other ferrous metals. There is a low heat input to the part, and penetrations are only about 3 mm.

The environmental impacts of oxyfuel gas welding include the generation of combustion products, which have to be *scrubbed* before release to the atmosphere, and the production of slag and waste metal that must be safely disposed.

Arc-welding Processes. Arc-welding processes employ the basic circuit shown in Figure 8. Welding currents typically vary from 100 to 1000 amps, with voltages in the range from 20 to 50 volts.

In one type of arc-welding process, the electrode is consumed and thus supplies the molten metal. A second process utilizes a nonconsumable tungsten electrode, which requires a separate metal wire to supply the molten metal. Filler materials must be selected to be compatible with the mating surfaces being welded. In applications where a close fit is required between mating parts, gas-tungsten arc welding can produce high-quality, nearly invisible welds.

In *plasma arc welding* an arc is maintained between a nonconsumable electrode and the workpiece in such a way as to force the arc to be contained within a small-diameter nozzle, with an inert gas forced through the stricture. Plasma-arc welding is characterized by extremely high (30,000°F) temperatures, which offers very high welding speeds and hence high production rates.

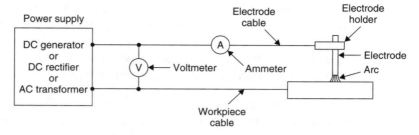

Figure 8 Basic circuit for the arc-welding process.

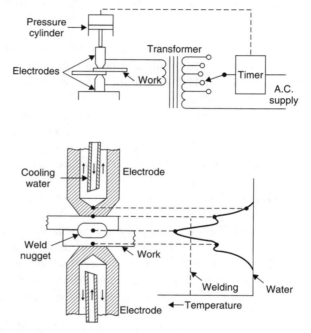

Figure 9 A typical resistance-welding circuit and configuration.

The environmental impacts of arc-welding processes include the generation of metal waste and the requirement for relatively high power.

Resistance Welding Processes. In *resistance welding,* both heat and mechanical pressure are used to induce coalescence. Electrodes are placed in contact with the material, and electrical resistance heating is utilized to raise the temperatures of the workpieces and the space between them. These same electrodes also supply the mechanical pressure that holds the workpieces in contact. When the desired temperature has been achieved, the pressure exerted by the electrode is increased to induce coalescence. Figure 9 illustrates a typical resistance welding circuit. It is important to note that the workpieces actually form part of the electrical circuit, and that the total resistance between the electrodes consists of three distinct components: (1) the resistance of the workpieces; (2) the contact resistance between the electrodes and the workpieces; and (3) the resistance between the surfaces to be joined.

The most important environmental consideration in resistance welding is the electrical power consumed.

Electron-Beam Welding. **Electron-beam welding** is a fusion welding process which utilizes the heating resulting from the impingement of a beam of high-velocity electrons on the metal parts to be welded. The electron optical system for the electron-beam welding process is shown in Figure 10. An electrical current heats a tungsten filament to about 4,000°F, causing it to emit a stream of electrons by *thermal emission*. Focusing coils are employed to concentrate the electrons into a beam, accelerate them, and direct them to a focused spot that is

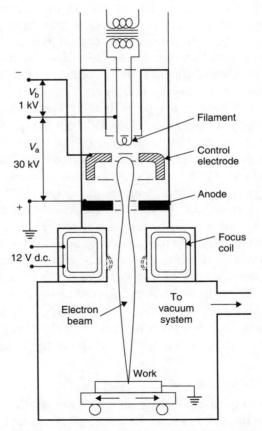

Figure 10 Schematic diagram for the electron-beam welding process.

between 0.8 and 3.2 mm in diameter. Since the electrons, which are accelerated at 150 kV, achieve velocities near two-thirds the speed of light, intense heat is generated. Since the beam is composed of charged particles, it can be positioned by electromagnetic lenses. To be effective as a welding heat source, the electron beam must be generated and focused in a high vacuum, typically at pressures as low as 0.01 Pa.

Almost any metal can be welded by the electron-beam process, including those that are very difficult to weld by any other process, including tungsten, zirconium, and beryllium. Heat-sensitive metals can be welded without damage to the base metal.

From an environmental standpoint, the absence of shielding gases, fluxes, or filler materials means that the waste material produced by the process is negligible. Only the high power requirements stand as a problem.

Brazing and Soldering.
Brazing is the permanent joining of similar or dissimilar metals through the application of heat and a filler material. Filler metals melt at temperatures as low as 800°F, typically much lower than those of the base metals, which makes brazing a useful joining process for dissimilar metals (ferrous to nonferrous metals, metals with different melting points, or even metal to ceramic). Strong permanent joints are formed by brazing.

Soldering is a type of brazing operation in which the filler material has a melting temperature below 850°F. It is typically used for connecting thin metal pieces, connecting electronic components, joining metals while avoiding high temperatures, and filling surface flaws and defects in metal parts. Soldering can be used to join a wide variety of shapes, sizes, and thicknesses, and is widely used to provide electrical coupling or airtight seals. The primary means of heating the filler material is to apply an electrically heated iron rod to melt the filler metal and position it in the proper location on the workpiece. Soldering filler materials are typically low melting temperature metals such as lead, tin, bismuth, indium, cadmium, silver, gold, and germanium. Because of their low cost and favorable properties, alloys of tin and lead are most commonly used.

The environmental impacts of brazing and soldering trace to the filler materials used in their application. Since 1988, the use of lead and lead alloys in drinking water lines has been prohibited in the United States. Japan and the European Union prohibit the use of lead in electronic applications.

4.5 Plastic Injection Molding

The injection-molding process involves the rapid pressure filling of a shape-specific mold cavity with a fluid plastic material, followed by the solidification of the material into a product. The process is used for thermoplastics, thermosetting resins, and rubbers.

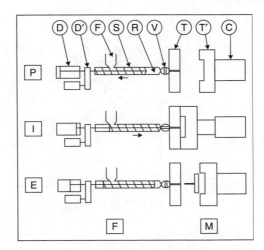

Figure 11 The principle of injection molding. (From Ref. 6.)

Principle of Injection Molding

The injection molding of thermoplastics can be subdivided into several stages as illustrated in Figure 11. At the plastication stage P, the feed unit F operates in much the same way as an extruder, melting and homogenizing the material in the screw/barrel system. The screw, however, is allowed to retract, to make room for the molten material in a space at the cylinder head, referred to here as the material *reservoir*, between the screw tip and a closed valve or an obstruction of solidified material from the previous shot. At the injection stage I, the screw is used as a ram (piston) for the rapid transfer of the molten material from the reservoir to the cavity between the two halves (T and T′) of the closed mold. Since the mold is kept at a temperature below the solidification temperature of the material, it is essential to inject the molten material rapidly to ensure complete filling of the cavity. A high holding or packing pressure (10,000 to 30,000 psi) is normally exerted to partially compensate for the thermal contraction (shrinkage) of the material upon cooling. The cooling of the material in the mold is often the limiting time factor in injection molding because of the low thermal conductivity of polymers. After the cooling stage, the mold can be opened and the solid product removed.

Equipment

Injection molding machines are now most commonly of the reciprocating screw type, as illustrated in Figure 12. Two distinct units, referred to as the feed unit F and the mold unit M, are mounted on a frame (F′). The feed unit F consists of the plastication/injection cylinder (screw, barrel, and feed hopper), the axial screw drive, and the rotation screw drive.

Although injection-molding machines may occasionally be dedicated to the molding of a single product, a machine is normally used with a variety of tools

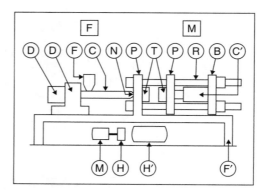

Figure 12 The injection molding machine. (From Ref. 6.)

(molds), which may imply frequent mold changes and the associated costly set-up period. Injection-molding machines are available in a broad range of sizes. They are normally rated by their maximum clamping force, with normal ranges from about 25 to 150 tons for "small" machines, 150 to 70 tons for "medium-sized" machines, and 750 to 5,000 tons for "large" machines; the current maximum is 10,000 tons.

Tooling

The interchangeable injection-molding tool, the *mold*, must (1) provide a cavity corresponding to the geometry of the product and (2) allow the ejection of the product after its solidification. Primary mold opening is achieved by fastening one-half of the mold to the stationary platen (T), as shown in Figure 12, and the other half to the moving platen (T′). The stationary mold half is sometimes referred to as the *front, cavity,* or *negative block*, and the moving mold half as the *rear, force,* or *positive block*. The removal of a product from a cavity surface requires, in addition to an ejection system, a suitable surface finish and an appropriate taper or draft. It need not require a mold release agent.

During injection, the material flows from the nozzle at the tip of the injection unit to the single cavity, or to each of several cavities, through what is referred to here as the *feed system*, generally comprising sprues, runners, and gates. In most cases, injection-molded products need to be removed from one mold half by an ejection (knockout, stripping) device. This device is normally incorporated in the moving mold half. Retractable secondary mold sections may be required when products feature undercuts, reentrant shapes, internal or external threads, and so on.

Runners are machined in mold halves, next to the parting surface. One solution, applicable to chemically stable thermoplastics, consists of having large runners cooled in such a way that a sleeve of insulating solid plastic forms around a molten core, where the intermittent injection flow takes place; this method is referred to as *insulated* or *Canadian* runner molding. Another solution, referred to as *hot runner molding*, involves a heated runner, or manifold block, and is often used in conjunction with valve gating. *Gates* serve several purposes in injection molding. Their easily altered, smaller cross-section permits a convenient control of the flow of the molten material, the rapid freezing of the material to shut off the cavity after injection, and the easy separation of the products from the feed appendage (de-gating). Important savings can be made by using hot runners.

The maximum pressure in injection molds is normally in the range of 4,000 to 12,000 psi corresponding to a clamping force per unit projected area of cavity and feed system in the range of 2–6 tons/square inch. The construction of injection molds requires materials with a combination of good thermal conductivity and resistance to mechanical wear and abrasion. Prototype molds can be cast from low-melting alloys. For short-run molds (about 10,000 to 100,000 moldings), tool steel is normally used. Long runs involving millions of moldings require special hardened and chrome-plated steels.

A variety of techniques are used to form mold cavities: machining of a solid block, computer-aided machining (CAM) centers, hobbing (cold forming), electrochemical machining (ECM), electrical discharge machining (EDM), or spark erosion, electroforming, plating, and etching. For short runs—fewer than 10,000 parts—a mold cavity can be fabricated using a selective-laser-sintering rapid prototyping process to build a copper-infiltrated iron part.

Auxiliaries

Many thermoplastic resins require thorough drying prior to molding, to avoid the formation of voids or a degradation of the material at molding temperatures. Mold temperature control is often achieved by the circulation of a fluid through a separate heater/chiller device. With increased interest in automation, robots have been introduced for the removal of products and feed appendages from open molds, and for separation (degating) and sorting. Feed appendages, startup scrap, and occasional production scrap are normally reground in granulators and recycled as a fraction of the feed material.

Materials

All thermoplastics are, in principle, suitable for injection molding, but since fast flow rates are needed, grades with good fluidity (high melt index) are normally preferable.

Products

A major advantage of injection-molded products is the incorporation of fine details such as bosses, locating pins, mounting holes, bushings, ribs, flanges, and so on, which normally eliminates assembly and finishing operations. Thermosetting resin systems, such as phenolics (PF); or unsaturated polyester (UP), often used with fillers or reinforcements, are increasingly injection molded at relatively high speeds. Curing, which involves chemical reactions, takes generally much longer than the injection, and multimold machines are thus often used with shuttle or rotary systems. Injection molding is increasingly used for producing relatively small rubber products significantly faster than by compression molding and, normally, with a smaller amount of scrap and a better dimensional accuracy. As in the case of thermosetting resins, a heated mold is needed for vulcanization (curing).

Environmental Analysis of Injection Molding Processes

Plastic components are major parts in electrical and electronic (E&E) products. About 8.5 percent of the plastic parts produced are for these products. Large number of plastic parts used for automobile industry. Some 33 percent of all small house appliances incorporate plastic components, and about 42 percent of all plastic materials are used in the manufacture of toys.

This environmental analysis of injection molding highlights a few important points. The type of injection molding machine (hydraulic, hybrid, or electric) has a large impact on energy consumption. Table 6 shows the energy related emissions for the injection molding process, including the compounder stage. Table 7 gives the total annual production of injection molded plastics. Table 8 gives the total annual energy consumption associated with the production of injection molded plastics. The impact of injection molding on the environment may seem benign, but it can be significant. We must take into consideration energy consumption, manufacturing process, and raw material usage. The product life cycle is important because it affects the production, energy, and raw material. The majority of plastic parts that are used in electrical and electronic products are parts made through the injection-molding process. Injection molding involves melting polymer resin together with additives and then injecting the melt into the mold to make the final products. This process may have an impact on the environment, but we have to reduce the effect of this process and make it benign as much as possible.

Table 6 Energy-related Air Emissions for the Compounder Stage and the Injection Molder Stage

Stage	SEC (MJ/Kg)	Energy Related Emissions				
		CO_2 g	SO_2 g	NO_x g	CH_4 g	Hg mg
Compounder	5.51	284.25	1.26	0.51	10.32	0.01
Injection molder						
Hydraulic	13.08	674.82	2.98	1.22	24.29	0.01
Hybrid	7.35	379.33	1.68	0.68	13.77	0.01
All-electric	6.68	344.57	1.52	0.62	12.50	0.01

Table 7 Injection-molded Polymer Totals in kg/year

	Injection Molded—Million kg/yr	
	U.S. Only	Global
Six main thermoplastics	5,571	23,899
All plastics	12,031	38,961

Note: The subdivision "6 main thermoplastics" refers to HDPE, LDPE, LLDPE, PP, PS and PVC.

Table 8 Total Energy Used in Injection Molding

Compounder and Injection Molder	U.S. GJ/year	Global GJ/year
Six main thermoplastics	9.34E + 07	4.01E + 08
All plastics	2.06E + 08	6.68E + 08

Note: The subdivision "6 main thermoplastics" refers to HDPE, LDPE, LLDPE, PP, PS and PVC.

Injection molding is used primarily to produce plastic parts with specific geometrics. The process starts by mixing polymer resin with additives that are specific to the part to gain desired properties such as increased strength. The mix of polymer resin and additives is also combined with colorants if needed at this point and is stored in a hopper. The material is gravity fed into a feeding tube that has screws to push material forward. When in the screws, the material is melted and mixed. The material is fed into the die that will shape the material to the desired part. During the fill stage, hydraulic clamps hold the two ends of the die together until all the necessary material has entered the die and cooled to the desired temperature for removal. Then the clamps release the part and it is removed from the die.

Life Cycle of a Plastic Product

When tracing the life cycle of the process to the beginning we need to look at how the polymer pellets are manufactured. In injection molding the overall process starts at production stage. This stage takes raw materials from the earth and transforms them with addition of energy into polymers. The raw polymer is shipped in bulk to the compounder, which mixes it with additives in order to give it required properties for application. The polymer is shipped to the injection molder, which transforms the polymer into finished products. The injection molder might add some additives in the process, such as coloring. After being injection molded and packaged, the product is ready for the consumer. When trying to develop the polymer resin used in the injection mold process, the manufacturer uses large amounts of petroleum, and large energy costs are associated with the production of the material. The additives added to the polymer base can be hazardous in large concentrations. The majority of the byproducts to the process can be hazardous and are not biodegradable.

Environmental Impact of Plastics Injection Molding. When considering the life cycle of a plastic-based product, it is important to understand the emissions that come from the polymer production stage. The emissions can be divided into energy-related emission and processing emission. Processing emission at the site is small compare to energy-related ones. It should be noted that plastics don't break down in landfills. Two solutions have been used over the last few years. The first is to burn the plastic that leads to toxic material into the air. This method is most commonly used today because plastics are petroleum based and have high heating properties. Countries like Japan and England have laws limiting the amount of petroleum-based products that can be incinerated and are moving toward more methods that recycle the product. Due to this trend, more effort is placed in the design phase of projects to ensure the correct mixture of recycled plastics, new polymer material, and additives for product performance. The second method is to recycle the plastic and make it into other products. This second solution can be used only for one of the two types of plastics. Thermoplastics can be melted, while thermoset plastics cannot be melted and have to be scrapped if a product is defective or at its end-of-life cycle. One area that has large opportunity for recycling is the plastic in automobiles. Current U.S. methods of recycling cars focus only on reusing the metal components. The plastic products are considered scrap and sent to landfills.

If we compare injection molding to other conventional manufacturing processes, injection molding appears to be on the same order of magnitude in term of energy consumption. For example, of processes such as sand and die casting have similar energy requirements (11–15 MJ/kg). However, when compared to processes used in the semiconductor industry, the impact of injection molding seems significant. But in order to understand this point, we have to understand the product's widespread effect on the economy. Injection-molding processes are more widely used and are growing in countries like China and India.

Although waste material is low and low levels of coolant are used in the process, the amount of energy used in the process has resulted in the research and development of ways to make the process more benign. It is critical to continue to improve the efficiency of the process in order to reduce the impact on the environment. It is essential to make a process that uses less energy, especially at this time when energy prices continue to rise.

5 THE MANUFACTURED PRODUCT

Most of the discussion in this chapter has focused on ways to ensure that manufacturing processes are environmentally benign. Any company that is morally and ethically committed to the goals of environmentally benign manufacturing cannot scrutinize its manufacturing processes without first giving due consideration to the manufactured product itself. It could legitimately be argued that the energy expenditure of certain products will easily surpass any savings in environmental impact achieved through optimally designed manufacturing processes very early in the product life cycle. An example is the large gas-guzzling truck or automobile, which is manufactured with the quaint notion of "bowing to customer demand" for large vehicles despite their poor fuel mileage performance.

It is curious, though, that a considerable marketing budget is expended to cultivate this customer demand. It is also curious that an automobile manufacturer recently withdrew from the marketplace a plug-in, all-electric vehicle that had managed to gain a great deal of approval from its customers. Yet, manufacturers offer the excuse that they cannot act unilaterally without suffering competitively in the marketplace. Lawmakers, too, are prone to succumb to the notion that "people should be free to buy the products they want." Where does that leave the premise, or promise, of environmentally benign manufacturing? And where does that leave future generations, who are predestined to live in the environment we leave them?

REFERENCES CITED

1. M. E. Zohdi, W. E. Biles, and D. B. Webster, "Production Processes and Equipment," Chapter 5 in *Mechanical Engineers' Handbook: Book 3—Manufacturing and Management*, M. Kutz, ed., John Wiley, New York, 2006, pp 173–244.
2. M. E. Zohdi, and W. E. Biles, "Metal Forming, Shaping and Casting," Chapter 6 in *Mechanical Engineers' Handbook: Book 3—Manufacturing and Management*, M. Kutz, ed., John Wiley, New York, 2006, pp 245–285.
3. Javed, S., C. W. Lovell, and L. E. Wood. "Waste Foundry Sand in Asphalt Concrete," *Transportation Research Record 1437*. Transportation Research Board, Washington, D.C. (1994)
4. American Foundry Society, "The AFS Teams with DOE, DOD, EPA, and DOT to Deliver Results for America," AFS, Schaumburg, IL (1999)
5. E. P. DeGarmo, J T. Black, R. A. Kohser, and B. E. Klamecki, "Materials and Processes in Manufacturing," 9th ed., John Wiley, New York, 2003, pp 920–998.
6. J.-M. Charrier, *Polymeric Materials and Processing: Plastics, Elastomers and Composites*, Hanser Publishers, Munich, 1990.

REFERENCES

C. J. Backhouse, A. J. Clegg, and T. Staikos, "Reducing the Environmental Impacts of Metal Castings through Life-cycle Management," Wolfson School of Mechanical and Manufacturing Engineering, Loughborough University, Loughborough, Leicester, LE11 3TU, UK *Progress in Industrial Ecology*, **1** (1–3) (2004).

W. E. Biles, "Plastics Parts Processing Part I," Chapter 32, in *Handbook of Materials Selection*, M. Kutz, ed., John Wiley, New York, 2002, pp 969–992.

S. Dalquist and T. Gutowski, "Life-Cycle Analysis of Conventional Manufacturing Techniques," Massachusetts Institute of Technology Proceedings of IMECE2004: 2004 ASME International Mechanical Engineering Congress & Exposition, Anaheim, California, November 13–19, 2004.

T. Gutowski, "Casting," www.geocities.com/fimutp/casting.pdf.

D. O. Harper, "Plastics Parts Processing Part II," Chapter 33, in *Handbook of Materials Selection*, M. Kutz, ed., John Wiley, New York, 2002, pp 993–1036.

D. W. Richerson, "The Metal Casting Industry," Chapter 8 in www.ms.ornl.gov/programs/energyeff/cfcc/iof/chap8.pdf.

CHAPTER 2

DESIGN FOR THE ENVIRONMENT

J. Jeswiet
Mechanical and Materials Engineering, Queen's University Kingston, Ontario, Canada

1	INTRODUCTION	29	4	USEFUL WEB SITES FOR DfE	40
	1.1 The Need for DfE	29			
	1.2 The Structure behind DfE	32	5	SOFTWARE	42
	1.3 Application of DfE	34	6	SUMMARY	42
2	LIFE-CYCLE ANALYSIS	37			
3	APPLYING DfE	37			

1 INTRODUCTION

Design for the environment has the commonly known acronym DfE and can be viewed in different ways. As the acronym implies, DfE is about products, their design, and keeping environmental impact to a minimum. Because it is about the impact of product design on the environment, it is primarily concerned with industry. Therefore, DfE is about industry improving and optimizing the environmental performance of products, impacts on human health, associated risks, and product and process costs.

1.1 The Need for DfE

Although the need for concern about environmental impacts is discussed elsewhere in this book, it is helpful to remind ourselves about why this is needed. First, DfE is concerned with product design and hence the markets where products are sold. In a recent paper the following observation was made: in modern markets, products will continue to be demanded, designed and made.[1] These products will have environmental impacts that must be kept to a minimum.

Ten points arise from the foregoing:

1. Environmental problems occur because people want and need products.
2. Consumer numbers will not reduce and their habits will not change dramatically.

3. In modern society, mass production is the norm.
4. All products are designed.
5. All designs are manufactured.
6. There is an environmental impact for all product designs.
7. There is an environmental impact for all manufacturing processes.
8. It is important to get the design right at the beginning.
9. The designer must think about potential environmental impacts at all design stages.
10. Treat nature fairly or suffer the consequences.

These points are supported by the following: Designers know that 70 percent of product costs are decided early in the design stage.[2] This can be extended to environmental impacts, where it can be observed that if we get the design right, at the beginning, environmental impacts can be reduced by an estimated 70 percent.[3] Therefore, the environmental impact of a product must be addressed at the early stages of design.

A product consists of one or more parts. Components and products are designed and then manufactured. Products and their components include a wide range, from screws to fuel cells:

- Single components (nail, bolt, fork, coat hanger, etc.)
- Assemblies of many components (ballpoint pens, automobiles, washing machines, etc.)

New-product innovation and change are occurring at a fast pace in design and manufacturing. The ability to manufacture at much smaller scales will create new opportunities in design. This will also create new environmental challenges in the future. As shown in Table 1, more complicated products have more parts. Also, the number of components per product has changed dramatically with time as shown in Figure 1.[4,5]

The airplane is the perfect example of how products have increased in complexity, with concomitant increase in the number of parts. Included in Figure 1

Table 1 Examples of the Number of Parts in Some Products, circa 2000

Product	Number of Parts
Rotary lawnmower	300
Grand piano	12,000
Automobile ca. 2000	15,000
C-5A transport plane	>4 million
Boeing 747-400	>6 million

Note: From Ref. 4.

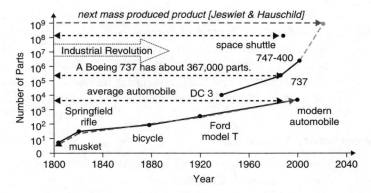

Figure 1 The increase in the number of parts per product with time. (From Ref. 5.)

is the prediction that, with the introduction of nanotechnology, there will be an order of magnitude increase in the number of parts per product.[5] This can be illustrated by considering the environmental effects that nanotechnology may have upon the environment; they are unknown. For example, the sizes of nanotechnology particles are those of blood cells.[6] At this scale, the effects of many elements and chemicals are unknown, and tests are not conducted.

With the changes occurring at a fast pace, and with the need for environmental compliance, companies are moving increasingly toward tools such as DfE. This section gives a few examples.

There is now concern about environmental image and how this will affect sales of a brand. Branding has become an important tool for companies, as indicated by *Forbes* magazine in its 2004 ratings of company branding.[7] One result is that, in many cases, positive action is taken to reduce environmental impacts.[8] In addition, it has been predicted that the cost of doing business will increase. For instance, banks might charge higher interest rates when environmental programs are not in place. In one example,[9] an oil company held secret meetings with environmental groups worldwide in an effort to change its hard-nosed public image on the environment. Critics claim the company has played a major role in the fight against the Kyoto Protocol. Accordingly, competitors have a softer public relations image as a result. Some energy companies have gone ahead and made it part of their mandate to decrease their production of greenhouse gases. For instance, in 1997 one oil company recognized the global warming problem and announced target reductions of 10 percent compared to its 1990 levels by 2010, but actually was able to do this by 2001.[10] Its net savings were $650,000. In two cases energy companies have become involved in solar energy research.[10,11]

Legal jurisdictions are creating regulations that are the main drivers in most cases. One of the more well-known cases is the EU directive on end of life for vehicles.[12]

We can see there is a great deal of activity with respect to *sustainability* as defined by the Bruntland committee.[13] Many engineering organizations also are becoming concerned with environmental work. Two examples are the American Society of Mechanical Engineers, which has formed a panel to look at product sustainability,[14] and the Engineering Institute of Canada, which held a conference recently to address climate change and contributing factors.[15]

1.2 The Structure behind DfE

Current activities on environmental design strategies, methodologies, and tools/techniques can be summarized into six categories:[16]

1. *Frameworks:* These contain general ideas about what should guide the product development process, including toolkits and strategic guidelines, including DfE.
2. *Checklists and guidelines:* Both are used as qualitative tools listing issues to consider in the development process, including DFE.
3. *Rating and ranking tools:* These are simple qualitative tools, typically using a specified scale for assessment.
4. *Analytical tools:* Comprehensive quantitative tools can be used for evaluating and measuring environmental performance, such as LCA.
5. *Software and expert systems:* The systems are designed to systemize and handle large amounts of information.
6. *Organizing tools:* They are important for giving direction on how to organize the process of tools, such as DfE, inside and outside the company.

What's in a name? Lagerstedt identified many of the existing methods concerned with design and environment as having the common goal of measuring and describing the environmental impacts of products and services.[17] Examples include life-cycle engineering (LCE), design for the environment (DfE), green design, environmentally conscious design (ECD), life-cycle design (LCD), sustainable design (SD), environmentally conscious design and manufacture (ECDM), industrial ecology (IE), design for recycling (DfR), and ecodesign. Although the acronyms and labels may differ, the people applying the labels all generally have the same goal—to reduce or eliminate environmental impacts of products. *Life-cycle engineering* (LCE) serves as a catchall for the many areas concerned with environmental impacts; see Figure 2.[18] This figure shows design for the environment as one facet of the life cycle of a product. As already shown, it must be included in the early stages of design.

Focus on LCE started in the 1990s. Over the last decade, it has prompted the following actions:[19]

- Legislation such as the EU take-back directive is now in place.[12]
- Environmental issues are being used in marketing to enhance brand names.

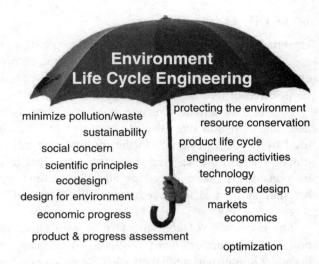

Figure 2 The many facets of life-cycle engineering. (From Ref. 18.)

- The cost of borrowing money is lower for environmentally responsible groups.
- Techniques such as DfE, ecodesign and DfD are in their infancy. They have yet to be adopted as a matter of doing business.

Products continue to increase in complexity, with concomitant, potential, environmental impacts.[5] For that reason, the environmental methodologies now being developed must play an important role in reducing those impacts. The likely side effect—and one that will spur LCE efforts—is that it will increase companies' bottom lines as efficiencies are found and waste is eliminated. There are many techniques available to industry in the environmental toolbox.

DfE is viewed in various ways, depending on the context. Usually it is applied as an extensive checklist at some stage in the design process. In a study of 28 companies, the list of methods and instruments for DfE included:[20]

- A negative list for harmful products
- A list of guidelines
- Company standards
- A positive list for environmentally friendly products
- Procedure descriptions in an environmental handbook
- List of compatible materials
- Examples of what competitors do
- Disassembly structure plans

These all appear in most DfE lists.

1.3 Application of DfE

Design for the Environment (DfE) and ecodesign are used synonymously. Both are about the impact of product design on the environment, and both use similar checklists. When DfE is used here, it should be understood that ecodesign can be inserted in its place.

DfE is a systematic consideration of design issues related to environmental and human health over the life cycle of the product.[21] It covers the approach used by product designers in the multifaceted field of product design and manufacture. DfE assumes that the effect a product has on the environment should be considered and reduced at all stages along the product life cycle, hence the inclusion in DfE procedures of life-cycle analysis (LCA), which has five stages.

The usual way of applying DfE is with checklists. Lists of design rules are popular, and there are many of them—which is appropriate? For instance, Anastas and Zimmerman published a list for green engineering,[22] but in this particular case it is more applicable to chemical engineering. In the case of product design, there are simple rules that can be followed. It should be noted that DfE is still young, and integrating it into design practice is not yet common; hence, checklists in existence are not standard and are continually evolving.[19,20,23]

One example is a list of eight self-evident axioms for DfE:[24]

1. Manufacture without producing hazardous waste.
2. Use clean technologies.
3. Reduce product chemical emissions.
4. Reduce product energy consumption.
5. Use nonhazardous recyclable materials.
6. Use recycled material and reused components.
7. Design for ease of disassembly.
8. Focus on product reuse or recycling at end of life.

Five other axioms appear in this list:[25]

1. Select low environmental impact materials.
2. Avoid toxic or hazardous materials.
3. Choose cleaner production processes.
4. Maximize energy and water efficiencies.
5. Design for waste minimization.

In these two lists, each principle is associated with at least one method used by designers. For instance, design for ease of disassembly (DfD) and manufacturing without producing hazardous waste belong with environmentally benign manufacturing (EBM).

A more detailed list called the "10 Golden Rules of EcoDesign" can be integrated into the foregoing DfE axioms to give a more detailed approach:[26]

1. Do not use toxic substances, but use closed loops when necessary to do so.
2. Minimize energy and resource consumption in production and transportation by striving for efficiency.
3. Minimize energy and resource consumption in the use phase, especially for products that have their most significant environmental aspects in the use phase.
4. Promote repair and upgrading (maintenance), especially for system-dependent products.
5. Promote durability for products with significant environmental aspects outside the use phase.
6. Use structural features and high-quality materials, to minimize weight, without interfering with necessary flexibility, impact strength, or functional properties.
7. Use better materials, surface treatments, or structural arrangements to protect products from dirt, corrosion, and wear.
8. Arrange in advance for upgrading, repair, and recycling through good access, labeling, modules and breakpoints, and provide good manuals.
9. Promote upgrading, repair, and recycling by using few, simple, recycled, unblended materials.
10. Use the minimum joining elements possible, and use screws, adhesives, welding, snap fits, geometric locking, and so on according to life-cycle guidelines.

Kaldjian has a similar list but with two additions:[27]

11. Adjust product design to reduce packaging.
12. Use recyclable materials when possible.

There are many *design for* topics, which can be summed up as DfX's. DfE encompasses many issues including design for disassembly (DfD), design for recycling (DfR) and design for remanufacturing. Designers must consider the utilization and the reuse of full products, subassemblies, and components in connection with a recycling process. They must design for ease of remanufacturing or recycling. Therefore, strategies, rules, and aids explained in VDI-guidelines 2243 are used in DfE.[28] The importance of designing for disassembly (DfD) and its aspects were discussed by Jovane et al. in a review of research into DfD.[29] The need for DfD techniques became apparent as recovering parts and materials from end-of-life products increased in popularity. There are a number of benefits of achieving efficient disassembly of products as opposed to recycling a product by shredding:

- Components that are of adequate quality can be refurbished or reused.
- Metallic parts can be separated easily into categories that increase their recycling value.
- Disassembled plastic parts can be easily removed and recycled.
- Parts made from other material such as glass or hazardous material can easily be separated and reprocessed.

These points can be incorporated into a DfE checklist, to make it more complete. The DfE checklist can be divided into three design modules:[30]

1. Product design
2. Material design
3. Energy consumption

This type of breakdown is fairly common and can be found in other techniques such as streamlined life-cycle analysis (SLCA).[31] Process design focuses on the reduction of energy consumption and the minimization of wastes and pollution processes. Material design is concerned with the selection and use of raw materials to minimize hazardous wastes, amount and type of pollution emitted, and total amount of materials required. Design for energy consumption is the selection of materials and processes that result in a reduction of the product's energy requirement for manufacture or use. Energy required for product production and use lead to the production of greenhouse gases; hence, the designer must be cognizant of both the production and use phase.

The philosophy is that DfE products are flexible, reliable, durable, adaptable, modular, dematerialized and reusable. In addition to proving economical feasibility and social compatibility, products must be ecologically viable. It is important to remember that the integration of environmental considerations must find its place among the many other priorities considered in the development of a new product.[17,19] Even from an environmental point of view, the weight given to the environmental performance of the product should not be higher than that which gives the strongest competitive ability of the product. If the product supersedes other less environmentally sound products, but does not perform well in the marketplace, no reduction in the load on environment is obtained. Often, however, environmental improvement can easily be attained without impairing other important performance parameters of the product. On the contrary, the work on optimization of the environmental performance often spurs creativity and brings innovative design solutions into the product development process, which allows improvement of the overall performance, together with the environmental performance—at least the first time a product is undergoing design for environment.[19]

2 LIFE-CYCLE ANALYSIS

Life-cycle analysis (LCA) in basic to most environmental work, and DfE is no exception. Life-cycle analysis consists of five stages. They include the extraction of the raw materials, the manufacturing of the product, its marketing and distribution, its use, and, finally, the disposal of a product. These are the same as the five stages used in streamlined life-cycle analysis (SLCA).[31] The stressors used in the SLCA matrix are materials choice, energy use, solid residues, liquid residues, and gaseous residues.

LCA is not a one-time effort; it is a repetitive procedure. Because DfE is strongly associated with LCA, it too is iterative. Figure 3 is an illustration of how LCA is an iterative process once goals and scope have been set and the DfE process has been started. Here, one can see that if the design is correct at the beginning, then the iterations become easier each time. Hence the incentive to "get it right the first time."

3 APPLYING DfE

Among the multitude of existing DfE tools, it is important to choose the right one for the intended purpose. Hauschild et al. present a hierarchy to be followed when using DfE:[19]

1. Determine whether the intended product is the right one to produce in the long term.
2. Identify the "environmental hot spots" in the life cycle of the product.
3. Select the DfE tool that supports optimization of the product by reducing these hot spots (see Figure 4).

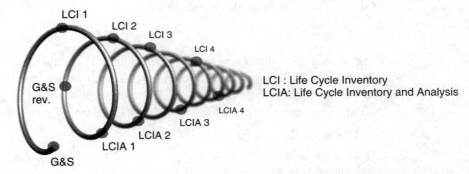

Figure 3 Life-cycle assessment is performed as an iterative exercise, and each phase may be revisited several times. After each iteration, the uncertainty is reduced, and the assessment is completed when the results are sufficiently certain to adequately answer the questions that were posed at in the goal and scope definition (G&S) at the beginning. (From Ref. 19.)

Figure 4 The corona of the life cycle represents the magnitude of a specific environmental impact or an aggregated measure of all environmental impacts at different points in the product's life cycle. The systematic identification of the environmental hot spots and improvement potentials in the life cycle of the product guarantees the relevant focus of the product development process. In this way, the right DfE tools can be chosen and applied to give the new product an optimal environmental performance, trimming the largest environmental impacts of the product's life cycle. (From Ref. 19.)

Graedel suggests three stages for environmental management: regulatory compliance (first stage), pollution prevention (second stage), and DfE (third stage). For the third stage, a similar set of guidelines can be used.[32] Combining these two approaches, we obtain the following list for initial steps:

1. Check for regulatory compliance.
2. Determine whether the intended product is the right one to produce in the long term.
3. Identify the environmental hot spots in the life cycle of the product.
4. Select the DfE tool that supports optimization of the product by reducing these hot spots (see Figure 4).

Once these steps have been completed, the following consolidated module checklists can be applied. This checklist is a composite of previous lists discussed in this chapter.

Product and Material Design

- Minimize waste.
- Use recycled material and reused components.
- Use few, simple, recycled, unblended materials.
- Use nonhazardous recyclable material.
- Design for ease of disassembly.
- Make remanufacturing a possibility.
- Reuse or recycle products at the end of life.
- Make maintenance easy, especially for system-dependent products.
- Design for upgrading, repair, and recycling through good access, labeling, modules, and breakpoints, and provide good manuals.
- Make products durable when there are significant environmental aspects outside the use phase.
- Use structural features and high-quality materials to minimize weight without interfering with necessary flexibility, impact strength, or functional properties.
- Use better materials, surface treatments, or structural arrangements to protect products from dirt, corrosion, and wear.
- Use the minimum joining elements possible (a DfD principle) and use screws, adhesives, welding, snap fits, geometric locking, and so on according to DfD guidelines.
- Components that are of adequate quality can be refurbished or reused.
- Separate metallic parts into categories that increase their recycling value.
- Disassembled plastic parts should be easily removed and recycled.
- Design to reduce packaging.

Process and Manufacture

- Manufacture without producing hazardous waste.
- Do not use toxic substances, but use closed loops when it is necessary to use them.
- Use clean technologies.
- Keep product chemical emissions to a bare minimum.

Design for Minimum Energy Consumption

- Minimize energy and resource consumption in production and transportation by striving for efficiency.

- Minimize energy and resource consumption in the use phase, especially for products that have their most significant environmental aspects in the use phase.

This list will serve most product designers very well, and is recommended for most people working in product design.

4 USEFUL WEB SITES FOR DfE

Two useful sites for information about DfE can be found at the U.S. Environmental Protection Agency (U.S. EPA)[33] and the National Research Council of Canada (NRC).[34] Both are government agencies and both are encouraging industry to address environmental issues. The U.S. EPA has a long list of industrial partnerships for a large variety of industries, and NRC has a set of useful checklists for industries interested in DfE.

The U.S. EPA states the following about DfE: "The DfE program is one of EPA's premier partnership programs, working with individual industry sectors to compare and improve the performance and human health and environmental risks and costs of existing and alternative products, processes, and practices. DfE partnership projects promote integrating cleaner, cheaper, and smarter solutions into everyday business practices."[33]

NRC states that "companies that apply DFE find it:

- Reduces environmental impact of products/processes.
- Optimizes raw material consumption and energy use.
- Improves waste management/pollution prevention systems.
- Encourages good design and drives innovation.
- Cuts costs.
- Meets user needs/wants by exceeding current expectations for price, performance and quality.
- Increases product marketability."[34]

DfE can also provide a means for establishing a long-term strategic vision of a company's future products and operations. The iterative aspect of LCA is part if this.

At the NRC Website the following can be found:[34]

Life-cycle Stage 1: Production & Supply of Materials and Components

- What problems can arise in the production and supply of materials and components?
- How much, and what types of, plastic and rubber are used?
- How much, and what types of, additives are used?
- How much, and what types of, metals are used?

- How much, and what other types of, materials (glass, ceramics, etc.) are used?
- How much, and which type of, surface treatment is used?
- What is the environmental profile of the components?
- How much energy is required to transport the components and materials?

Life-cycle Stage 2: In-house Production

- What problems can arise in the production process in your own company?
- How many and what types of production processes are used (including connections, surface treatments, printing, and labeling)?
- How much and what types of auxiliary materials are needed?
- How high is the energy consumption?
- How much waste is generated?
- How many products don't meet the required quality norms?

Life-cycle Stage 3: Distribution

- What problems arise in the distribution of the product to the customer?
- What kind of transport packaging, bulk packaging, and retail packaging are used (volumes, weights, materials, reusability)?
- Which means of transport are used?
- Is transport efficiently organized?

Life-cycle Stage 4: Product Use

- What problems arise when using, operating, servicing, and repairing the product?
- How much, and what type of, energy is required, direct, or indirect?
- How much, and what kind of, consumables are needed?
- What and how much auxiliary materials and energy are required for operating, servicing, and repair?
- Can the product be disassembled by a layman?
- Are those parts often requiring replacement detachable?

Life-cycle Stage 5: Recovery and Disposal

- What problems can arise in the recovery and disposal of the product?
- How is the product currently disposed of?
- Are components or materials being reused?
- What components could be reused?
- Can the components be disassembled without damage?
- What materials are recyclable?
- Are the materials identifiable?

- Can they be detached quickly?
- Are any incompatible inks, surface treatments, or stickers used?
- Are any hazardous components easily detachable?
- Do problems occur while incinerating nonreusable product parts?

The NRC *Design for Environment Guide* is a combination of the SLCA stages and the checklists developed earlier.

5 SOFTWARE

Two software packages that have DfE components are Simapro and GaBi. These can be found at the following Web sites:

- Simapro, http://www.pre.nl/simapro/default.htm
- GaBi, http://www.gabi-software.com/

Both software packages are useful, but they take a long time to complete a full analysis and the price of these packages can be prohibitive.

6 SUMMARY

This chapter created a composite checklist for product designers who wish to apply DfE. The list is extensive, and when applied with other tools such as DfD or design for remanufacturing, a comprehensive design analysis is possible. The checklist given in the chapter is similar to those by national environmental agencies. The chapter briefly touches on two software packages with DfE components.

REFERENCES

1. J. Jeswiet, "Reasons and Objectives for Life Cycle Engineering at an Undergraduate Level," LCE 2006, 13th CIRP International Conference on Life Cycle Engineering, Leuven, Belgium, June 2006; pp. 173–176.
2. G. Boothroyd, P. Dewhurst and W. Knight, *Product Design and Manufacture for Assembly*, Marcel Dekker, 1994.
3. T. E. Graedel and B. R. Allenby, *Industrial Ecology*. Pearson Education Inc., 2003, Upper Saddle River, New Jersey.
4. S. Kalpakjian and S. Schmid, *Manufacturing Engineering and Technology*, 5th ed., Pearson Prentice-Hall, 2006, Upper Saddle River, New Jersey.
5. J. Jeswiet and M. Hauschild, "Ecodesign and Future Environmental Impacts," *Journal of Materials & Design*, **26**(7), 629–634 (2005).
6. "The Current," Canadian Broadcasting Corporation [CBC], December 28, 2005.
7. "How companies manage intangibles like brands...is tough to quantify. Until now," *Forbes* (2004), www.forbes.com, accessed July 1, 2005.
8. Australian Broadcasting Corporation radio interview, 0815, June 1, 2004.

9. T. Macalister, "Exxon Seeks to Clean up Its Image as Global Villain," *Guardian Weekly* (October 16–22, 2003), 7.
10. BP 2003 Sustainability Report.
11. "Profits and Principles, Is There a Choice?" Shell Oil advertising, The Economist (October 2003).
12. "EU automobile take back directive;" Directive 2000/53/EC, September 18, 2000, on the EOL of vehicles.
13. G. H. Brundtland, "Our Common Future," The World Commission on Environment and Development (WCED), 42nd session of UN General Assembly, August 4, 1987.
14. ASME Panel on Sustainability—panel session on November 7, 2005. ASME Winter Annual Meeting; ASME BRTD (Board Research Task Development) committee on Products and Processes for Sustainability.
15. EIC Climate Change Conference, Ottawa, May 11, 2006.
16. H. Bauman, F. Boons, and A. Bragt, "Mapping the Green Development Field, Engineering Policy and Business Perspectives," *Journal of Cleaner Production*, **10**, 409–425 (2002).
17. J. Lagerstedt, KTH, "Functional and Environmental Factors in Early Phases of Product Development—Eco Functional Matrix," Ph.D. thesis, January 24, 2003.
18. J. Jeswiet, "A Definition for Life Cycle Engineering," 36th International seminar on Manufacturing Systems. Saarbrucken Germany. Plenary Speech, pp. 17–20, June 3, 2003.
19. M. Hauschild, J. Jeswiet, and L. Alting, "From Life Cycle Assessment to Sustainable Production: Status and perspectives," *Annals of CIRP*, **54**, 70–87 (February 2005).
20. C. Gruner, F. Dannheim, and H. Birkhofer, "Integration of EMS and DFE: Current Practice and Future Trends," CIRP 6th International Seminar on Life-cycle Engineering, p. 131, June 21, 1999, ISSN 1561-9265.
21. J. Fiksel, *Design for the Environment: Creating Eco-Efficient Products and Processes*. McGraw Hill, New York, 1995.
22. P. T. Anastas and J. B. Zimmerman, "Design through the Twelve Principles of Green Engineering," *Environmental Science Technology* (2003).
23. W. Dewulf, B. Willems, and J. Duflou, "Estimating the Environmental Profile of Early Design Concepts for an Electric Juicer Using Eco-Pas Methodology," 12th International Seminar on Life Cycle Engineering, Grenoble, 2005, pp. 321–324.
24. C. A. Catania, "Biodegradable Plastics for Hygienic Disposable Products. *Design and Manufacture for Sustainable Development 2003*," The University of Cambridge, p. 278, September 3, 2003, ISBN1 86058 427 6.
25. B. Hill, "Industry's Integration of Environmental Product Design," 1993 IEEE International Symposium on Electronics and the Environment.
26. C. Luttropp and R. Zust, "Ecoeffective Products from a Holistic Point of View," 5th International CIRP Seminar on Life-cycle Engineering, Stockholm, Sweden, 1998, p. 105.
27. J. Kaldjian, "Ecological Design: Source Reduction, Recycling, and LCA. *Innovation*, **11**(3): 11–13, 1992.
28. VDI 2243, "Konstruieren Recyclingerechter Technischer Produkte," *Dusseldorf*, VDI (1991).
29. F. Jovane, L. Alting, A. Armillotta, W. Eversheim, K. Feldmann, G. Seliger, and N. Roth, "A Key Issue in Product Life Cycle: Disassembly," *Annals of CIRP*, **42**(2), 651(1993).

30. H. Wenzel, M. Z. Hauschild, and L. Alting, "Environmental Assessment of Products," Vol. 1, *Methodology, Tools, Techniques and Case Studies*, Chapman & Hall, United Kingdom, Kluwer Academic Publishers, Hingham, MA, 1997.
31. T. E. Graedel, *Streamlined Life Cycle Assessment*, Bell Laboratories, Lucent Technologies, Prentice-Hall, 1998.
32. T. E. Graedel and B. R. Allenby, *Industrial Ecology*, AT&T, published by Prentice Hall, NJ, 2003.
33. U.S. EPA (U.S. Environmental Protection agency) Web site: http://www.epa.gov/dfe/.
34. NRC (National Research Council of Canada) Web site: http://dfe-sce.nrc-cnrc.gc.ca/overview/benefits_e.html.

Other Useful Web Sites for DfE

√ University of Washington Web site: http://faculty.washington.edu/cooperjs/.
√ TURI (Toxic Use reduction Institute) Web site: http://www.turi.org/content/content/view/full/1230/.

Additional Reading

B. R. Allenby and A. Fullerton, "Design for Environment: A New Strategy for Environmental Management," *Pollution Prevention Review*, pp. 51–61 (Winter 1991–92).

T. A. Bhamra, S. Evans, T. C. McAloone, M. Simon, S. Poole and A. Sweatman, "Integrating Environmental Decisions into the Product Development Process: Part 1—The Early Stages," in *Ecodesign '99. 1st International Symposium on Environmentally Conscious Design and Inverse Manufacturing* conference proceedings, Waseda University International Conference Center, Tokyo, Japan, 1999.

H. Brezet, T. van der Horst, H. te Riele, G. Duijf, S. Haffmans, H. Böttcher, S. de Hoo, A. Zweers, and H. Verkooyen, "Guideline for Environmental Product Design," SDU Uitgeverij, 's-Gravenhave, The Netherlands (in Dutch, later revised and issued in English as: H. Brezet and C. van Hemel, (principal editors and authors), Ecodesign, a promising approach to sustainable production and consumption, UNEP, Paris, 1997).

M. Z. Hauschild, and H. Wenzel, "Environmental assessment of products," Vol. 2, *Scientific Background* Chapman & Hall, United Kingdom, Kluwer Academic Publishers, Hingham, MA, 1998.

CHAPTER 3

ORGANIZATION, MANAGEMENT, AND IMPROVEMENT OF MANUFACTURING SYSTEMS

Keith M. Gardiner
Lehigh University, Bethlehem, Pennsylvania

1	INTRODUCTION: WHAT IS THIS CHAPTER ABOUT?	46	7	IMPROVEMENT, PROBLEM SOLVING, AND SYSTEMS DESIGN: *AN ALL-EMBRACING RECYCLING, REPEATING, SPIRALING CREATIVE PROCESS* 59
2	THE NATURE OF THE MANUFACTURING SYSTEM: WHAT IS THE ARENA FOR OUR IMPROVEMENT?	47	8	WORKFORCE CONSIDERATIONS: *SOCIAL ENGINEERING, THE DIFFICULT PART* 61
3	THE EVOLUTION OF LEADERSHIP AND MANAGEMENT: THE HANDICAP OF HIERARCHIES	49	9	ENVIRONMENTAL CONSCIOUSNESS: *MANUFACTURING EMBEDDED IN THE SOCIETY* 64
4	ORGANIZATIONAL BEHAVIORS, CHANGE, AND SPORTS: A FRUITLESS QUEST FOR STABILITY	52		9.1 Sustainability 64 9.2 Principles for Environmentally Conscious Design 66
5	A SYSTEM OF MEASUREMENT AND ORGANIZATION: STIMULATING CHANGE	54	10	IMPLEMENTATION: *CONSIDERATIONS AND EXAMPLES FOR COMPANIES OF ALL SIZES* 67
6	COMPONENTS OF THE MANUFACTURING SYSTEM: A SIMPLIFIED WAY OF LOOKING AT THE SYSTEM	57		10.1 Vertical Integration 67 10.2 Real-World Examples 68 10.3 Education Programs 70 10.4 Measuring Results 71
			11	A LOOK TO THE FUTURE 72

1 INTRODUCTION: WHAT IS THIS CHAPTER ABOUT?

There are many books, pricey consultants, guides, expensive courses, and magazine articles telling us how to improve. Improvers tell us how to do everything from diet, exercise, staying healthy, relaxing, sleeping, investing, fixing our homes, and growing vegetables to bringing up our children—there are recommended fixes available for every human condition! This trend is nowhere more prevalent than in business and industry, and most especially in manufacturing. The challenge for this chapter is to deliver meaningful content that, if applied diligently, will enable readers to improve their manufacturing systems.

We must go beyond the acronyms and buzzwords, and here there are strong parallels with self-improvement. To be successful, self-improvement and a diet or exercise regime first requires admission, recognition, and consciousness of the necessity for improvement. The next step required is to realize that improvement is possible; then there must be a willingness and eager enthusiasm to meet the challenges and commence the task, or tasks; this can be very difficult. It is too easy for managers, or erstwhile change-agents, to place placards by the coffee and soda machines and in the cafeteria with messages like "Learn today and be here tomorrow." Inspirational posters, T-shirts, and baseball caps with logos and slogans are often made available as promotional incentives. This is ignorant folly, and can rapidly turn any improvement project into a cliché and workplace joke.

A leading slogan (maybe some slogans are unavoidable) is continuous improvement. Here the models from sports or the arts are appropriate. Athletes and musicians practice, learn, and train, almost as a way of life. Similar approaches and habits must be introduced to the manufacturing regime. Here, management must lead by example and act as coaches, while at the same time accepting that they also must be engaged in continuing endeavors to improve. Commitment and enthusiasm of management, accompanied by visible participation, are essential. In fact, no improvement initiative should be launched without a prior thoroughgoing and preferably independent objective analysis to assess the morale of the whole operation, or enterprise. Incorrect assumptions by leadership will result in poor planning, possibly inappropriate emphasis and ineffective implementation. As a consequence there could be negative effects on workplace morale, and the initiative could be destined for failure.

Beyond this it is wise to recognize that any initiative will inevitably have a life cycle.[1] Thus, planning and implementation must be very careful and deliberate. Initiatives of this nature should not be considered as *once and done*. There must be long-range plans for continuation, revitalization, and refreshment. To be successful, the improvement initiative(s) must become embedded into the culture and practices of the enterprise. It must become a habit, and resources must be allocated to support successful implementation and on-going maintenance.

Improvement can be an abstract notion, but any improvement must be accompanied by a thorough analysis and understanding of exactly what is to be improved. An athlete has many performance metrics, such as resting pulse, heart and lung capacities, treadmill and weight performances, times for standard tests and ultimately, of course, competitive results. Practice and training regimens are developed to focus on areas of weakness and to develop greater capabilities in zones of opportunity. Time is spent in counseling, measuring, and planning with development of very specific exercises on a continuing basis. It is rare to discover this kind of detailed attention being paid to the improvement of individuals, teams, or their performance in manufacturing enterprises.

2 THE NATURE OF THE MANUFACTURING SYSTEM: WHAT IS THE ARENA FOR OUR IMPROVEMENT?

Systems for manufacture, or production, have evolved appreciably in the last 4,000 or so years. The achievements of the Egyptians, Persians, Greeks, Romans, and others must not be ignored. They were able to leave us countless superbly manufactured artifacts and equip their military as efficient conquerors. It is interesting and worthwhile to define the production or manufacturing system in this context. Our system can be viewed as "a system whereby resources (including materials and energy) are transformed to produce goods (and/or services) with generation of wealth."[2] Our current systems, recent developments, and, particularly, our prejudices can be best appreciated and understood by taking a brief glance back in time to review the nature, management, and characteristics of some of these early production systems.

Most early systems were directed and under the control of local rulers. In many locations these pharaohs, princes, chieftains or tribal leaders levied taxes for defense and other purposes of state and also to support their military, social, and manufacturing systems. In Europe after the fall of the Roman Empire, a distributed regional, state, or manorial system arose that was hierarchical. The local earls, dukes, princes, or lords of the manor owed allegiance and paid taxes to the next levels, the church, and/or threatening despots. This manorial system relied on a tiered dependent and subservient vassal or peasant society. The manor, district, or local manager (or seigneur) gave protection and loans of land to the vassals proportional to perceptions of their contribution to the unit.[3] Products required for daily living, agriculture, clothing, food, meat, and fuel were produced as ordered, assuming weather and other conditions were satisfactory.

Major large-scale projects to meet architectural, marine, defense, societal, and funereal purposes (harbors, fortifications, aqueducts, and memorial structures) involved substantial mobilization of resources and possibly the use of slaves captured in wars. Smaller artifacts were made by single artisans, or by small groups working collectively; agricultural production was also relatively small scale and primarily for local markets. In these early days the idea of *an enterprise* was synonymous with the city, or city-state itself. When the armies needed equipment, swords, and armor, orders were posted and groups of artisans worked to fill them. Organization during these periods was hierarchical and devolved around the state and a ruling class. Religion also played a major role in structuring the lives of the populace.

The artisan groups organized themselves into guilds establishing standards for their craft, together with differentiation, fellowship, and support for those admitted to full membership. There was training for apprentices and aid for widows and orphans when a member died. Guilds participated actively in the religious life of the community, built almshouses, and did charitable works.[4] It can be surmised that guild leaders of the miners in Saxony, for example, would have the power, experience, and qualifications to negotiate working conditions with the lord of the manor or leader of the principality and mine owner. The guild would also claim some share in the revenues of the mining and metal winning operations. Mining and manufacturing operations in Saxony were described extensively in *De Re Metallica*, a notable text by Agricola in 1556 translated into English by the Hoovers.[5]

The guild workplaces, mines, smelters, waterwheel-powered forges, hammers (described by Agricola), grist mills, and the like were the early factories. The existence of a water-powered paper mill in England is recorded as early as 1494. The printing operations of Gutenberg in what was to become Germany, and of Caxton in England in 1454 and 1474, respectively, were small factories. Early armorers must have worked in groups supported by cupolas, furnaces, hearths, and power systems. A most renowned early factory was the Arsenale (arsenal) in Venice. This was a dockyard operated by the city–state that opened around the eighth century, with major new structures (Arsenale Nuovo) started in 1320. At its height in the sixteenth century, the arsenal was capable of producing one ship per day using an assembly line with mass-production methods, prefabrication of standardized parts, division of labor, and specialization.[6] Power sources during these periods were limited to levers, winches, and cranes driven by human or animal power, wind, or water. To a large extent these systems were reasonably sustainable, but were vulnerable to unpredictable social, climatic, or other disasters.

During the period marked as the Industrial Revolution, available power densities increased markedly. Improvements in engineering and materials increased the efficiency and size of waterwheels and their associated transmission systems. There is a tendency, certainly in the United Kingdom and United States, to mark the improvement of the steam engine by Boulton and Watt, and the discussions of the Lunar Society, as the inception of the Industrial Revolution.[7] In fact, effective production systems were already extant and evolving as the result of global influences. The scale and scope increased as result of this major change in available power density. Factories grew up around sources of power, materials, and potential employees.

3 THE EVOLUTION OF LEADERSHIP AND MANAGEMENT: THE HANDICAP OF HIERARCHIES

History has given us effective models for the organization of our manufacturing systems. The notion of the paid worker as a vassal has tended to predominate, notwithstanding the wise thoughts of Adam Smith,[8] predating W. Edwards Deming by almost 200 years.[9] He expressed the need for the workforce to be positively integrated as a factor engaged in the furtherance of the objectives of the manufacturing system as follows:

> *But what improves the circumstances of the greater part can never be regarded as an inconvenience to the whole. No society can surely be flourishing and happy, of which the far greater part of the members are poor and miserable. It is but equity, besides, that they who feed, clothe, and lodge the whole body of the people, should have such a share of the produce of their own labor as to be themselves tolerably well fed, clothed, and lodged. ... The liberal reward of labor, as it encourages the propagation, so it increases the industry of the common people. The wages of labor are the encouragement of industry, which, like every other human quality, improves in proportion to the encouragement it receives. A plentiful subsistence increases the bodily strength of the laborer, and the comfortable hope of bettering his condition, and of ending his days perhaps in ease and plenty, animates him to exert that strength to the utmost. Where wages are high, accordingly, we shall always find the workmen more active, diligent, and expeditious than where they are low.*[9]

It is clear that an understanding of physical, economic, social, organizational, and behavioral processes are an important aspect for the whole manufacturing or production enterprise.

And, of course, if we combed the words of Machiavelli in *The Prince,* or Sun Tzu, *The Art of War,* we would find that the idea of treating workers with care and respect is not original.[10,11] Management, to be effective, must also comprise leadership. Frederick Taylor in his work on *The Principles of Scientific Management,* brought important attention to the importance of *managing the numbers* but also took care to mention that the workers should earn a share of the prosperity resulting from improving the efficiency of their labors.[12] Henry Ford is remembered for his drive for the efficiencies of mass production, and his groundbreaking $5-a-day announcement in 1914 that aimed to enable his employees to acquire their own vehicles.[13] The worst and—unfortunately—most remembered aspects of using a moving production line and managing the numbers were first described graphically in 1906 by Upton Sinclair in his book *The Jungle,* about the meat-packing industry.[14] Hounshell's work *From the American System to Mass Production 1800–1932* provides an excellent account of the development of these early manufacturing systems.[15]

The styles of management that developed fertilized the growth of the union movement and an inimical separation between workers and management. The unions did to some extent follow the pattern of the earlier guilds in providing qualification metrics and welfare for their members, but a principal role was as negotiators with management. A further unfortunate consequence was a proliferation of job descriptions that later inhibited cross-training, job sharing, and worker transfer. The leadership and management of any enterprise wishing to succeed must take note of the historical and linguistic baggage accompanying the words like *management* and *workers* and develop alternatives. Today, *associate* is a popular synonym for employee or worker.

In the second half of the last century a majority of the U.S. workforce enjoyed tremendous prosperity by comparison with workers in war-ravaged Europe and Asia. Nevertheless, there were strikes, hard negotiations, and, more latterly, waves of downsizings and reengineering causing lost jobs as foreign competitors grew more aggressive. However, the economy was generally robust, and some current opinions suggest that U.S. consumers were held to ransom as both management and their workforce gained large pay and benefit packages. This was sustainable when the United States possessed a quasi-island economy, importing and exporting almost at will and with a positive balance of trade. As the economies, productivity, efficiency, and manufacturing prowess of competitor nations grew, conditions became arduous. Now major union tasks are to negotiate reduced salaries and lay-off conditions as their membership has declined from 20 percent of the workforce in 1983 to 12.5 percent, or 15.7 million members in 2005.[16] It is likely that union affiliations and power will continue to decrease. More workers are being *empowered,* and given opportunities to become increasingly multiskilled. Additionally, the vision of lifelong employment—doing one task serving one enterprise—has faded as a result of the need for greater flexibility and responsiveness.

Traditionally, we are accustomed to large hierarchical operations with relatively specialized division of labor, and aggregation into functional groups for purposes of command, communication, control, and planning. These organizational structures took advantage of ideas of process simplification and the division of labor. They enabled effective production and had minimal requirements for development of the skill base of the employees. In a general sense, the skills were embedded in the tooling, and in the fitters who set up the tools. This system was, perhaps, far from optimum, but based on the theories of the time, skills available, social needs, and economics, it generated a reasonable level of prosperity. In a comparative sense, the long era of this style of mass production brought higher levels of wealth and prosperity to many more people and societies than any previous system.[15]

In the latter part of the twentieth century it became obvious that large hierarchical structures were a great hindrance to decision processes. There are many conflicts and appreciable difficulties in handling innovative ideas. Certain modifications were adapted from the military practice of creating special task forces, or teams with specific focused missions, operating outside the traditional reporting structures and management envelope. The "success" of task forces led to the adoption of many variations of matrix structures, disposing employees from differing functional groupings into project- or program-focused teams. These matrix methods are contrasted with functional groupings in numerous treatises dealing with management. Groover provides a succinct analysis of matrix management methods in a recent text on work systems.[17]

Most large organizations are unavoidably dyslexic; they become bureaucratic and fossilized. Any organization eventually develops to preserve forms, stabilize activities, and provide secure protocols for our interpersonal behavior. Organizations of their nature inhibit change and restrict the development of ideas leading to continuous improvement. To be successful in the future, organizations must be structured with a recognition of the ineluctable life cycle of inception, growth, and maturation, with a, perhaps, evanescent stability preceding the inevitable decline. A similar cycle is shared by every process, product, and individual associated with an enterprise, although with varying time constants. Organizations must be structured (and restructured) with a facility to accept and adapt to continuous and often unpredictable change.[1] Fresh paradigms must be evaluated and welcomed continually. There is need to create a pervasive awareness that stability is unwelcome.

In developing our ideal organization structure that is accepting of change and improvement, it must be recognized that the success of the earlier hierarchical pyramids was associated to a great extent with the co-location of individuals with similar affinities. Cross-disciplinary or matrixed cross-functional teams are a wonderful idea, but it is important to recognize that few individuals choose their career paths and disciplines by accident. These choices are related to their own social or psychological attributes. The most successful individuals, it can be assumed, are those who attain the closest match between their internal psyche's and their professional activity. For example, there are appreciable differences in the communication and perceptual skills of many electrical and mechanical engineers. Such contrasts and potentials for conflict and team disruption become even greater as the needs of a team call for involvement from additional disciplines, such as accounting, economics, ergonomics, finance, industrial design, manufacturing engineering, marketing, materials management, safety, waste management, and the like. These interpersonal factors are exacerbated when different divisions of any large enterprise must collaborate, or when international cultures are represented. All individuals have differing interpretations of the world, their own responsibilities to the enterprise, and to the project at hand. The integration, management, and leadership of diverse multifunction teams require skills equal to those of the best counselors and therapists.[18]

4 ORGANIZATIONAL BEHAVIORS, CHANGE, AND SPORTS: A FRUITLESS QUEST FOR STABILITY

It seems implicit in the human psyche that we assume tomorrow will be a close approximation of our "ordinary" yesterday. Both as individuals and as groups in organizations, we assume that "if only we can get over this workload hump, or this crisis, and past the next checkpoint and deadline, then we will enter a domain of calm and a plateau of stability." In the main, our organization structures, measurements, and expectations are based on this idea that stability is an attainable, and virtuous, state. In the affairs of man this is patently untrue. At no time has history been free of change, and of concerns for the unstable future. Explaining and forecasting this future occupies many economists. Kondratieff produced his ideas of waves following innovations or major changes in 1924. Joseph Schumpeter followed in 1942 with the idea of the *creative destruction* of older systems as result of new methods or technologies.

Notwithstanding these ideas about change, it is clear that from the earliest of times the human race has endeavored to organize itself to achieve surprise-free environments. We tend to gravitate to those groups that we know; where we will be safe, sheltered, understood, and free of surprises. In general, both individuals and organizations shun change. Enterprises create organizations to prosecute their objectives and to advance their interests. Every organization, if it embodies more than a few people is compelled to develop bureaucratic structures to handle routine matters uniformly and expeditiously. Organizations of their nature strive to create surprise-free environments for their customers and employees. Thus, we see that people and the organizations in which they arrange themselves are highly change resistant.[1]

Studies exist that demonstrate extraordinary productivity results when people are placed in self-managed teams with significant challenges in highly constrained environments. An idea and personnel are isolated and left alone and brilliance emerges, notwithstanding an awful environment and severe constraints. This has been called a *mushroom effect* because spores, or ideas, are left in a dark corner on a pile of metaphorical horse manure and almost forgotten. There is substantial literature relating tales of bandit or pirate operations working against impossible deadlines with minimal resources, thereby becoming extraordinarily motivated and sometimes flouting the expectations of a mature parent organization. Stories of the success of small entrepreneurial endeavors abound, but there are many failures. Some of these projects are poorly structured but, nevertheless, succeeded as result of the personalities of the leaders. Memorable examples have been excellently described by Kidder in *The Soul of a New Machine*, a book about the development of a new Data General computer model, and Guterl with his Apple Macintosh design case history.[19,20]

Many enterprises recognize that major improvements, such as accelerated new product development and introduction, require a different organization. They attempt to accomplish this by embedding specially assembled project groups within an existing but already archaic hierarchic framework. The transfer, or loan, of individuals with special skills into special quality circle task forces, early manufacturing involvement (EMI), or concurrent engineering teams is often an effective solution to overcome the dyslexic characteristics of an historic organization structure. However, it can be postulated that any success may be wholly due to the close attention that "special" projects receive from senior executives, and is likely to be transient. It is difficult to evolve special teams into an ongoing search for continual improvement. It can be observed that these special high-profile teams lose their adrenalin fairly rapidly, and a string of *me-too results* follows. Ideally any major changes, new processes, or new product developments should be accompanied by a reconfigured organization. Special measures and personnel rotations are needed to ensure refreshment, revitalization, continual organizational evolution, and renewal.

When we compare practices in the arts and sports with those of industry, we can see many parallels. Clearly extraordinary performance can be generated by organizations that may be perceived as almost anarchist in character (c.f., jazz groups). However, some form is detectable by the team members. Many leaders talk of teams and imply analogies with sports activities; others use the arts, and Drucker speaks of *orchestral management*. [21,22] In many team sports the emphasis is often placed on moving a ball effectively. Aficionados of each different sport know exactly what is effective in their context. In most cases, the specialties of the players rest on either particular hitting skills or handling skills. In some cases, there are special positions on the field or pitch with a subsidiary requirement for either hitting or delivery. For the handlers, delivery becomes everything. They specialize; they practice; they examine every move in slow motion; they visit psychologists, chiropractors, and frequently specialist surgeons to improve and maintain their skills. They are rested, rotated and measured with great refinement. Their rewards are public record; and they are accorded the esteem of their peers. Even with the *star* systems, most individuals and their management recognize the interdependencies of an effective team.

In team sports that do not involve a ball or puck, the measure of final excellence or speed may be easier, but integration of the individuals can be more difficult. Rowing, for example, requires great individual ability; but this is worthless in a four or eight, unless the output of the whole team is synchronous. The bobsled event looks like the application of brute force with pure gravity, and the margins are remarkably tight. To the nonexpert, the contest results almost appear random. However, there is a regularity and consistence expressed in hundredths of seconds that demonstrates the excellence of the best teams. Measurements for attaining team excellence are demonstrably much more than just the assembly of the fastest pushers. The ability to think and act with one's fellows and get onto the sled at the last possible moment also play a great part, and cannot be measured by singular tests. However, the measure of integrated team performance is conclusive.

5 A SYSTEM OF MEASUREMENT AND ORGANIZATION: STIMULATING CHANGE

Building on the sports analogy, an enterprise wishing to improve must consider itself as engaged in some cosmic league of global proportion. Although continuous improvement and high productivity are abstract concepts, they must be understood and defined in the context of the organization seeking to excel. There must be benchmarks; some "stake in the ground" must be established. A product cycle can be judged against historic comparisons or competitive benchmarks, and the time to initial generation of profits can be contrasted with earlier products. A higher-productivity product cycle will reach the breakeven point faster, and with less trauma within the organization. Institutional learning or human resource development should be an additional measure, as this has strong correlation with future prosperity.

Clearly, customers, shareholders, employees, and other stakeholders are continually measuring the attributes of the enterprise with which they are involved. The sum of these measures could be said to be the value placed on the enterprise by both the engaged communities and by the stock market. This aggregate value is a composite measure of management competence, adherence to targets, efficiency of resource utilization, customer satisfaction, and product/process elegance. *Elegance* is a subjective measure that could be assessed from reviews of industry consultants, or *experts*. It may also be inferred from customer experiences, warranty claims, life cycle costs, and level of engineering change orders, or equivalent measures in service industries. The successful implementation of the Malcolm Baldrige awards shows that it is possible to make useful measurements of intangibles in business environments.[23] Such measures can readily be adapted for individuals and teams, as well as organizations. Criteria for the Malcolm Baldrige awards are presented in Figure 1.

Once there is a measure of the enterprise, it is relatively simple to decompose this and abstract a measure for every division, site, or department in the organization. This may well relate to long-term revenue projections, short-term profitability, or to volumes, new-product introductions, market share, or global rankings; the organization measure adopted is a strategic issue for the enterprise. Any sports team or arts group possesses some intrinsic ability to judge its standing in whichever league it chooses to play. Ultimately, this becomes a numerical tabulation and is a measurement of organizational effectiveness in competing in the chosen market. The measurement intervals used must relate to the life cycle or time constants associated with the product cycles and the overall rate of change within the industry.

Further decomposition can be undertaken to evaluate each team and the individuals therein. Individuals making contributions to several teams will carry assigned proportions from every team evaluation. Individual evaluations (and rewards) should include recognition of all contributions to each team with which

The following categories of activities are examined:

1. *Leadership* — how senior executives guide the organization, and how the organization addresses its responsibilities to the public and practices good citizenship.
2. *Strategic planning* — how the organization sets strategic directions and how it determines key action plans.
3. *Customer and market focus* — how the organization determines requirements and expectations of customers and markets; builds relationships with customers; and acquires, satisfies, and retains customers.
4. *Measurement, analysis, and knowledge management* — the management, effective use, analysis, and improvement of data and information to support key organization processes and the organization's performance management system.
5. *Human resource focus* — how the organization enables its workforce to develop its full potential and how the workforce is aligned with the organization's objectives.
6. *Process management* — aspects of how key production/delivery and support processes are designed, managed, and improved.
7. *Business results* — the organization's performance and improvement in its key business areas: customer satisfaction, financial and marketplace performance, human resources, supplier and partner performance, operational performance, and governance and social responsibility. The category also examines how the organization performs relative to competitors.

Figure 1 Malcolm Baldrige Award Criteria for Performance Excellence.

the individual was engaged. There should also be components acknowledging creativity, innovation, extraordinary contributions, an ability to integrate, and development of future potential. A valuable contribution to performance measurement can be gained by seeking reviews from the team colleagues, managers, and technical coordinators or leaders that work with the individual being assessed. There are a variety of ways to administer these 360-degree reviews and it is important that they are treated seriously and confidentially as a potential aid for improving performance. Each employee (or associate) may nominate colleagues for including in her/his survey with the concurrence of the primary supervisor. The review process must be based on data from several sources and should be dealt with one on one as a coaching session. There should be no surprises (or fear) because all contributors to a well-managed, continuously improving operation should have been encouraged to acquire superior levels of consciousness in their relationships with other team members and leadership. Measurement schemes must stimulate continuous lifelong learning and professional growth. After all, the human resources of any enterprise are avowedly the most potent and responsive resource available for enhancing quality, productivity and continuous improvement.

In larger organizations during recent decades there have been sufficient turbulence, internal rearrangement, and reorganization, with reassignments to new programs such that hardly anyone had an opportunity to attain stability. Some of this churn was not productive for the enterprise overall, although there was appreciable, often involuntary, vitality added to the careers of affected personnel. Our new evaluation processes must recognize the life cycles of the organization, teams, and individuals. Change must be deliberate and planned. It should not necessarily be assumed that any individuals should stay with a project through the whole life cycle. There should be changes on some planned matrix, relating to the performance and developing (or declining) capabilities and interests of each employee, the needs of the project, and the requirements arising elsewhere within the organization. It is essential for the prosperity and success of the enterprise that any battles for resources, headcount, and budget allocation details between different departments, functions, and divisions are dealt with swiftly so that they do not impact morale and responsiveness. Musicians and athletes change teams or move on to different activities. Similar career styles in engineering should be anticipated, encouraged, and promoted by the measurement schemes adopted in all organizations that aim for continual improvement. There is need for circumspection when there are excellent contributions by departments, teams, and individuals to projects that fail or are canceled. Clearly, some rewards may be merited, but only if there was useful learning consistent with the longer-term interests of the enterprise.

Organizational maturity implies a tendency toward a stability that can impede change and improvement. Therefore, it is essential to create measuring and management strategies that discourage the onset of maturity. There is a clear need for the stimulation and excitement occasioned by a degree of metastability. However, there is a contrasting need for security, stability, and confidence in the enterprise to enable creative individuals to interact in relatively nonthreatening environments. We are reminded of Deming's concern for the abolition of fear—this must be balanced by a strong touch of paranoia about competition, the onset of process or product obsolescence, changing technology, and other factors expressed so well by Grove.[24] There should be expanding horizons and opportunities for individuals within every section in the enterprise, accessible to all the employees. Total quality objectives, improvement, and high productivity can only be approached when all individuals gain in stature and opportunity as tasks are integrated or eliminated. In quasi-stable or service industries, there must be anticipation of new markets as resources are released by productivity improvements.

Organization structures and measurement intervals must relate directly to product/customer needs. Recognition of suitable organizational time constants is an essential concomitant to delivery of well-designed products into the marketplace, with a timely flow and continuous improvement. The management structure that is likely to evolve from the use of these types of measurement schemes will have some orchestral or sports characteristics. There will be teams, project leaders, specialists, conductors, coaches, and the inevitable front office. The relationships between different teams with alternate priorities may resemble that between chamber, woodwind, and string or jazz ensembles in our orchestra. The imposition of the rotation requirements, the time constants, will cause these almost cellular arrays to grow, modify, evolve, and shrink in organic fashion responding to the demands and pressures of an environment. The most responsive organization will accumulate skills and experience in the manner of some learning neural network, and an organization diagram may possess somewhat similar form.

6 COMPONENTS OF THE MANUFACTURING SYSTEM: A SIMPLIFIED WAY OF LOOKING AT THE SYSTEM

The manufacturing system provides concept implementation from design through realization of a product and completion of the life cycle to satisfy the customer and society. The manufacturing system can be said to exist for generating wealth in a societal sense.[2] It is useful from a design, planning, and improvement viewpoint to break down the internal aspects of the manufacturing system by contemplating the interactions of six major components: materials, process, equipment, facilities, logistics, and people. These components and their integration form the system, and their organization is affected by factors external to the system.

The manufacturing system transforms *materials* into products and consumes materiel resources such as energy in doing this. There are also waste products and eventual recycling to consider. This component embraces all physical input to the system and resulting material outputs.

Materials are transformed by a *process*; this defines chemical, physical, mechanical, and thermal conditions and rates for transformations. If properly understood, the process component is amenable to application of computer technology for sensing, feedback, modeling, interpretation, and control.

The processes require *equipment* or tooling. The equipment must possess the capability for applying process with appropriate precision on suitable volumes, or pieces, of material requiring transformation at the required rates. The equipment must be intrinsically safe, environmentally benign, and reliable. Today most equipment is electronically controlled, and there may be advantages gained by interfacing with other tools through a factory network to facilitate communications. (Feed-forward of process data can permit yield and quality enhancements in subsequent processes if they are designed to be adaptive.)

Process equipment requires an appropriate environment and services to maintain proper functionality; it may also be integrated with material-handling systems and other pieces of equipment. There are special requirements for provision of utilities, contamination control, waste management, access for materials input, and output, which must be addressed under the category *facilities*.

These components are integrated and deployed by *logistics*. The logistics comprise product, process, and systems design data; forecasts; development schedules; materials management; accounting, business, financial, marketing, and distribution arrangements; maintenance; and service, including eventual recycling requirements. This component is information-rich and of similar nature to *process*, only in a more macro sense. These are factors that are subject to change while designs are being carried out; they are also liable to suffer dramatic instabilities after the system is brought online. The logistics component comprises a most fruitful area for research and innovative strategies, which can be a significant commercial advantage over the systems of competitors. There are several notable enterprises, such as Amazon, Dell, Federal Express, Lands' End, and others, whose core competencies are primarily logistical rather than focused on technological differentiation.[25]

The whole system requires operating agents or *people*. A system is dependent on people as employees, customers, stockholders, or owners; as suppliers or subcontractors; and as stakeholders residing in communities affected by the system. There are, again, many unpredictable factors involving all aspects of human behavior.

There are significant human resource, leadership, management, recognition, and reward issues internal to the system. These also become a reflection of the expectations of the external society that accommodates the system. All people variously seek stability, with secure horizons, and shelter from turbulent times; however, in the new industrial society there can be no stability. Stability means no growth—and eventual decline. There must be pervasive quest for continuous improvement, with lifelong learning. Some social parity must be equally accessible to all who make contributions. These ideas raise questions with regard to equality of opportunities for contributing to increasingly technological endeavors. Drucker[26] postulated a population of *knowledge workers* in the 1994 Edwin L. Godkin lecture "Knowledge Work and Knowledge Society—The Social Transformations of this Century," at Harvard. His model of the future is certainly credible, and it places heavy responsibility on educational systems to equip individuals for this future. Investment in *human capital* is an essential aspect of all future planning. All these matters come down to how whole societies are organized, how expectations are developed, and to the development of concomitant reward structures. These factors have great impact on improvement efforts and productivity, and there are significant differences across different regions and cultures. The tasks of inspiring collaboration and continued workforce enthusiasm present greater problems than the tasks of acquiring and deploying available technologies.

Although the classification into six components aids the internal aspects of the design, many constraints to the processes and choices for the components and their integration derive from the relationship of the new system to its environment. These systems are not closed, and they are subject to perturbations that affect economies, social groups, nations, and continents.

7 IMPROVEMENT, PROBLEM SOLVING, AND SYSTEMS DESIGN: *AN ALL-EMBRACING RECYCLING, REPEATING, SPIRALING CREATIVE PROCESS*

When improving and reconfiguring the manufacturing system, it is advisable to have a final future vision in mind. The characteristics of a globally ideal future manufacturing system must be founded on sound principles of thermodynamics and design. Entropy conservation and minimization of trauma must be the governing rules, both for systems design and for associated organizational and social structures.[2] Significant emphasis must be given to quality of working life and conservation of resources. For such systems to prevail and be successful, environmentally acceptable, and sustainable, there must be recognition of the global commons, as espoused by Hardin,[27] the Greenpeace organization, and green enthusiasts. The systems must aim to be environmentally benign while providing useful products that satisfy human needs and solve human problems, meanwhile affording employment with wealth generation for the host communities and all stakeholders. To meet the competition, the systems must be able to handle frequently changing customer needs. This calls for fast design cycles, minimum inventories, and short cycle times to afford maximum flexibility and responsiveness at least cost.

The improvement, problem solving, and design activity must recognize responsibility for the whole system whereby a design is to be realized; design is holistic and must be total. It is not reasonable to design or improve products, or processes, independently of the system for realization and eventual revenue generation. Equally so, the whole manufacturing system must be consciously integrated with the needs of both the enterprise, customers, and host communities. It should be noted that few products are everlasting, and neither are the organizations that strive to produce them. Organizations and their structures must be designed so that they adapt with comparable life cycles to the products that they aim to generate.

Any improvement program must be regarded as a pervasive activity embracing such divisions of labor as research, development, process planning, manufacturing, assembly, packaging, distribution, marketing, and so on, and including an appreciation for integrating the activity with the whole environment in which it will be implemented. There must be a thorough consciousness of all the likely interactions, both internal and external to the enterprise. Because this can become such a vast activity, it becomes a problem how it may be best organized for outcomes with the least trauma (and delay). There are now systems and software available for life-cycle management (LCM); these require large-capacity servers that may be beyond the ambitions of smaller enterprises. However, smaller business operations can now 'rent' access to powerful servers and appropriate software from major corporate subcontractors. Collaborations in this field between industry, universities, and research centers are being stimulated through governmentally funded centers.[28]

Designing any improvement activity must involve planning, interpretation of needs, assessment and ordering of priorities, and a definition and selection from choices. There are measurements, and some degree of organization of resources is implicit. Design is an art of selecting and integrating resources using diverse tactics to address problems with consciously optimized degrees of success. It is important to gain a full appreciation of the problem; today this is emphasized as paying attention to the needs of the customer. It should be noted that future manufacturing systems will have many customers—and not just those purchasing the products that are generated. To some extent, all those involved with and affected by the systems should be regarded as customers who must be satisfied. There are both internal and external customers, the next worker down the line, or the assembly operations across the country or ocean, and then the end-use purchaser. This is consistent with the most recent thrusts emphasizing quality. The measurements of success may be objective (such as revenue or profit, increased throughput, higher quality), or subjective (such as elegance). In general, history shows that elegant but simple and economical solutions will deliver satisfaction.

By considering customers and the problem definition or statement of requirements, possible measurement strategies can be derived. The idea of success affords the converse opportunity of failure and implies a gradation of performance levels. Problems can be defined (however metaphysical and obscure), and the level of success can be estimated. The measurements may be totally subjective, absolutely commercial, or physical, like durability, size, weight, and so on. Nevertheless, once there is a specified problem and an indication of measurements for a successful solution, then problem-solving design activities can proceed. Brainstorming through this matrix will result in an improved problem definition and a superior measurement structure. Later, this measurement structure will support the evolution of organizational and administrative arrangements, resource allocation, scheduling, and so on. There are a wide range of quality tools that may be applied to aid the analysis. Most of these have greatest value for mediating discussions and interactions among the improvement team. For smaller-scale matters, Pareto plots and Ishikawa, or *fishbone* diagrams, may be sufficient.[9,23] Then at a more complex level *quality function deployment* (QFD) (otherwise known as *house of quality*) techniques can be valuable, or Taguchi principles can be applied to analyze *system noise*.[9,23] *Smoothness* and placid but rapid flow without churn are good indicators of solutions that are likely to lead to success.

At this point, it is important to gain an overall appreciation for the whole environment in which the problem of improvement is being addressed. The environment can be considered as that which cannot be changed. It is there, it is unavoidable, and it must be recognized and dealt with while undertaking the design. Next, there must be an appraisal of the schedule and resources, both materiel and personnel. These are all primary regulators of the quality levels attainable for the results of the project and have a substantial impact on final costs and eventual consequences. Meticulous attention to early organizational details, responsibilities, and communications ensures cost-efficient decisions as a project accelerates and as the rates of investment and sensitivities to risks increase. These considerations are not necessarily prescriptive, serial, or sequential, and many may be revisited repeatedly as the project progresses. The closest analogy for these procedures is to the helical design of the chambered nautilus. There is spiraling recycling of problem-solving processes with continual accretion, growth, and accumulation of learning as the final solution is approached.

Design or problem solving is like a journey and, just as there are many adequate alternate routes to a destination, there is a possibility of many different improvements or implementation schemes of equivalent merit. If the measurement system does not give a clear answer, then some measurement at a deeper level should be developed. The measurements should be as unambiguous as possible, or there is danger to the morale of the team. There must be single choices for serious focus and further development, or the explosion of options becomes too large to handle. Something that can be of assistance here is a search for analogs, a comparison with other systems, benchmarking, and analysis of the competition. Aspects that may be intangible from a viewpoint of strict functionality can have valuable impact in terms of brand, or corporate identification, ease of customer association, and so on. Such factors are all part of the team responsibility, and ultimately they can be measured, although subjectively. The process must proceed simultaneously (concurrently) with the development of any necessary changes in manufacturing systems infrastructure. Additionally, designers should contemplate a risk analysis with best- and worst-case scenarios to cover either phenomenal success at start-up or utter failure. It is also wise to assess market volatilities and dependencies on unforeseen influences and competitive responses. These can range, for example, from drastic economic shifts due to oil embargoes and energy crises through environmental regulation changes eliminating materials and processes, which could damage the ozone layer. The marketplace has many vagaries that must be considered. The life cycle and eventual retirement/reclamation of the product/system must also be considered as part of the project responsibility.

8 WORKFORCE CONSIDERATIONS: *SOCIAL ENGINEERING, THE DIFFICULT PART*

A key requirement of a system that desires continuous improvement is that the workforce is empowered and capable—that is, encouraged by continuous learning and with adaptability and enthusiasm for change. Here the ideas of Deming and other quality experts are indispensable.[9,23] The ideas of total quality management (TQM), quality circles, and self-directed teams are valuable tools for operational improvement. In the case of Six Sigma implementation, it is a requirement that the procedures are learned and introduced by senior management. Each management level is required to be fully engaged and to collaborate in the training of the workforce. Several levels of accomplishment and attainment are recognized by colored belts (a judo analogy).

Empowerment has many faces and can be very threatening to established bureaucracies. Teams must be empowered if they are to operate effectively, and they must be accorded authority to match the responsibilities of the problem(s) they have selected or been assigned. It is therefore mandatory to provide appropriate education and training opportunities for the whole workforce to ensure development of capabilities matching these responsibilities. Thoroughgoing empowerment should flatten organizations, eliminating many levels of management. Faster decision making, and on the spot improvements enhance enthusiasm and participation of the workforce. Success engenders success. Increasing efficiencies are accompanied by quality improvements, reduced costs, improved throughput and higher output. There are thereby opportunities for reducing manpower while maintaining steady production levels, alternatively pricing structures can be changed aiming to deliver greater volumes and increase market share. Ideally, surplus manpower could be diverted to the development and introduction of new products. More customarily we hear phrases like *down-sizing* or *right-sizing*, and *reengineering*. The improvement ideas just given are as attractive as they are simple, but thorough implementation is a severe test for the leadership and management of any organization that genuinely wants to promote change and improvement. To maintain these ideas requires the dedication and concentration usually reserved for sports teams and their coaches. In sports there is continual measurement, evaluation, rotation, and renewals. This will undoubtedly apply in successful industries in the future, and societies may well be compelled to adapt to continually changing and impermanent employment prospects.

There are two main motivations for workforce reduction, the initial one being cost reduction and the other being productivity improvements. Less labor reduces total costs, but recent analysis shows that direct labor costs rarely exceed 10 percent of the cost of manufacturing. Thus, any reduction in labor costs has only fractional impact; this measure similarly implies that relocation to countries with cheaper labor may be a false economy. The other, more persuasive reasons for involving fewer people all relate to productivity: quality, throughput, cycle times, flexibility, and responsiveness. The only work that should be done is that which adds value for the customer or ensures learning and future process improvement. When value engineering or reengineering is applied to customary organizations, it is found that there are many people keeping each other busy performing obsolescent tasks. Although technology is occasionally misapplied or is overly sophisticated, when used intelligently it enables workforce reduction and productivity improvements that afford cost-performance benefits. There are also opportunities to improve the quality of working life as a corollary.

As the workforce is reduced, the remaining employees become capable of working better, faster, and more effectively as tasks are redefined. The rewards and recognition for tasks well done are easier to determine. The elimination of layers of management, supervisors, and other titles boosts morale and improves communications, team spirit, and the rate and quality of results. However, depending on the levels of threat that created the need for change, it may take some time and care to raise morale. Simultaneously, individual responsibility levels and stress may increase as the nature of work and contribution in the workplace change totally. Workers deserve a level of security, support, and confidence if they are to take risks in making decisions for themselves. Further education and training becomes a continuing life-long requirement. Often, the entire involvement with work may be forced to become a stressful integral part of the life of each individual. As in preindustrial times, the needs of society for production become inseparable from the life in the community, although machines and the intensity of international competition dictate what may be a less optional and crueler pace.

It is easiest to introduce new systems of these types as response to severe external threats, or in times of great and imposed changes. There is instability, fear, and paranoia and, hence, a great appetite for new solutions. However, this still requires leadership, courage, and determination. Additionally, threats of bankruptcy, or similar trauma, for example, may make conservation of resources a more urgent priority. Thus, it may be more vital to deal with the threat immediately to gain breathing space for later improvements and reorganizations. Gerstner gives an excellent account of his early days at IBM when securing some stability was more important than spending time developing a vision.[29] Nevertheless, his confident actions suggest that he did have his own nascent if selfish vision! An entirely new organization and approach with a clean slate stands the best chance for survival and prosperity. Due to prevailing conditions, Gerstner was compelled to transform the IBM organization and restore profitability somewhat stealthily.

Several methods have been described for developing team spirit and breaking down disciplinary and professional barriers that inhibit effective integration.[30] The General Motors Saturn Project broke new ground in organization structure and development of employee commitment.[31] The Project started with a small, carefully selected cadre of employees, future associates, representing all levels of workers. This team was given responsibility of working with the architects and factory designers to resolve issues of equipment and facilities layout, break area, and restroom locations. They were given coaching sessions to introduce architectural concepts and notions of scale and space, safety and OSHA requirements etc. so everyone was on the same page. The initial team was then charged with interviewing and recommending hiring of their future colleagues. When Saturn vehicles were first introduced, they were a well-received and successful team-based product. There were many new and different organizational features that carried through from design right onto the floor of the showroom. Customer satisfaction was high, for a time sales were limited by factory capacity problems, and expansion was debated. Subsequently, and most unfortunately, this grand idealistic experiment was found to be unaffordable, and Saturn was folded back into the GM organization.

9 ENVIRONMENTAL CONSCIOUSNESS: *MANUFACTURING EMBEDDED IN THE SOCIETY*

It is clear that manufacturing has changed profoundly in the last few decades. Formerly manufacturing more closely approached the model developed from the blacksmith pounding hot metals on an anvil. Ultimately, these and the accompanying assembly processes were automated and became the mass production manufacturing lines of the last century. Now it is reasonable to regard manufacturing more holistically. The success of the enterprise requires that management satisfy the demands of the marketplace and meet the competition. These objectives no longer mean simply the generation of revenues, profits, minimized costs, and a stream of new products, with ever better and more comprehensive services to satisfy the customers (and stockholders). They also require sustainability with respect to the environment. Both require respect for the long-term needs of the global community. A manufacturing system can be defined as a means of transforming resources, including materials and energy, into products and/or services to satisfy customer needs and concurrently generate wealth for society without trauma and waste.[2]

9.1 Sustainability

The definition of *sustain* is *to keep in existence; maintain*.[32] Today, increasing attention is being paid to the whole concept of *sustainability*. In earlier times there was little thought given to the ideas of life-cycle engineering. Products were designed for an estimated, or forecast, life span, and there was no consideration of subsequent retirement, reclamation, recycling, or eventual disposal. The idea that product design must cater for the whole life cycle of any product, from the concept out to end-of-life and safe disposal, or reconfiguration for further use, is relatively new. This idea also encompasses the safe disposal and/or reuse of process residues, and elimination, or significant reduction of emissions into the atmosphere, or into the environment, ground and water. In a manufacturing systems context, sustainability has the implication that any system should be designed and implemented according to certain principles with no negative effects, present or future.[2] Additionally, the products should be as benign as possible in use, and they should be readily capable of reuse, renewal, reclamation, recycling, or safe conversion and disposal. If these ideas had even been contemplated a couple of centuries, or even longer, ago, the Industrial

Revolution would have been stillborn. The countries and economies that we now know as the developed world would still be reliant upon agriculture and crafts. The movement from agrarian and peasant-based economies to mercantile capitalism and globalization would have been slowed appreciably.

Aspects of design for sustainability are now receiving wide attention. Here it is appropriate to encapsulate the four key strategies defined by Hawken and others.[33]

Effective Use of Resources

Hawken's group espouses amplification of efforts to use all resources more effectively. Their major and first aims are for radical resource productivity, reduced rates of depletion, lower pollution, and creation of jobs. The metrics implied are not only the immediate bottom line, but also the generation of wealth for all stakeholders, and the entire host community without traumas (present or future). This is consistent with precepts discussed in publications by this author.[2]

Biomimicry

A second strategy is that of *biomimicry*. Here the aim is to copy natural systems of biological design and reuse (phytoremediation). In the forest, trees grow up, mature, weaken, fall down, and rot, and the decomposition gives rise to many reproducing and multiplying organisms. Ultimately, more trees sprout up in a repeating and enduring sustainable cycle. It is worthy of recall that the first notions of sustainability were associated with crop rotation in medieval agricultural practice. Wilson, in *The Future of Life*, commends a case study in Guatemala where a small local population was enabled to live sustainably by marketing natural products from a stable rainforest, as opposed to selling the timber and creating farms.[34] Ben & Jerry's ice cream company originated with these kinds of principles. Its product remains moderately successful among a growing group of other environmentally conscious offerings. Its practices follow the concept of the *Commons* espoused by Hardin in 1986.[27] This is in marked contrast to U.S. agri-business practices in the fast-food supply chain for beef, chickens, hogs, and potatoes described by Schlosser.[35] The idea of encouraging small, flexible, individually oriented and entrepreneurial operations is intrinsically attractive. For people with the right skills it is possible to launch and then expand uniquely innovative firms using the Web. Nanotechnologies adopt a somewhat parallel approach in that atoms or molecules are experimentally configured into the form desired. Recently announced processes involve the precise patterning of an appropriate surface and atoms, or molecules, are then caused to deposit or grow preferentially in the presence of a catalyst to produce specifically oriented arrays with extraordinary properties. These extremely small-scale processes could have application in future products. However, although these processes *grow the product* in a quasi-organic sense, at this point these are not truly sustainable. On the human level, analogies based on studying the behavior of ants and bees are being discussed.

Extending the Manufacturers' Responsibility

Hawken's third strategy reminds us that we exist in a *service and flow* economy. This revalues relationships between commercial operations and their customers. The manufacturer's or product originator's task is not completed when the customer walks out of the store, showroom, or sales office; there are responsibilities beyond the instant exchange of cash. This aligns with the idea of *extended products*, which implies taking a lifetime product support perspective and includes all services to support the product in addition to the manufacturing of the product itself, and then retirement of the product. This may require that intelligence be embedded in the product. Accompanying this is the idea of providing an ongoing relationship that gives customers solutions, experiences and delight. There is less emphasis on the product itself.[36,37] Both tangible and intangible products and services can be included. It is a novel challenge to develop support structures for these extended products.

Conservation and Restoration

The fourth principle emphasizes the importance of conservation of everything, and restoration of the original status wherever possible.

9.2 Principles for Environmentally Conscious Design

There are several corollaries for systems design that expand on these strategies, not necessarily in rank order. There is some inevitable overlap and redundancy:

1. There are no negative effects on the environment.
2. Conservation, elimination of waste or scrap, capability for reuse, or remanufacture—product can readily be made equivalent to new (ETN).
3. Products, or services, are benign in use.
4. System should accommodate global concerns for energy, fuel, natural resources and degradation—preserves balance.
5. The future system life cycle is considered.
6. Quality-of-life of users (a "necessity factor") is considered.
7. The development and future horizons of the workforce are important.
8. Ergonomics and heath concerns are factors.
9. The effect of the system on host community must be determined.
10. Local, regional, national, and global ecological, economic, and financial matters are considered. The design embraces global warming, pollution, entrained diseases, and other factors.
11. System should be based on smooth flows—evolutionary, organic, and nondisruptive.
12. The design must respect people as individuals—customers, employees, stakeholders.

13. There must be only positive long-term impacts.
14. The system overall must improve the common weal.

10 IMPLEMENTATION: *CONSIDERATIONS AND EXAMPLES FOR COMPANIES OF ALL SIZES*

Probably at no previous moment in world history have commerce and industry been so complex, extensive, and globally interrelated. In the twenty-first century, almost all of our commercial activities have global connotations and are increasingly digitally based. The growth of information technology (IT) has accelerated a shift from the simple exchange of cash for products or services to entire manufacturing or commercial systems that market solutions enabling customer success, provide experiences, and build continuing relationships. Process execution is critical; the process is continual and not intermittent, as it was in the past with discrete transactions. The provider entity must learn metaphorically to walk continually "in the moccasins" of the customer. *Customer delight* is the major metric for attention, but at the same time the needs of the community must be recognized. Strategies that emphasize customer success are imperative; these may often dictate collaboration with former, and even present, competitors.[36] In the twenty-first century successful enterprises must recognize the power of their customers (and the market) and must endeavor to establish long-term relationships based on providing solutions, service and a successful ongoing experience that replaces the earlier tradition of product delivery and cash transfer.[37]

10.1 Vertical Integration

Digital business (or *e-business*[29]) affords a radical new way of operation, and the factor of *speed to customer delight* challenges industry strategists to incorporate many new approaches and activities. The most successful enterprises have reorganized and refocused their activities. Vertical integration with control of all the materials, processes, and operations that go into producing a product is no longer preferred. This was the model adopted long ago by Henry Ford at his Rouge plant.[15] It is now important to focus on core competencies, activities that depend on special skills or advantages that the enterprise possesses or has developed. These are the processes that represent the *intellectual property* or *crown jewels* of the organization and generally provide the best opportunities for revenue maximization. Subsidiary operations can be delegated to specialist subcontractors. (Referring back to our athletic models, it can be observed that all-rounders do not often win Olympic medals). These tiers of subcontractors are organized into supply chains and are managed electronically. Excellence in communications, delivery, reliability, and quality are all key requirements to support just-in-time (JIT) manufacturing systems.

There are risks and some disadvantages with the dependencies occasioned by lengthy supply chains. Larger enterprises frequently insist that their suppliers build factories and/or warehouses nearby final assembly plants. Some enterprises have specialized contractors designing, developing, and manufacturing substantial proportions of major components. Boeing is a key exponent of this technique. Its Web site lists worldwide contributors to the 787 Dreamliner project (http://www.boeing.com/commercial/787family/). There are multiple reasons for delegating production activities, expertise, and focus are primary. Also there is cost; large organizations may carry very high overheads due to research and development, infrastructure, and legacy costs. When these are written-off against the manufacture of noncore components, parts, or subassemblies, then the raw cost of these items becomes unsupportable. For example, in the early 1980s when IBM introduced its personal computer, the internal pricing of its world-class hard drives, memory, and microprocessor chips would have made realistic market pricing impractical. IBM relied on vendors for all these components, the plastic housing, and peripherals.

10.2 Real-World Examples

Dell affords one of the best models for reliance on supply chains. Its core competency is giving customers exactly what they wish for just as quickly as Dell can have it assembled and shipped. It uses an array of suppliers to fulfill carefully figured inventory needs for just a few days production. Computers and peripherals are custom-built through final assembly in direct response to orders. Some bare chassis may receive standard boards and components, and special models may be finished for promotional uses. The history of its direct-to-customer *build-to-order* process is described in a book by Michael Dell.[25] Their revenues for 2005 totaled $55.9 billion, a 14 percent increase over the prior year. There are three Golden Rules: (1) disdain inventory, (2) always listen to the customer, and (3) never sell indirect. Dell also segments so that each business unit stays "small and focused on the needs of specific sets of customers."[25] Business information and plans are shared with all employees. Dell is virtually integrated through the purchasing arms of its large account customers. It continually strives to step back to examine the whole context and environment, and avoid "breathing your own exhaust."[25]

One can hardly discuss manufacturing system improvement without mentioning Toyota (see Table 1). The Toyota Production System (TPS) has been discussed and written about at length, and many Japanese phrases have passed into the manufacturing lexicon. *Kaizen* is a watchword, and the principles of *Lean* are covered comprehensively in other texts. Expressing it simply, *lean* involves only doing things that add value for the eventual customer, eliminating waste in all its forms, and not building inventory between operations, or at the end of the manufacturing line. Theoretically, a request by a customer must pull products out of final assembly or processing, and once that product is on the way, then the upstream processes must be instantly ready to fill the void (just in time). In the

10 Implementation: *Considerations and Examples for Companies of All Sizes* 69

Table 1 Toyota Precepts

1. Fostering an atmosphere of continuous improvement and learning
2. Satisfying customers (and eliminating waste at the same time)
3. Getting quality right the first time
4. Grooming leaders from within rather than recruiting them from the outside
5. Teaching all employees to become problem solvers
6. Growing together with suppliers and partners for mutual benefit

idealized case, every internal or external unit in the supply chain must respond identically. The flow may be controlled and maintained by exchange of signals, or *kan-bans,* up and down the delivery chain.[38]

When we think of manufacturing systems, we tend to think of Boeing, Ford, General Motors, IBM, Johnson & Johnson, Pfizer, and similar giant enterprises with multiple and globally distributed sites. Their actions and strategies are often drivers of the economy, but they would be powerless without the contributions, endeavors, and support of the smaller units tiered in the supply chain. In fact, a very large proportion of all manufacturing is undertaken by much smaller firms. It is reported anecdotally that 99 percent of U.S. manufacturers have fewer than 500 employees and they produce some 55 percent of the value-added output while affording 66 percent of all manufacturing employment. Data from 2001 show that only 27 percent of the U.S. workforce serve companies with more than 10,000 employees; there were a mere 930 firms of this size across all U.S. industry. More than 4 million (77%) of the 5.66 million firms in the country have fewer than 20 employees, although some may just be P.O. boxes and may not have any kind of manufacturing facilities! About half a million have between 20 and 100 employees, representing 18 percent of the workforce. There are 85,000 firms employing between 100 and 500 individuals, or 14 percent of the workforce. Twenty six million workers, or 23 percent, are in the 500 to 10,000 group. The total workforce accounted for in this tabulation was 115 million.[39] Smaller firms suffer difficulties with respect to acquisition of the newest technologies and may be forced into bidding wars with OEMs that further constrict their resources and development budgets. On the other hand they have the potential advantages of less bureaucracy, faster decision making and greater flexibility. The case of Liberty Brass in New York City is a model of the advantages that accrue to a smaller nimble company with courage and imagination.[40] This is a 40-employee machine shop that with adroit equipment choices reduced set-up times to turn around smaller and special orders responsively. It improved planning and estimating using the latest software so that quotes and eventual deliveries were accelerated, and outsourced suitable jobs to specialists. Essentially, it focused on core competencies and developing

Table 2 Essential Ingredients for Improvement

1. Vision: Aspirations and dreams
2. Mission, goals, and objectives (metrics)
3. Personnel and their motivation (empowerment & teams)
4. Strategies and tactics (core competencies, cooperation and partnerships)
5. Customer focus (experiences, relationships and solutions)
6. Communicate and listen to all stakeholders
7. Priorities and resources: Consistency and persistence

relationships for speedy customer service. Its prices are higher than Chinese suppliers, but its quality, reliability, responsiveness, and lower shipping costs gain Liberty Brass repeat orders. There are similar cases of quite small enterprises taking profitable advantage of improvement programs stimulated by small business initiatives.

10.3 Education Programs

Improvement systems cannot be implemented without education and training. Any program must have wholehearted commitment and demonstrated support by senior management (as required by Six Sigma). There must be not only commitment but also time spent, either in welcoming employee students to the first sessions and introducing the objectives, or as a visiting speaker for an executive overview, luncheon, or coffee break. The program must be accessible to all qualified or selected employees without undue conflict with regular responsibilities. There should not be any work penalty associated with attendance. All participation for work-related technical vitality, and improvement programs should be on the clock and during working hours, unless there are special exceptions. When Harley-Davidson wished to transform its operations, its analysis showed that it needed to hire a few additional people so that everyone could be scheduled for up to one hour per week of education or training time.

The long-range objectives to be accomplished must be clearly enunciated. Preferably, common administrative phrases and buzz words should be avoided in all announcements and program descriptions. Words worth avoiding include business conduct, continuous improvement, diversity, ethics, lean, quality or quality circles, safety, sexual harassment, and teamwork—these are important topics that should be embedded into programs with newer, different, and more invigorating titles.

Ideally, educational activities should be accessible to the whole workforce, including contract workers, everyone in the workplace should be on the same *playing field*. Special programs are appropriate to certain workers, levels, and technical areas. There should be some base program for everyone on a rotating basis, with time included in regular work assignments (maybe up to 40 minutes a week, or every couple of weeks) for training. Consistency and regularity are important; employee stimulation must endure after the first banners in the cafeteria fade. Training should cover vital topics such as quality tools, assessments and appraisals, career planning, latest trends in teamwork, personal development, and investments and retirement. *State of business* sessions with local executives or politicians are a useful change of pace. Special cross-training needs can be accommodated using in-plant experts. Mixing individuals from different departments and functions in every group, and having employees introduce their new-found friends, encourages interactions. The program must be perceived as worthwhile and not a time sink. Sessions should be short, frequent, and regular, with occasional promotional items, badges, ribbons, or certificates. Above all, they must not be boring. The occasional use of an outside facilitator or tapes by speakers such as John Cleese or Tom Peters can be very worthwhile. Employees should be entertained so that different learning styles can be accommodated. Make sure there is lots of action, vitality, and stimulation with immediate value, discussion time, coffee breaks, graduation breakfasts, and similar events. Finally, measure everything to support continuous improvement of the program. Useful books for handouts are available in bulk from suppliers such as www.goalqpc.com, a good source for memory joggers and Six Sigma materials, and Price Pritchett @ http://topics.practical.org/browse/Price_Pritchett.

10.4 Measuring Results

Successful implementation of improvements can be verified with traditional production metrics. The most convincing metrics are with regard to costs. Improvements and quality are not exactly free. Investment of resources and time are required to accompany, continue, and maintain appropriate schemes. Customary accounting methods can needlessly introduce deceptive chimera into cost analysis. For example, a small, extraordinarily responsive company making custom products to order is likely to have low inventories, a small amount of work in progress, and a limited backlog. Notwithstanding excellent revenues and decent profits, if such a company wishes to expand and seeks additional capital, it possesses few assets and apparently limited prospects. Any banker would be skeptical of approving loans for equipment that would facilitate growth, increased market share, or entry into a related market. Inventory and materials, or partially finished goods in the pipeline, may be counted traditionally as assets, whereas in the prevailing dynamic conditions they are a practical liability that must be sold off before newer, more current products can be introduced. In fact, to the loan officer a poorly managed operation loaded with inventory and WIP could be regarded more favorably than a bare-bones fast-turnaround highly responsive operation. As a result, when measuring the fiscal value of improvement schemes, it is vitally important to secure the costs and savings numbers that truly reflect the performance of the new or improved activities (Table 2). This technique is known as *activity-based costing*. Once adopted, it leads directly to *activity-based management* (ABC/ABM).[41]

Costs are very sensitive to the methods used for sharing the overhead burden. The ABC analysis looks at the costs that are directly attributable to the component, part, or assembly being manufactured—in other words, the direct cost of production. Analyses using this approach can reveal surprises when comparing manufacturing costs for specials against high-volume regular offerings. ABC analysis makes it possible to decide which of several ranges of products show the best cost to revenue ratios. Traditional accounting methodology does not do this effectively. Explorations of this type are essential when comparing the costs of a part made at the home location against the final delivered cost of the same part shipped from abroad. Companies must factor in costs of goods in the lengthy pipeline and possible risk exposures when deciding whether to make or buy.

11 A LOOK TO THE FUTURE

The ideal manufacturing system of the future, no matter how large, must be flexible and responsive and simulate the performance of a small operation. It must be focused on the best and most effective methods of bringing its core competencies to the service of customers, and on maintaining relationships with these customers Excellent relationships, collaborations, and partnerships must be established with subcontractors, suppliers, and associated groups for mutual benefit. Teams will comprise individuals within an enterprise, and also employees of partner enterprises. Virtual and global relationships and dependencies will become customary, and even essential, as a response to unpredictable and urgent challenges and opportunities. The employees of manufacturing organizations that wish to survive and prosper in the future must be well-trained, knowledgeable, and empowered team players with excellent communication skills and empathy for different cultures. Management and continuous improvement of these operations calls for vision, enthusiasm, with excellent leadership skills. Table 2 highlights important ingredients.

There are significant dichotomies as all these concepts are developed and integrated. There will always be economies of scale, but penalties of increased impedance, noise, and reduced quality will always accompany bureaucracies and size. By increasing degrees of order, it becomes easier to perform standard operations, but by employing total centralization, degrees of freedom dwindle, there is less variety, innovation is suppressed, boredom grows, and systems degrade.[29] This paradox applies universally and requires that leadership of large organizations acquires great skills of diplomacy and humility; very difficult and balanced judgments must be sold to diverse constituencies. Modern management requires walking a tightrope and balancing the priorities for survival and prosperity of the enterprise against the desires of the stakeholders and host communities. Some short-range suffering may be inevitable, as skills, abilities, needs, economics, and manufacturing strategies are rebalanced in response to global challenges. As Professor Joseph Stiglitz has observed in relation to globalization trends and lean implementations—"Short-term losers are concentrated, often very noisy, and their pain can be considerable." "The average worker in rich countries may actually be getting worse off, and there are probably more losers than winners." "There is need for a strong social safety net ... with more progressive taxation." [42] Ideally, local rearrangements can be deployed in response to the most punishing inequities. Re-education programs, community improvement projects, and expanded leisure opportunities, with adequate funding, are possible alternatives to early retirement or unemployment. It is important to remember that all resources, whether human or materiel, are finite and should not be wasted. All opportunities should be grasped effectively, and new means of securing revenues may require some creativity.

It is a dream that the manufacturing systems of the future will be not unlike a drive-up fast-food operation, or the food court of a major shopping mall. There will be arrays of small factories that each can accommodate multiple needs but represent specialized sets of skills (or core competencies) materials, processes, assembly and fabrication techniques, and finished items. There will also be series of agents, product integrators, advisors and consultants, or sales counselors, their function being to advise and act for customers as an interface with the various manufacturing facilities required to procure the desired custom product. Possibly all these functions could be handled through networks or catalogs in some *virtual* manner, as often already applies today. Customers will design, construct, buy, and equip their own houses, kitchens, computer systems, meals in restaurants, and other goods. Behind the retail suppliers are arrays of agents, distributors, wholesalers, and ranges and sizes of subcontractors and manufacturers. The future will be some spectrum of all these possibilities, with an extraordinarily wide range of time constants affecting the acceptance of new technologies across industries, regions, countries, and continents. Even in the presence of new technologies, there is little doubt that examples of past practice will continue to add value effectively in certain applications. There will always be the need for highly skilled craft workers to develop, fit, and maintain new tooling and, although overall the numbers of old-style tool-and-die makers has diminished, there will be niche areas for specialists in activities regarded as obsolescent.

It will always be prudent and economically reasonable to assemble many products in proximity to the final customers. This shortens delivery, reduces inventories, increases supplier responsiveness, and accelerates the concept-to-cash metric. Large enterprises are now reconfiguring their factories and replacing dedicated 'hard' tooling with flexible tooling capable of fabricating a wider range of products. These factories are also being located proximate to centers of population nationally and worldwide. Where the large enterprises go, their suppliers follow. The former mega-factories with several thousands of employees have been displaced by smaller distributed units that are simpler to manage and control. This trend is accompanied by an increased focus on the continuing education and technical vitality of the workforce. Greater numbers of distributed factories serving customers will entrain more jobs, and there will also be service, maintenance, education, and healthcare needs to be satisfied, thus, generating prosperity, and consequently more customers. As the population ages, provided they are allowed to continue earning, and a community remains prosperous, there will be an increase in discretionary expenditures for recreation and travel, and again, more jobs. There is need for global acceptance of the cycle that jobs engender customers and that prosperous customers fund more jobs, and then those jobs empower more potential customers. As customers of these systems ourselves, we must consider whether these notions are satisfying or efficient. Such questions can only be resolved by the system of measurements that must be implicitly agreed as we almost accidentally embark on improvements socially and professionally. The important thing is to start the journey knowingly. We may enjoy and learn from our nostalgia for the past, but we must anticipate the future eagerly.

REFERENCES

This chapter owes much to substantially amended, modified, and updated rewritten materials extracted from prior works by the author. The cooperation of the publishers of these items is gratefully acknowledged.

Author's Works

"Design: Organization and Measurement," chapter in the *Handbook of Design Management*, Basil Blackwell, Oxford, 1990, pp. 155–166.

"Management Structures for Realization of High Productivity," Proceedings, Int. Conf. on Productivity & Quality Research (ICPQR-93), Vol. 1, Industrial Eng., & Management Press (IIE), 1993, pp. 237–246.

"Globally Ideal Manufacturing Systems: Characteristics & Requirements," Flexible Automation and Integrated Manufacturing (FAIM'93), Conference Proceedings, Part II, Computer Integrated Manufacturing Systems, MIA: CIM Strategy, pp. 67–77, CRC Press, 1993.

"An Integrated Design Strategy for Future Manufacturing Systems," *J. Manufacturing Systems*, **15** (1), 52–61 (January 1996).

"Management Strategies for Sustainable Manufacturing," in Proceedings of the Seventh International Pacific Conference on Manufacturing, Bangkok, Thailand, Volume 1, pp. 13–22, November 27–29, 2002.

"The Evolving Production Enterprise," in Proceedings of the 18th International Conference on Production Research, Salerno, Italy, International Foundation for Production Research (CD), 2005.

"Manufacturing: The Future," material presented at 16th National Manufacturing Week, Chicago, Session 2D21, Reed Exhibitions and ASME: http://www.reedshows.com/nmw/handouts/2D21.doc.

Works Cited

1. K. M. Gardiner, "Management Structures for Realization of High Productivity," Proceedings, Int. Conf. on Productivity & Quality Research (ICPQR-93), vol. 1, Industrial Engineering & Management Press (IIE), pp. 237–246, 1993.
2. K. M. Gardiner, "An Integrated Design Strategy for Future Manufacturing Systems," *J. Manufacturing Systems*, **15** (1), 52–61 (January 1996).
3. The Columbia Encyclopedia, 6th ed., 2001 (online).
4. E. Burton and P. J. Marique, *The Catholic Encyclopedia*, vol. VII, Robert Appleton Co., online edition, 2003, http://www.newadvent.org/cathen/07066c.htm.
5. G. Agricola, De Re Metallica, translated from the first Latin edition of 1556 by Herbert C. and Lou H. Hoover, Dover Publications, Inc., New York, 1950.
6. http://www.nationmaster.com/encyclopedia/Venice-Arsenal and Wikipedia at http://answers.com/topic/factory.
7. D. Cardwell, *The Fontana History of Technology,* Fontana Press, London, 1994.
8. Adam Smith, "An Inquiry into the Nature and Causes of the Wealth of Nations," 2 vols., *Everyman's Library*, Dent & Sons, London, 1904, vol. I.
9. W. Edward, Deming, every text on quality offers ample descriptions of his principles. See, for example, J. R. Evans and W. M. Lindsay, "The Management and Control of Quality," 6th ed., South-Western College, Mason, Ohio, 2004.
10. Niccolo Machiavelli, *The Prince,* many editions; for example, Bantam Classics, New York, 1984.
11. Sun Tzu, *The Art of War,* many editions; for example, Dover Publications, New York, 2002.
12. F. W. Taylor, *The Principles of Scientific Management,* unabridged republication, Dover Publications, 1998; original, Harper & Brothers Publishers, New York 1919.
13. Special to the *New York Times,* "Gives $10,000,000 to 26,000 employees—Minimum Wage $5 a Day," January 6, 1914.
14. Upton Sinclair, *The Jungle,* several editions, including *The Uncensored Original Edition,* See Sharp Press; New edition (April 1, 2003), ISBN: 1884365302.
15. D. A. Hounshell, *From the American System to Mass Production 1800–1932—The Development of Manufacturing Technology in the United States,* Johns Hopkins University Press, Baltimore, MD, 1984.
16. Bureau of Labor Statistics, Press Release on Union Membership, January 20, 2006, available at http://www.bls.gov/cps/.
17. M. P. Groover, "Work Systems and the Methods, Measurement, and Management of Work," Pearson Prentice Hall, Upper Saddle River, NJ, 2007.
18. K. M. Gardiner, T. E. Schlie, and N. L. Webster, "Manufacturing Globalization and Human Factors," Proceedings, 16th Int. Conf. on Flexible Automation & Intelligent Manufacturing, Limerick University, Ireland, June 2006.
19. T. Kidder, *The Soul of a New Machine,* Atlantic-Little, Brown and Company, Boston, 1981.
20. F. Guterl, Design Case History: "Apple's MacIntosh," *IEEE Spectrum,* **21**(12), 34–44 (December 1984).

21. P. F. Drucker, "The Coming of the New Organization," *Harvard Business Review,* **66**(1), 45–53 (January–February 1988).
22. P. F. Drucker, "The Emerging Theory of Manufacturing," *Harvard Business Review,* **68**(3), 94–102 (May–June, 1990).
23. Several texts refer to the Baldrige Awards, including J. E. Ross, *Total Quality Management,* 3rd ed., St. Lucie Press, CRC Press, Boca Raton, Florida, 1999.
24. A. S. Grove, *Only the Paranoid Survive,* Currency, Doubleday, New York, 1996.
25. M. Dell, *Direct from Dell: Strategies that Revolutionized an Industry,* HarperCollins; Paperback edition, New York, 2000.
26. P. F. Drucker, "Knowledge Work and Knowledge Society—The Social Transformations of this Century," Edwin L. Godkin lecture, Harvard University, May 4, 1994.
27. G. Hardin, *Filters Against Folly: How to Survive Despite Ecologists, Economists, and the Merely Eloquent,* Viking Penguin, New York, 1986.
28. For further explanations on Life-Cycle Management (LCM) see Wikipedia at http://en.wikipedia.org/wiki/Product_life_cycle_management or check a vendor site at http://www-03.ibm.com/solutions/plm/index.jsp
29. L. V. Gerstner, "Who Says Elephants Can't Dance? Inside IBM's Historic Turnaround," *HarperBusiness,* 2002.
30. See, for example: "Teamwork—The Team Member Handbook," Pritchett & Associates, http://www.PritchettNet.com; or "The Team Memory Jogger," GOAL/QPC and Orial Inc., Salem, New Hampshire, 1995, http://www.goalqpc.com.
31. See C. Clipson, "Design as a Business Strategy," chapter in the *Handbook of Design Management,* Basil Blackwell, Oxford, 1990, pp. 96–105; supplemented by private communication concerning Saturn Project.
32. *The American Heritage Dictionary,* Second College Edition, Houghton Mifflin, Boston, 1982.
33. P. Hawken et al., *Natural Capitalism: Creating the Next Industrial Revolution,* 1st ed., Back Bay Books, 2000.
34. E. O. Wilson, *The Future of Life,* Knopf, New York, 2002.
35. E. Schlosser, *Fast Food Nation—The dark side of the all-American meal,* Perrenial, HarperCollins, New York, 2002.
36. S. L. Goldman, R. N. Nagel, and K. Preiss, *Agile Competitors and Virtual Organizations—Strategies for Enriching the Customer,* Van Nostrand Reinhold, Princeton, NJ, 1994.
37. R. Dove, *Response Ability—The Language, Structure, and Culture of the Agile Enterprise,* Wiley, New York, 2001.
38. J. Liker, *The Toyota Way: 14 Management Principles from the World's Greatest Manufacturer,* McGraw-Hill, New York, 2003.
39. Multiple U.S. government sites offer many statistics: www.census.gov and www.bls.gov etc. This data originates at www.census.gov/epcd/susb/1999/us/US—. HTM.
40. A. S. Brown, "Staying Alive—Forget competing with China on Price. These U.S. Manufacturers have Found Ways to Earn Their Bread," *Mechanical Engineering,* **128**(1), 22–26 (January 2006).
41. See, for example, P. B. B. Turney, "Common Cents—The ABC of Performance Breakthrough," *Cost Technology* (1991).

42. J. Stiglitz, "... benefits of globalization are unevenly spread," quoted in *The Guardian Weekly* (March 10–16, 2006).

Biography

The author is director of the Center for Manufacturing Systems Engineering and has been a professor at Lehigh University, Bethlehem, Pennsylvania, since 1987. He spent 21 years with IBM in semiconductor manufacturing and with the Corporate Manufacturing Technology Institute. Prior to this, he worked on manufacturing methods for gas turbines with Rolls-Royce, and on the development of nuclear fuel elements with English Electric. He has degrees in metallurgy from the University of Manchester in England, and is a registered professional engineer (CA). He is a member of the College of Fellows of the Society of Manufacturing Engineers, past member of the SME board of directors, where he was secretary-treasurer in 2000. He received the Joseph A. Siegel Service Award in 2003. His affiliations include ASME, Sigma Xi, and the Engineers Club of the Lehigh Valley. Cycling, photography, and industrial archeology are among his hobbies. He teaches courses on organizational planning and control, and manufacturing management for the department of Industrial and Systems Engineering, and is associated with a team-taught graduate class on agile organizations and manufacturing systems as a part of the cross-disciplinary MS in MSE program, which he directs. In addition he coordinates and gives lectures for a first year introduction to engineering practice course and laboratory sequence.

CHAPTER 4
MANUFACTURING SYSTEMS EVALUATION

Walter W. Olson, Ph.D., P.E.
University of Toledo, Toledo, Ohio

1	INTRODUCTION 79	4	SYSTEM EFFECTS ON ECM 85
2	COMPONENTS OF ENVIRONMENTALLY CONSCIOUS MANUFACTURING 80	5	ASSESSMENT 86
			5.1 Assessment Planning 87
			5.2 Data Collection 89
3	MANUFACTURING SYSTEMS 81		5.3 Site Visit and Inspection 90
	3.1 Levels of Manufacturing Systems 82		5.4 Reporting and Project Formulation 91
	3.2 The Plan, Do, Check, and Act Cycle 83	6	SUMMARY 92

1 INTRODUCTION

Environmental conscious manufacturing (ECM) is the production of products using processes and techniques selected to be both economically viable and have the least impact on the environment. Process selection criteria include minimum waste, minimum use of hazardous materials and minimum use of energy in addition to the production goals. Products produced in this manner are often more competitive in the marketplace because these criteria result in cost reduction, cost avoidance, and increased appeal to the consumer.

This chapter discusses several techniques for assessing and evaluating environmentally conscious manufacturing performance of manufacturing systems. These techniques begin with an analysis of the paperwork/data followed by a visit of the plant floor and conducting interviews. This provides the basis for making evaluations of improvement areas and performing the tasks necessary to formulate improvement projects. The emphasis here is on improving the manufacturing systems; however, the benefits occur in production.

This chapter focuses on very practical and fundamental issues in manufacturing. As a result, it is not readily applicable to a firm seeking ISO 14001 accreditation. Nor is it meant to be. Whereas ISO 14001 is an environmental management system, this chapter is about manufacturing systems and their evaluation. Ghisellinia and Thurston identified several decision traps that ISO 14001 imposes "the 'management' nature of the standard, failure to identity a rigorous environmental baseline, misconception of pollution prevention, inordinate emphasis on short-term goals, focusing on regulatory compliance, and diversion of EMS resources to the documentation system."[1] The processes here seek to avoid these traps. Therefore, there will be little reference to ISO 14001 or life-cycle assessment (LCA) in the following pages.

Manufacturing systems are the planning, communications, coordination, monitoring, and management aspects of manufacturing. Industrial engineers often perform these tasks within the overall manufacturing system. The goal of this chapter is to provide guidance to assist the industrial engineer in finding better methods and techniques to make the overall manufacturing system more responsive to the goals of environmentally conscious manufacturing.

2 COMPONENTS OF ENVIRONMENTALLY CONSCIOUS MANUFACTURING

Although one can reduce ECM in a number of specific details, all of which seem independent of each other, the engineer would do well to remember three major points:

1. Reduce waste.
2. Reduce hazardous materials and processes.
3. Reduce energy.

Almost every other ECM detail relates to these simple three objectives.

Waste is probably the single most import element. *Wastes are expenditures of resources that are not incorporated into the product.* Resources used in manufacturing are time, money, capital, labor, and materials. Wastes are characterized by emissions, machining offal, cutting fluids, person hours expended in waiting, buffering of items waiting for the next process, and machine downtime due to change over or maintenance, for example. When waste occurs, the product is more expensive.

A major division of manufacturing processes results in material separations. It is important to realize that it may be necessary to have waste in order to produce a product (see Figure 1). However, the material lost should be minimized. A systematic approach to identifying and eliminating wastes results in increased ECM while also increasing profitability. Often trade-offs must be made. For example, to be economically viable a process may require the use of a more wasteful material than one where material waste can essentially be eliminated but at a cost that prevents economic production of the product. Although this is an extreme example that is rarely encountered, effective ECM requires that several objectives be met while producing a product that can be marketed successfully.

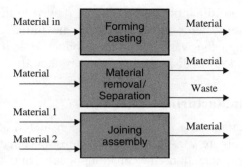

Figure 1 Classification of manufacturing processes.

The greatest waste occurs when a product was produced in quantity that failed to meet market requirements. Thus, it is important to identify wastes and know exactly where wastes are occurring, but it is even more important to know how to reduce waste without incurring greater costs.

Hazardous materials are defined as materials that by their nature, chemistry or conversion, result in reduced health of those exposed to them, degraded safety, or environmental risks produced by their uncontrolled release. These include toxic, mutagenic, radioactive, inflammable, and explosive substances. In addition to substances, hazards can include such things as noise, compressed fluids, and dusts. In many cases, specific controls and permissible levels are associated to these substances imposed by governments. Because of the additional controls and the additional increased handling awareness, use of these materials increases manufacturing costs.

Every industrial engineer in a manufacturing plant is intimately familiar with the cost of energy. However, they are not necessarily familiar with the environmental damage that energy production creates. Regardless of the form of energy or the source of energy, all energy has associated environmental damage. For example, direct solar conversion, which many vaunt as clean energy, requires the use of cadmium, selenium, and tellurium—all of which are harmful to the environment. Therefore, the reduction of energy requirements not only reduces manufacturing costs, but also reduces environmental damage. Thus, reduction of energy is also an objective of ECM.

3 MANUFACTURING SYSTEMS

Systems are required to produce products. A system is an organized whole consisting of subsystems that receive inputs from the external environment, transforming the inputs to outputs.[2] Many manufacturing activities required well-defined, systematic approaches to ensure efficiency. Manufacturing requires close coordination of labor, tools, materials, and information. If the systems that ensure this coordination are missing or flawed in their performance, waste is produced. These systems must start with the initial concept of the product and extend through the warranty of the product after it is in use. In some cases, this also includes the disposition of the product after the usage phase.

3.1 Levels of Manufacturing Systems

Manufacturing systems can be considered at four different levels of refinement (see Figure 2). At the top level, loosely called the *metasystem*, are decisions about the organization, the subcomponents, and the interactions of the components within the overall manufacturing systems. Factors that influence the metasystem are the top decisions as to what the foci of the businesses are, whether to produce to order or produce to stock, the complexity of the product, and the economic responsiveness needed. These will determine the components of the manufacturing system and establish the outputs of the components. The planning horizon for these decisions is the expected life of the corporation. It ranges from years to decades.

At the *strategic* level, forecasting and capacity planning (including facility planning and workforce planning) are the major subsystems. They are taken together with an output of the aggregate plan. The aggregate plan is based on the capacity of the manufacturing firm and includes the labor polices to drive the resources at the firm level. Aggregate planning horizons typically range from three months to five years, depending on the size of the firm, the product complexity, and the commitment levels of resources needed. Since it might take two

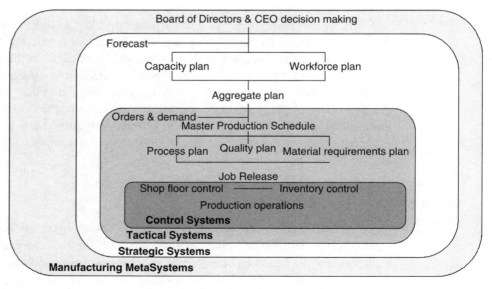

Figure 2 Manufacturing system levels.

years or more to bring a new factory online, aggregate plans need to incorporate the long-term economic outlook of the corporation and be consonant with the corporation strategic plan. In fact, one could argue that the aggregate plan is the operational statement of the strategic plan.

The next level down is the *tactical* level of systems. Here, process plans, quality plans and production scheduling need to be developed. Actual customer orders enter through a demand management system. Depending on decisions made at the metasystem level, orders are met from stock or by introduction of new production orders to the operations. Critical decisions at the tactical level determine what tools, what level of quality, and what response times are needed to meet the aggregate plan. The major output of the tactical systems is the production master schedule. At the tactical level, planning horizons range from one week to more than a year.

At the lowest system level are *control* systems. Shop floor control, inventory control, quality control, and maintenance control systems are the major components. These controls are driven by the production master schedule and direct and monitor resources to meet the production master schedule. Planning horizons may involve activities of a few seconds up to three months.

Decisions that have the most far-reaching effects occur at the metasystem level. As decisions are made at lower levels, they decrease in scope and reach but become more detailed. Thus, decisions at high levels involve more movement of resources and have greater potential for waste reduction than do decisions at the lowest levels.

Although various forms of manufacturing systems may have different names for the systems above and stipulate set of interactions between the systems, any complete manufacturing system will have subsystems performing the functions just described. Essentially, these systems relate real-world factors to models that can be used to direct future manufacturing activities. In this context, it is essential that all systems implement a *plan, do, check and act (PDCA) cycle*. This is illustrated in Figure 3.

3.2 The Plan, Do, Check, and Act Cycle

The PDCA cycle, sometimes called the Deming cycle, expresses a relationship between planning, operations/production, evaluations of outcome and management that are essential in a well organized and profitable manufacturing system. The "Plan" phase of the cycle prepares the system to meet its operational objectives and goals. In this phase, the events necessary for performance of the system are organized and the resources needed for production are scheduled. Planning is fed by the action phase which expresses management intent. The "Do" phase is the expression of the system performance. In most cases, the system is named after the "Do" phase as this is where the system produces its product and outputs. The "Check" phase is the assessment and evaluation of the do and plan phases.

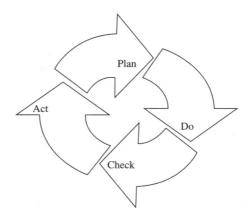

Figure 3 Plan–do–check–act cycle.

In the "Check" phase, it important to observe "what went right" as well as "what went wrong." This is often the role of quality control within the system. Finally, the "Act" phase is the role of management in the system. Leaders and managers set goals and objectives based on the reports from the check phase and their observations of the do and plan phases. Furthermore, they provide the directive action to correct and improve the system for the next cycle of PDCA.

Within manufacturing systems, the majority of methods to detect system wastes are based on measuring the system variability. In general, if a system is highly variable, the system is not performing well. The major causes of system variability are insufficient resources to perform the assigned tasks, insufficient capacity to handle the workload and insufficient time for planning and execution. Often times, these problems are exacerbated by failure of higher-level systems that provide inputs to the system in question. For example, aggregate plan changes frequently raises havoc with production scheduling and material ordering systems in the attempt to keep the master production schedule responsive. As results, long lead-time materials may be ordered, labor hired, and plant resources designated that are either too much (hence, wasteful,) or too little (again, wasteful of time) when actual production is executed.

Overall, the manufacturing systems organize production in an orderly way from high-level goals to the very detailed and mundane activities. Inefficient manufacturing systems results in inefficient production operations. Thus, observations of inefficiency at the operational level must be tracked to the level of system where the inefficiency first was manifested. This process reveals improvement opportunities. Thus far, these systems have been considered independently of environmentally conscious manufacturing. In the next section, it will be shown how these systems affect ECM.

4 SYSTEM EFFECTS ON ECM

Manufacturing systems are used to plan manufacturing, develop all of the subsystems, and then control the activities of manufacturing. They have a great impact on ECM. In fact, it is doubtful that ECM could be accomplished without active manufacturing-system participation. At the metasystem levels, there must be an overt commitment to ECM. A company is driven by risk reduction, corporate image, and economic objectives. The leaders of the corporation must understand and accept that ECM can reduce risk, improve corporate image, and improve the economics of the firm.[5] Without such an understanding at the top decision-making levels, ECM stands little chance of being fully implemented. A company pursuing ECM should state that ECM is part of the core business structure. This has the influence of directing the subordinate-level activities toward choices that support ECM.

In addition, the *make to order* and the *make to stock* decisions influence the amount of material that a firm must hold to support its operations. Because of external constraints on certain materials and lead times for certain components, there will be differences in what manufacturing strategy a corporation will take. However, anytime that a company must store raw materials, components, or finished product, there is an increased opportunity for waste to occur. If these materials also include hazardous materials, handling awareness also increases costs.

At the strategic level, ECM is most affected by capacity and facility planning. Facilities determine the processes available to the firm. They also determine the capacity. Process selection for a facility has the greatest influence on whether ECM can be achieved in a given facility. Any process that requires auxiliary materials to achieve an objective is subject for consideration for improvement under ECM. For example, a process that requires hydrochloric acid to clean a metal surface of rust and preservatives must be carefully examined. There are alternative processes that do not require the acid. However, once the process is selected and built into a facility, cost of replacement, space limitations, and a host of other factors may render it impossible to replace the process and therefore limit the effectiveness subsequent ECM attempts.

Additionally, at the strategic level, choices are made on the layout of a given facility. Traditionally, these are transfer line, cell or flexible machine layouts, colony or job shop process layout, and (in the most extreme case) stationary product layout. This choice must satisfy the given production requirements that include product, quality, and process flow. Choice of the wrong layout strategy results in increased energy consumption, increased work in progress, increased internal plant transportation, and increased waste.

The second greatest influence on ECM occurs at the tactical level in process planning. Process planning is the conversion of the engineering design to definition of the detailed processes that will enable creation of the product. Process planning can be considered at two levels:

1. The *macro level* where decisions are made as to what process of the facility will be used, in what order
2. The *micro level*, where all of the details of machines, individual tools, and instructions are created

Because of the influence of process planning on the individual details of creating a product, choices here directly influence waste, use of hazardous materials, and energy consumption. As the old cliché goes, "The devil is in the details!" Process planners intimate with the facility can greatly reduce waste and energy consumption by choice of energy saving and near net shape processes. An additional benefit for this is that the cost of production is also less.

Another tactical decision that can have an influence on ECM is production scheduling. Long batch-run schedules tend to waste less and consume less energy. Process setups and change-overs are notoriously wasteful and costly. In addition, much of the quality control wastage occurs during the early part of new batch runs. However, the same benefits can often be achieved through planning the facility for flexibility at the strategic level.

At the lowest levels, controls have the least impact on ECM, although significant improvements are still possible. For example, early detection of a faulty process can prevent excess material and energy use. This is even more critical if the process in question is near the beginning of the product's manufacturing.

The Controls are often the best indicators of inadequate ECM decisions made at higher levels, even though they have the least impact. For example, while monitoring the production process, shop floor foremen should be aware of places where waste collects and should report this to the process planners and process engineers. Machine maintenance is often an indicator: frequent stoppages and unplanned maintenance are an indication of an improper process and of waste. Thus, assessment activities focus on data and observations at these levels. But where problems exist, oftentimes the improvement will be in a higher-level system.

This section discussed illustrations of systematic effects on ECM. The next sections focuses on assessment and evaluation and improvement methods for achieving ECM.

5 ASSESSMENT

Assessment as used here is the examination and evaluation of the effectiveness of a manufacturing facility and its systems in meeting the objectives of ECM. Assessment is properly the *check* phase of PDCA. In assessment we examine the processes that are occurring and determine how well they function. The purpose of assessment is to benchmark the current state of operations against an ideal state of operations. Plants will have both exemplary operations and operations that do not meet desired standards. It is essential to recognize the exemplary operations for several purposes: exemplary practices should be copied where feasible across the corporation. In addition, recognition of what is being performed right is important to encouraging and rewarding the plant management for making improvements.

There are three parts to assessment: data collection, evaluation, and reporting. During data collection, information is gathered and analyzed to determine how the plant is performing over an extended period, usually a year or more. The evaluation is conducted on site and is a snapshot of where the plant is at the time of assessment. Reporting is essential in documenting the assessment procedures, the assessment recommendations, and the plant exemplary practices so that they can be communicated appropriately.

5.1 Assessment Planning

The time and effort required for assessment will vary, depending on the size of the organization being evaluated. Typically, an assessment team will consist of a team leader and three to five team members. The team leader should have familiarity with the plant and its operations. However, the team leader and its members must be independent of the plant in the corporation organization. It is most common to select members of an industrial staff from another plant to perform the assessment if no formal organization exists in the corporate structure. For small firms with only one plant, these requirements cannot be met; however, assessments can still be performed using either outside resources or by subdividing the plant and limiting the assessments to subunits. An excellent program to assist small companies can be referenced at http://www1.eere.energy.gov/industry/bestpractices/iacs.html.[3] Often the team leader is a senior industrial engineer. The team members are normally engineers and planners for the organization.

A typical assessment will require about a week to perform, provided the team is trained and has performed previous assessments. This will be far less for small plants and firms having only one plant, but the proportion of time planned will be approximately the same. The first three days of the assessment concentrates on data collection and analysis. The fourth day is the onsite plant visit. The fifth day is the onsite evaluation and report writing.

It is not uncommon and in some cases is desired that the first three days be separated from the plant visit by a time period of up to a month. This is to allow more time for analysis and to ensure that the correct data have been provided. It is desirable to have actual data, not summaries. However, the manufacturing unit will often try to provide summary only: this must be resisted. The actual data will have timing information and anomalies that are usually absent from summarized data. Where summaries have been furnished, time will be needed to make requests for the actual data and to receive it in time for the analysis to be completed before the site visit.

The first part of the assessment is analysis of the documentation of a plant. Depending on the plant, this may be performed at the plant location or at another corporate location. Off-site evaluation is preferred to on-site evaluation to reduce potential interference with plant operations, reduce assessment cost, and provide the team with the most time to analyze the data. On-site visits are usually less productive because of the need for the plant management to perform briefings, set up work areas and pre-interview team members from the concerned staff. During the first two hours of the first day, the team leader assigns team members to analyze specific data. One team member should focus on plant processes, quality, and scheduling. Another team member should focus on plant purchases and inventory. Yet another team member should concentrate on plant emissions and hazardous substances. A team member should be assigned to studying energy, water, and solid wastes for the plant. This analysis will focus on specific manufacturing systems improvement opportunities.

At the end of the third day, the team meets for approximately two hours to discuss findings, compare notes, and establish what areas are of concern for the on-site visit. Team members should report what they have analyzed, what were the results, and any major omissions that may have occurred. The team leader may ask certain team members to concentrate on certain parts of the facility, based on these findings.

In addition to accessing documents, data collection also includes visiting the processes or facilities being assessed and interviewing managers and employees. The visit is organized by arrival on the site at the beginning of the work day. A management briefing on the plant will review the layout, processes, and safety requirements for the plant. This is usually followed by a management-guided walkthrough of the plant. The management briefings and the management-guided walkthrough should be performed in approximately two hours. The team leader then assigns members areas of closer inspections and interviews in areas of interest.

At midday, the team should meet in a private location for the midday meal and should discuss any significant findings. Following this meal, the team leader may make additional assignments. The next three hours should be used to complete the site inspection and to perform any more interviews needed. In multishift operations, assessment team members may be required to attend those shifts for interviews and inspections. Often in a multishift operation, certain operations are only performed in the night shift, or the graveyard shift.

The first hour of the fifth day should be used to formulate any major findings that will be in the team report. At the end of this hour, the plant management is presented with these findings for their further comments. After hearing the plant management, the next two hours are used to discuss the formulation of recommended improvement projects. The remainder of the fifth day is used to formalize the team report. The formal report should completed within two weeks of the site visit.

5.2 Data Collection

During the data collection phase of assessment, an attempt is made to collect all of the documents relevant to the operating state. Usually, a year's worth of consecutive data is needed. Information needed relates to operations, material storage and use, identified hazardous materials, and energy usage. This would include production data in units of production, aggregate plans, production schedules, downtime reports, material safety data sheets (MSDS), quality defect reports (QDR), purchase orders for materials, shipping orders, inventory turnover reports, emission reports, and energy bills. In some cases, the records will be too voluminous to copy and the data may have to be examined at the plant location for the usage of the data. Most facilities now keep electronic data rather than paper records. If not, this may be a big area for improving ECM. In addition, if previous assessment reports exist, these should be incorporated into the data analysis and used for comparative purposes.

When electronic records exist, usually the ability to perform statistical analysis is much easier. Basic statistics such as means and variances should be computed and compared to benchmarking statistics if available. In addition, trend analysis may be useful to find processes that may be approaching a critical situation. Special care should be given to data that seem unusual when taken in the context of the complete data stream. Outliers are often improper measurements; however, they may also flag conditions that require further action.

Production schedules indicate how often the production is being changed. In addition, they will indicate by the plan date how much lead time is given to changing the schedule. Short planning lead times are an indicator of poor planning and waste. Downtime reports will provide information on the reliability of a process. Planned schedules should be compared to actual performance to find discrepancies. Where these exist, they should be investigated to determine the reason. QDRs are particularly important, as these will highlight problematic areas with the production line. Most firms today use some form of statistical quality control. If this is not in use, this will flag a very important area for improvement. Excessive QDRs often highlight process planning problems.

Material purchase orders are measures of the inputs that the system is actually consuming. This needs to be compared to shipping orders to determine whether the material ordered is being converted to product. It is particularly important in the collecting and analyzing of purchase orders and shipping orders to observe waste collection volume, type, and frequency. Inventory turnover reports will indicate materials that are stagnant within the system. A high percentage of non-rotating stockages indicates poor production scheduling, poor material planning, or an inability of managers to release one-time stocks once acquired. The result is waste in time, storage, and capital to support maintenance of this inventory.

MSDS will indicate the hazardous materials in use, as well as the handling requirements. The person(s) doing the assessment should observe in the site visit where the materials are and how they are being handled. The emission reports from the facility indicate what substances are being released into the atmosphere. Because of the cost of licensing and compliance, any savings here can have a big impact on ECM.

The energy bills will indicate how energy is used and what types of loads are being billed against. Computation of power (the ratio of real power to total power) and load factors (overall percentage use of electrical equipment) should be performed and graphed. Power factors below 95 percent should be flagged as opportunities for improvement for reducing energy costs. Load factors below 70 percent indicate that equipment is not being used near its peak operating requirements. When these data are taken in their entirety, a picture will form of the state of current operations. In addition, inappropriate load factors will be highlighted that are out of place and will become opportunities for improvements. More extensive energy auditing information is located at the Energy Star Web site, http://www.energystar.gov/, or the more specific link, http://www.energystar.gov/index.cfm?c=industry.bus_industry_plant_energy_auditing.[4]

5.3 Site Visit and Inspection

The site visit begins with a plant management briefing and plant management-guided walk through. During the management briefing, often in the question and answer period after the formal briefing, the assessment team should relate any significant findings that may have been noticed in the data and should ask questions regarding these. This discussion will prepare the management staff for the interviews to follow later in the day so they can gather the relevant information.

Information that was found in the documentation should be confirmed on the floor of the facility wherever possible. During the visit, carefully note the state of the operations. Is the facility adequately lighted? Is the facility operating in an orderly manner? Are the floors and the machines clean? What smells exist, and from where do they originate? Is work in progress entering the areas reserved for transportation and passage? Are you receiving the same messages from the employees and the managers when you interview them and discuss their operations?

In addition, the assessment team member should walk the outside of the facility. It is often surprising what is found on the perimeter of the property of a plant. The team member might spot excess materials, defective product, liquid wastes, and other indicators of ECM defects. In the past, these were often unreported, and they hid errors. These need to be investigated and cleaned up.

5.4 Reporting and Project Formulation

The report is formulated in the following outline:

1. Title page
   a. Name of the plant assessed
   b. Name and contact details of who the report was prepared for
   c. Assessment report number
   d. Date of the report
   e. Name and contact details of the assessment team leader
2. Executive summary
3. Assessment recommendations
   a. Title
   b. Observed problem
   c. Recommendation
   d. Estimated costs and benefits
4. Exemplary plant practices
   a. Title
   b. Observed practice
   c. Estimated benefits
   d. Contact information for more details
5. Synopsis of assessment
   a. Plant background
   b. General plant information
   c. Plant leadership
   d. Plant processes
   e. Description of wastes
   f. Description of energy usage
   g. Hazardous materials in use
6. Data analysis—synopsis only (details are in appendices)
7. Site visit observations
   a. What was observed
   b. Who observed it
   c. When was it observed
   d. Why is it significant
   e. Corrective action or plant response, if any

8. Findings and conclusions
9. Appendices as needed

The most important parts of the assessment report are the assessment recommendations and the exemplary plant practices. Each assessment recommendation (AR) should be itemized. The ARs provide the foundation for creating an improvement project. To do this, an AR should be titled in an active manner, starting with a action verb. An example is "Meter water use to reduce sewage charges." This then should be followed by a description of anticipated benefits and savings by performing the action. Where possible, this should cite both dollars in savings, as well as fundamental units such as kWh/yr or pounds CO_2. This is followed by who observed the operation and what exactly was observed. The team member should take or obtain pictures to illustrate the process recommended for change. Cost of the improvement project should be estimated if possible, with a short financial analysis of payback.

A key component of assessment is two-way communication between the assessment team and the plant management. In the process, there are ample opportunities for this to occur. On the fourth day, the plant management team provides information to the team and may wish certain areas to be emphasized in the study and may make certain recommendations for improvements. The team has discussed with the management staff the findings from the documentation. On the fifth day, the team reports its significant findings to the plant management team for comments.

Exemplary plant practices—or *best practices*, as they often titled—have a similar format. The title should indicate an action: "XYZ Plant uses ultrasonic tank cleaning." Benefits achieved are itemized. Then the report should give short description of the practice, followed by a contact person in the plant who can provide more information.

The final report must be timely. The report should be ready within two weeks of the site visit and should be provided to the plant management team and the corporate vice president for the division. This may result in an order for a follow-up assessment to ensure that the assessment recommendations are being implemented and corrective actions were taken where needed.

6 SUMMARY

Manufacturing system evaluation for environmentally conscious manufacturing is based on three overriding goals:

1. Reduce waste.
2. Reduce hazardous materials and processes.
3. Reduce energy.

These goals not only improve the responsiveness of the manufacturing unit to the environment, but they also result in significant cost savings and a better product when performed properly. Evaluation of manufacturing systems for environmentally conscious design begins with an assessment that highlights significant opportunities for improvement. Although the assessment focuses on actual operations, the solutions must be applied to the systems that created, support, and monitor the operations. Whereas ISO 14001 is a worthy goal to pursue and may be essential to the business practices of the firm, it is largely a paperwork exercise that does not solve common problems in manufacturing and does not necessarily lead to continuous improvement of the plant, the product, or the manufacturing systems. The evaluation process described in this chapter has been applied to several hundred different plants in the United States by a large number of auditors and assessors and has been shown to lead to real results in a timely fashion. It can also assist the ISO 14001 aspirations of the firm, if used properly.

REFERENCES

1. A. Ghisellinia and D. L. Thurston, "Decision Traps in ISO 14001 Implementation Process: Case Study Results from Illinois Certified Companies," *Journal of Cleaner Production*, **13**, 763–777 (2005).
2. W. W. Olson, "Systems Thinking," in *Sustainability Science and Engineering*, M. A. Abraham, ed., Elsevier, Amsterdam, 2006, p. 93.
3. U.S. Department of Energy Industrial Assessment Program, http://www1.eere.energy.gov/industry/bestpractices/iacs.html.
4. U.S. Energy Star, http://www.energystar.gov/ and http://www.energystar.gov/index.cfm?c=industry.bus_industry_plant_energy_auditing.

CHAPTER 5

PREVENTION OF METALWORKING FLUID POLLUTION: ENVIRONMENTALLY CONSCIOUS MANUFACTURING AT THE MACHINE TOOL

Steven J. Skerlos
The University of Michigan at Ann Arbor

1	INTRODUCTION	95
2	POLLUTION PREVENTION IN METAL-FABRICATED PRODUCTS MANUFACTURING	96
3	METALWORKING FLUIDS: USES AND CONCERNS	97
4	ENVIRONMENTAL IMPACT OF MWFs	99
	4.1 Hazardous Metal Carry-Off	99
	4.2 Hazardous Chemical Constituents	99
	4.3 Biochemical Oxygen Demand	100
	4.4 Oil and Grease	100
	4.5 Nutrients	100
5	HEALTH IMPACT OF MWFs	101
6	MWF POLLUTION PREVENTION THROUGH PROCESS PLANNING	102
7	MWF POLLUTION PREVENTION THROUGH PROCESS MODIFICATION	105
	7.1 Dry Machining	105
	7.2 Minimal MWF Application	106
	7.3 MWF Mist Reduction	107
	7.4 MWF Formulation Considerations	108
8	MWF POLLUTION PREVENTION THROUGH IN-PROCESS RECYCLING	109
	8.1 Phase Separation	110
	8.2 Magnetic Separation	113
	8.3 Filtration	114
	8.4 Biological Inhibition	117
	8.5 Conclusions: In-Process Recycling	118
9	CONCLUSION	119
10	ACKNOWLEDGMENTS	120

1 INTRODUCTION

Increasing environmental consciousness in society can be thought of as a reflection of numerous opportunities to improve the environmental performance of manufacturing. To some degree, these opportunities exist because their benefits cannot be captured as profit within the current accounting paradigm. Despite this, numerous opportunities currently exist to simultaneously improve environmental performance and profitability. Often these opportunities are overlooked due to conventional wisdom that suggests environmental improvement can occur only at the expense of profitability. Traditional end-of-pipe treatment technologies necessitated by reaction to environmental legislation have heavily contributed to this notion. But to the extent that environmentally focused manufacturing practices can bring along with them improved process efficiency and higher product quality, environmental improvement and profitability go hand in hand.

A window of opportunity currently exists to capitalize on environmental-based process efficiency improvements as a source of competitive advantage before either future legislation or competitors' actions require it. However, the most profitable environmental improvements usually require fundamental process understanding and change. This chapter illustrates this concept by detailing process issues associated with the environmental improvement of metalworking fluids used in metal-fabricated products manufacturing.

2 POLLUTION PREVENTION IN METAL-FABRICATED PRODUCTS MANUFACTURING

Metal-fabricated products manufacturing is critical to the aerospace, automotive, electronic, defense, furniture, and domestic appliance industries. Metal-fabricated products manufacturing includes metal shaping, surface preparation, and surface finishing. In 1992 the U.S. Environmental Protection Agency (EPA) produced a bibliographic report to create awareness within the metal fabricated products manufacturing industry regarding opportunities to prevent pollution.[1] The techniques offered in this report to prevent pollution are listed in Table 1.

Housekeeping, waste segregation, and training are the subset of pollution-prevention techniques that can be conducted with minimal change to basic manufacturing methods. By contrast, process planning, process modification, raw material substitution, and in-process recycling require full understanding of the environmental ramifications of a given manufacturing process and change driven by an understanding of the fundamental mechanisms driving the process itself. Data gaps in both environmental and process knowledge present the largest source of resistance to this second subset of pollution prevention techniques. However, it is typically these forms of pollution prevention that provide the greatest potential for improved efficiency and competitive advantage.

This is especially true for metalworking fluids (MWFs) used in the metal-fabricated products industry. Metalworking fluids have a variety of environmental liabilities associated with them that the industry is currently trying to reduce. Fundamental solutions to these environmental problems can take the form of MWF volume reduction, alternative MWF application strategies, MWF formulation changes, and MWF recycling technologies. However most of these solutions require fundamental understanding and change at the metalworking process level. Not surprisingly, the solutions that have been most readily adopted have been

Table 1 General Examples of Pollution-prevention Techniques

Harder to Implement: Greater Potential Gain	Easier to Implement: Smaller Potential Gain
• Process planning • Process modification • Material substitution • In-process recycling	• Waste segregation • Good housekeeping • Training

those involving the least amount of manufacturing change. This is due to large data and modeling gaps that exist with respect to MWF usage.

This chapter has three objectives: (1) describe the environmental and health issues associated with MWFs; (2) discuss established and emerging technologies that have been proposed to minimize environmental and health impacts of MWFs; and (3) highlight the data and modeling gaps at the process level that are required to achieve environmental improvement with respect to MWFs. Section 3 provides a brief introduction to MWFs. Sections 4 and 5 discuss environmental and health issues associated with MWFs. Section 6 details the data and models required for environmentally conscious process planning in MWFs used in metal-cutting operations. Section 7 describes process modifications that have been researched for minimizing MWF impact. Section 8 details recycling technologies that have been implemented to extend the life of MWFs.

3 METALWORKING FLUIDS: USES AND CONCERNS

Metalworking fluids benefit a variety of metal cutting and shaping processes by cooling and lubricating the workpiece and tool, transporting chips out of the cutting zone, and imparting corrosion protection. Approximately 1 billion gallons of MWF were used in 1998. MWF usage increased 5.3 percent between 1997 and 1998.[2]

Metalworking fluid chemistries are complex and vary significantly, depending on the manufacturing operation they are used in. By typical definitions, there are four categories of MWFs: straight oils, soluble oils, semi-synthetics, and synthetics. Straight oils consist of a petroleum or vegetable oil base with or without specialty additives. The remaining three types of MWFs are water-soluble and are classified by the ratio of water to mineral oil in their concentrated form. The concentrate is usually diluted in 80 to 95 percent water when used in process. Typically, soluble oils contain greater than 20 percent mineral oil in the concentrate, while a semi-synthetic will typically contain 5 to 20 percent. Synthetic metalworking fluids contain no mineral oil. Water-soluble metalworking fluids contain varying amounts of specialty additives including lubricants, corrosion

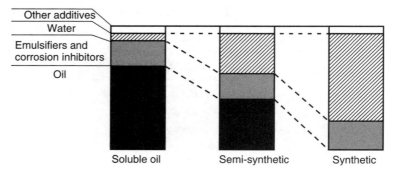

Figure 1 Relative proportion of water, oil, and additives in water-soluble MWFs. (Adapted from Ref. 3.)

inhibitors, emulsifiers, chelating agents, pH buffers, defoamers, and biocides. Figure 1 illustrates the relative percentage of oil, water, and additives found in water-soluble MWFs.[3]

Four major concerns have been raised about the state-of-the-art application of MWFs.[4] First, particulates, tramp oils, and bacteria are known to reduce the quality of metalworking operations over time.[5,6] Second, these contaminants eventually render the fluid ineffective for metalworking operations, creating significant acquisition and disposal costs that reduce profitability.[7,8] Third, the disposal of MWF places a significant burden on the environment.[9] And fourth, bacteria and the biocides used to control their growth in MWFs can be a significant health hazard.[10] These concerns are summarized in Figure 2. Sections 4

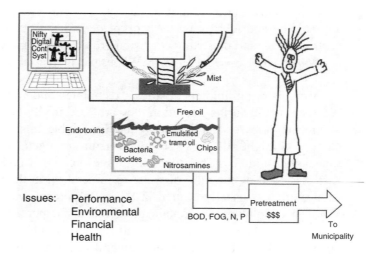

Figure 2 Illustration of contaminants and issues associated with MWF usage.

and 5 detail what is currently known about environmental and health impacts of MWF usage.

4 ENVIRONMENTAL IMPACT OF MWFs

Raw materials, energy, and water resources are consumed at each stage of the MWF life cycle. Reduction in resource consumption provides a clear opportunity to benefit the environment and improve process efficiency. The life-cycle stages that can be considered in this context include:

- Refining and production
- Transportation
- Preparation
- Use at the machine tool
- Recycling and maintenance
- Treatment
- Oil recovery
- Disposal

Of the life-cycle stages directly influenced by MWF end users, the disposal stage has the largest impact on the environment. MWF disposal results from a breakdown of the product by metal residuals, microorganisms, leak oils, and excesses in heat, water hardness, and water evaporation. MWF disposal can impact the environment due to hazardous metal carry-off, hazardous chemical constituents, oxygen depletion, oil content, and nutrient loading.[11]

4.1 Hazardous Metal Carry-Off

An otherwise nonhazardous MWF can become hazardous at the machine tool by carrying off particulate containing inclusions of heavy metals such as lead, chromium, or cadmium. Hazardous MWF disposal can exceed $250 per 55-gallon drum and is strictly regulated. There is a trade-off to consider between the benefit provided by hazardous metal inclusions found in machining workpieces and the environmental cost of disposing of the waste fluid once it becomes hazardous by chip contamination. Specifications requiring potentially hazardous metals should be reexamined from the context of disposal cost and environmental impact.

4.2 Hazardous Chemical Constituents

An MWF may contain known hazardous ingredients that are reported on its accompanying Material Safety Data Sheet (MSDS). Generally, hazardous ingredients found on an MSDS may have one or more of the following characteristics: combustibility, flammability, oxidativity, instability, reactivity, carcinogenicity, toxicity, or corrosivity. Besides being potentially hazardous to plant personnel when present beyond threshold levels, MWFs with hazardous ingredients can be a hazard in the environment if disposed of improperly. They can also be subject to special regulatory disposal requirements.

4.3 Biochemical Oxygen Demand

As MWF enters a receiving water, the bacterial population will begin breaking it down into metabolites, using up oxygen in the process. As the dissolved oxygen in the water is used up, the fish and plants that compete for oxygen will have less oxygen for respiration. If dissolved oxygen levels fall significantly, the organisms can no longer survive and the ecology of the receiving water will be altered. To account for this form of pollution, the biochemical oxygen demand (BOD) of a MWF is measured as a metric of its oxygen-depleting capacity. Maximum permissible levels of BOD (measured over 5 days) in sewered MWF are usually about 200 mg/l. Untreated MWFs may have BOD concentrations more than 20 times this limit.[8]

4.4 Oil and Grease

Fat, oil, and grease (FOG) contaminants in MWFs come from petroleum, mineral, animal, or vegetable sources. FOG can lead to taste, odor, and other aesthetic problems in water at relatively low concentrations. At elevated concentrations, FOG is toxic to aquatic life. In MWFs, FOG contaminants comprise a significant amount of the BOD, but FOG is regulated separately due to the significant impacts that FOG contaminants can have at low concentrations. FOG can also have detrimental impacts on treatment processes for BOD.

4.5 Nutrients

MWFs can also have elevated levels of nutrients such as nitrogen and phosphorous. Excessive nutrient concentrations can lead to ecosystem disruption in receiving waters. *Eutrophication*, for example, is the gentle aging process a lake goes through that transforms it to a marsh or bog. When excesses in nitrogen and phosphorous exist due to industrial, agricultural, or domestic influxes, this aging process profoundly accelerates, causing a premature death to the higher forms of aquatic life and to the lake itself.[12] Moreover, nitrates in drinking water have been suspected of contributing to incidence of blue baby syndrome.

EPA has become increasingly interested in the regulation of MWF-containing wastes. In 1995, the agency proposed effluent limitations guidelines and standards for the metal products and machinery industry. This proposed Metal Products and Machinery Rule would require oil and grease disposals below 17 mg/l.[13] Since untreated MWFs typically contain levels of oil and grease above 2000 mg/l, achieving this standard would increase on-site treatment and disposal costs significantly.[14]

5 HEALTH IMPACT OF MWFs

Before the 1970s, most metalworking operations employed either straight oils or soluble oils. Semi-synthetic and synthetic MWFs have gained market share since then because of the transition to higher-speed machining requiring more cooling in MWFs, increases in petroleum costs during the 1970s, and because synthetics are less susceptible to microbial attack than soluble oils and hence have a longer life. Straight oil consumption has decreased due to hazardous oil mists and inherent fire risks.

Water-soluble MWFs may also contain hazardous ingredients. Common MWFs used in the 1970s and 1980s were found to contain potentially carcinogenic nitrosamines.[5] The addition of nitrosamine-forming ingredients such as sodium nitrite has since been prohibited. Triethanolamine, a common ingredient in today's MWFs, has been associated with occupational asthma and is a suspected carcinogen.[10,15] Several other ingredients found in MWF formulations are listed as potential exposure hazards and must be listed on the MSDS provided with every MWF formulation.

Apart from inherent hazards that may be associated with MWF chemistry, in-plant hazards may exist due to microbial contamination. Bacterial populations of 10^5 to 10^9 organisms per milliliter of MWF are not uncommon in poorly maintained MWF systems and are the primary reason biocides are used in all water-soluble MWF formulations. Table 2 lists some typical metalworking fluid bacteria, their survivability in MWF, and their relative sizes.[16]

Microbial contamination has been recognized as a cause of deterioration of MWFs and has been studied extensively.[17,18] The primary reasons for the study of microbes in the past were directly concerned with reducing microbial metabolism of MWF ingredients. However, pathogenic or potentially pathogenic organisms such as *Pseudomonas aeruginosa, Escherichia coli*, and *Legionella sp.*[10] have been isolated from MWFs in recent years. This has raised concern about worker safety. Fortunately, infections directly attributable to microorganisms in MWFs appear to be rare. One notable exception is an outbreak of Pontiac fever in an engine manufacturing plant that was caused by *Legionella sp.*–contaminated MWF aerosol.[19]

Increasing attention has been paid to bacterial byproducts in MWFs. In the past decade, it has been discovered that bacterial and fungal spores and endotoxins can exist in MWF mists. Endotoxins are functional components of the cell wall of some bacteria found in Table 2. Endotoxins are generally released by microbial lysis or by destabilization of the bacterial cell wall. They can cause allergic responses in the upper respiratory tract, airways, or the distal portions of the lungs (e.g., HP also known as allergic alveolitis). HP has been associated with microbially contaminated MWFs in several recent studies, and more generally, heightened incidence of asthma and other disorders of the pulmonary airways have been reported in individuals exposed to MWFs.[10]

Table 2 Bacterial Species Found in MWFs and Their Survivability, Length, and Diameter

Species	Survival/ Growth	Length (um)
Pseudomonas aeruginosa	Excellent	1.5–3.0
Pseudomonas fluorescens	Good	2.3–2.8
Pseudomonas putida	Good	2.0–4.0
Proteus mirabilis	Excellent	1.0–3.0
Enterobacter cloacae	Very Good	1.2–3.0
Enterobacter agglomerans	Good	1.2–3.0
Citrobacter frueundii	Very Good	2.0–6.0
Escherichia coli	Good	2.0–6.0
Klebsiella pneumonia	Very Good	0.6–6.0
Desulfovibrio sp.	Fair–Good	2.5–10
Salmonella sp.	Fair	2.0–5.0
Legionella sp.	Not known	2–20
Mycobacterium sp.	Good	1.0–10

Biocides used to control bacterial populations in MWFs also have health risks associated with them. Biocides have been linked to allergic contact dermatitis.[5,20] Some biocides have also been linked to nitrosamine formation in MWFs.[21]

After a comprehensive review of respiratory disorders associated with MWF exposure, the National Institute for Occupational Safety and Health (NIOSH) has recommended that exposure to MWF aerosols be limited to 0.5 mg/m^3. This is an order of magnitude below the current standard. In addition, NIOSH has recommended that dermal exposure to MWF be limited where possible.

6 MWF POLLUTION PREVENTION THROUGH PROCESS PLANNING

Figure 3 illustrates a process-planning strategy for comprehensively integrating environmental considerations into traditional metal-cutting operations. The figure outlines the models required for environmentally conscious process planning and

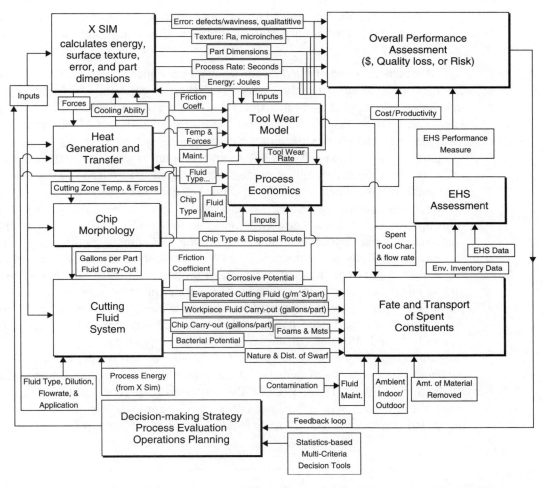

Figure 3 Conceptual process-planning strategy for environmentally conscious machining.

the detailed flow of information required between them. The models can be classified in five general categories:

1. Process simulation models for tool wear, part quality, heat generation, and chip morphology as a function of operating parameters, workpiece and tool materials, and MWF chemistry
2. An economic model to assess productivity and profitability of different sets of operating parameters, materials, and MWF
3. MWF and material transport models to elucidate potential receptors and exposure to the process
4. Environmental, health, and safety models to quantify health risks and environmental impact of process exposure

5. A decision-making model to optimize the metal-cutting operation with respect to profitability, productivity, health risks, and environmental impact

Unfortunately, most of the data and knowledge required to create these models are decentralized or currently nonexistent. Process modeling for a variety of machining processes has been extensively studied in the absence of MWF.[22] However, models that can account for the impact of different MWF application strategies and chemistries on tool wear, part quality, heat generation, and chip morphology do not exist. Work has recently begun to understand the transport of MWF from the metal-cutting process (see Section 7.3), but the environmental and health ramifications of MWF exposure are not fully understood.

The process-planning strategy outlined in Figure 3 demonstrates the relationship between process knowledge and the ability to improve environmental performance. For instance, deficiencies in the literature regarding part quality and productivity as a function of MWF chemistry and application volume are a major deterrent in identifying opportunities to modify MWF chemistry and application volume in a manner that might improve environmental performance. This example supports the general experience of many pollution-prevention researchers who have concluded that environmental improvement is generally bounded by the current state of process knowledge.

Some researchers have begun to bridge this gap by integrating environmental data and process performance data. One example of this is the Cutting Fluid Evaluation Software Testbed (CFEST), illustrated in Figure 4.[23] CFEST represents a framework within which MWFs can be compared in machining. The integrated computer model allows for the input of MWF data, process data, and

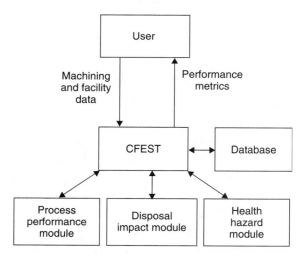

Figure 4 Cutting Fluid Evaluation Software Testbed (CFEST) concept.

site-specific data and returns metrics of MWF performance in terms of cutting forces, temperatures, treatment costs, mist droplet size, and relative health and safety. However, CFEST is limited by available data on MWF formulations, MWF transport, and human health impacts. Comprehensive process planning in the future will require tool wear, force, and workpiece quality models, coupled with this type of selection tool, to reveal exact function of MWF in a given operation and opportunities for dry machining, volume reduction, and recycling. Current research in these areas is discussed in Sections 7 and 8.

7 MWF POLLUTION PREVENTION THROUGH PROCESS MODIFICATION

A study in the German automotive industry revealed that workpiece-related manufacturing costs associated with MWFs (7–17%) were several times higher than tool costs (2–4%).[24] These costs would be eliminated if machining processes could be conducted without MWF. Environmental impact and health risks directly associated with MWFs would also be eliminated. These opportunities have encouraged research into machining without MWF (dry machining) and into minimizing MWF usage. Dry machining is discussed in Section 7.1. Minimal MWF application strategies are discussed in Section 7.2.

Other process modifications currently under active research aim to reduce worker exposure to MWF mist and to reduce the health risks and environmental impact of MWF chemistries. Section 7.3 describes current efforts to reduce worker exposure to mist. Section 7.4 describes environmental and health considerations in MWF formulation.

7.1 Dry Machining

Replacing the functions of MWF during dry machining has proven challenging. This is largely due to the fact that MWFs perform multiple functions simultaneously. These include lubricating the workpiece and tool, conducting heat from the cutting zone, inhibiting corrosion of the workpiece, cleaning the workpiece, and flushing chips from the cutting zone. The relative importance of these MWF functions depends on the operation, workpiece and tool material, cutting speed, feed rate, and depth of cut.

The absence of MWF can have negative impacts on machining. The absence of cooling and lubrication leads to a temperature increase due to friction. This can result in accelerated tool wear, residual stresses within the component being machined, dimensional errors, poor surface finish, and metal chip build-up on both the tool and workpiece. The absence of active chip removal from the cutting zone can also lead to high temperatures, as well as tool failure.

Several technologies have been researched to compensate for the cooling functions of cutting fluid during dry machining. Internal tool cooling, cryogenic systems, thermoelectric systems, and air cooling systems have all been tested for their ability to keep process temperatures low and reduce tool wear. Altering the geometry of the process to favor increased heat removal by chips ejected from the process has also been shown to keep process temperatures low in certain operations. Altering the tool geometry to reduce the amount of contact between chips and tool can have the same effect.[25]

Special tooling has been developed to replace lubrication and chip evacuation functions of cutting fluid during dry machining. Carbide, cubic boron nitride, silicon nitride, and diamond-coated tools all have some application in dry machining for this purpose. These materials all feature high-temperature hardness and wear resistance. Special tooling produced with advanced geometries can also perform chip control and chip-breaking functions, assisting chip-removal functions in the absence of cutting fluid. Vacuum and air-jet systems can also remove chips from the cutting zone without the need for cutting fluid.

Operating conditions and materials that tend to produce short chips, low cutting forces, and low temperatures are most amenable to dry material removal. Young et. al (1997) has ranked the technical capabilities of dry machining in different applications.[26] Table 3 lists operations that have been researched under dry conditions in order of increasing difficulty from milling to deep drilling. It also describes the current applicability of common workpiece materials in these operations. Cast iron materials have been traditionally used in turning and milling without cutting fluid. However dry drilling, reaming, and tapping cannot currently be conducted at profitable rates with cast iron.[24] The development of new tool coatings and geometries holds some promise to increase profitability and to extend the applicability of dry machining in these operations.[27] However, grinding, honing, and lapping will remain as even greater challenges to dry machining for some time.

7.2 Minimal MWF Application

Difficulty in operations such as dry grinding has lead to research of applications that can be described as "minimal MWF application strategies." Instead of

Table 3 Applicability of Dry Machining Operations to Different Workpiece Materials

Operation	Cast Iron	Non-Ferrous Alloys	Construction Steel	Heat Treated Steel	Cast Aluminum Alloy
Milling	●	●	●	●	●
Turning	●	●	●	●	●
Boring	●	●	⊙	⊙	⊙
Tapping	●	●	⊙	⊙	⊙
Reaming	⊙	●	⊙	⊙	⊙
Deep drilling	⊙	○	⊙	⊙	⊙

● Currently Possible ⊙ Currently Under Development ○ Currently Not Possible
Note: Adapted from Ref. 26.

flooding the tool and workpiece with MWF quantities between 5 and 40 gallons per minute, minimal application strategies attempt to apply MWF directly to the cutting zone in amounts no greater than actually required by the process. The benefits of minimal MWF usage include reduced acquisition, maintenance, recycling, and disposal costs. It also reduces operator exposure to mists.

Recent research has been directed toward determining the minimum amount of MWF required by a given process. For instance, it has been found that 40,000 times less MWF could be used in some turning operations compared to normal application rates without increasing friction on the tool and without decreasing tool life.[28] These results were a significant improvement over dry machining, which was found to reduce tool life in the same operations. In another application it was found that MWF usage in typical grinding operations of the German automotive industry could be reduced 90 percent without increases in cutting forces or grinding wheel wear.[7]

Despite the ability of minimal MWF strategies to extend the applicability of dry machining, these strategies ultimately have similar limitations to dry machining and are not currently effective in severe operations. This is because while the lubricating effect of MWF is maintained under minimal application strategies, cooling and chip-flushing functions of MWF are lost. Thus, minimal cutting-fluid applications must still utilize alternate technologies to cool and remove chips from the cutting zone (see Section 7.1). However, minimal cutting-fluid strategies do provide significant benefits over dry machining under conditions where extreme pressure additives can be most active to minimize heat generation and where chip flushing can be handled relatively easily by air or vacuum systems.

7.3 MWF Mist Reduction

The degree of respiratory hazard associated with MWFs is directly proportional to worker exposure to MWF mist. As mentioned earlier, this has motivated NIOSH to recommend that MWF aerosol standards be lowered an order of magnitude compared to current limits. Attempts to reduce in-plant mists generated by MWFs have taken three forms:

1. Improved enclosure and ventilation of machine tools
2. Adjustment of machining parameters to reduce mist formation
3. Formulation of MWFs with specialty additives to supress mist

First, improved ventilation systems and enclosed machine tools have been applied toward reducing mists from machining. This can be achieved reasonably well in many facilities and operations. However the costs can be significant and reducing operator access to the process is not always feasible.

Second, effort has begun to understand the fundamental mechanisms driving mist formation.[29] The ultimate object of this effort is to determine if operating parameters or the geometry of the process can be adjusted to reduce mist. Three primary mechanisms have been identified as causing mist in turning: high temperature in the cutting zone causing fluid evaporation, spin-off of fluid from the workpiece due to rotational motion, and the splash due to the kinetic energy of fluid-workpiece impact.[30] It has been found that reducing application volume, fluid velocity, rotation speed, and workpiece diameter can significantly reduce mist, but these variables are generally fixed in a given operation. However, it may be possible to modify the geometry of the process in certain situations to achieve mist reductions.

Third, MWF polymer additives have been developed that reduce worker exposure to mists.[31,32] These mist-suppressant additives encourage the formation of larger MWF droplet sizes, which have a lower residence time as mists in the air before contacting the floor. In-plant testing of one mist suppressant showed a decrease in mist levels over 40 percent for a period of up to two weeks. Potential drawbacks of mist suppressants include frequent replenishing and high cost.

7.4 MWF Formulation Considerations

The toxicological impact of a MWF depends on a variety of ingredients and additives. In some cases, the toxicity of a given MWF ingredient may be known in terms of a parameter called lethal dose 50 (abbreviated LD_{50}). LD_{50} is the single-exposure dose of substance required to cause death in 50 percent of test animals exposed. Toxicity data resulting from chronic exposure to substances at lower doses are more rare, even though chronic exposures can have hazardous ramifications at lower doses than the LD_{50}. Although some data exist regarding individual MWF ingredient toxicity, these data are not available for MWF mixtures. This fact, coupled with a general lack of performance data for MWF mixtures and the market-driven confidentiality of MWF formulations, severely limits the ability to select MWFs with the aim of reducing potentially toxic effects.

From the environmental perspective, MWFs formulated to be more amenable to recycling and waste-treatment are preferred. The proposed Metal Products and Machinery Rule would place higher demands on on-site waste treatment prior to municipal sewering and would likely encourage more MWF recycling. Generally, synthetic MWFs are the most difficult and costly to waste treat. This is due to the requirement of biological treatment to remove BOD and FOG associated with highly water-soluble ingredients. However synthetics generally have a longer life and greater bacterial resistance, and they are more amenable to recycling. The compatibility of common MWF ingredients with waste-treatment operations is discussed in detail by Childers et al.[33]

8 MWF POLLUTION PREVENTION THROUGH IN-PROCESS RECYCLING

Ideally, in-process recycling involves a perfect separation of contaminants from MWFs, returning the MWF to its initial state before it was exposed to metalworking processes. MWF contaminants to be removed by the recycling process include microorganisms, machining particulate, and oils. They lead to reduced quality of machining over time, costly disposal and acquisition, frequent loadings of environmental pollutants (e.g., BOD and FOG), and risks of respiratory illness in workers. Eliminating contaminants can extend the life of MWFs, reduce costs, improve the quality of manufacturing, reduce health risks, and decrease environmental loadings. However, contaminant removal cannot address issues such as evaporation, pH reduction due to microbial growth, and adsorptive loss of surfactants due to exposure to metal surfaces. Thus, even under ideal recycling scenarios, chemical maintenance is still required to maintain MWF functionality over extended periods.

Some form of MWF recycling is almost always performed, driven by costs of MWF acquisition and disposal. In practice, MWFs are used until either bacterial infection or performance has reached unacceptable levels. Thus, simply inhibiting biological growth, even without necessarily removing it, can be considered *recycling* since this alone can extend the life of MWFs in process. General principles and specific technologies that have been applied to MWF recycling in manufacturing are listed in Figure 5.

The following sections discuss the specific recycling technologies grouped by their primary principle of operation. Section 8.1 discusses phase-separation technologies. Section 8.2 describes magnetic separation. Section 8.3 details filtration technologies. Section 8.4 surveys biological inactivation techniques used for microbial reduction in MWFs.

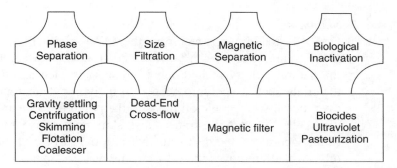

Figure 5 Technologies used for MWF recycling categorized by mode of operation.

8.1 Phase Separation

Although large particulate and free oil contaminants get mechanically mixed with MWFs, they do not dissolve in MWFs. Thus, if a contaminated MWF were left for some time in a large tank, the particulate (with density greater than the MWF) would eventually fall to the bottom of the tank and could be removed. This process is called *gravity settling*. During gravity settling, oil contamination also rises to the surface since oil is less dense than water. All oil and chip control techniques described in this section exploit this fundamental separation force between different phases (e.g., oil, chips, and MWF). Skimming, coalescing, and flotation supplement the gravitational separation force with the natural affinity of oil and chips to other hydrophobic substances such as air or plastic. Centrifuges and hydrocyclones replace the gravitational force with centrifugal force.

In considering the following separation technologies, an important distinction needs to be made between contaminant oils that exist in a free state and those that exist bound in a chemically emulsified state. Contaminant oils can become bound and stabilized in water due to emulsifiers common to MWF formulations. Emulsified contaminant oils (called *tramp oils*) are a major concern because they provide an excellent growth medium for bacteria, can reduce the cooling ability of the MWF, and can destabilize the MWF. Highly emulsified oils cannot be easily separated with any recycling technique.

Gravity Settling

Gravity settling is limited by the amount of time required for settling to occur, especially since MWFs are used in high volume in machining processes. The separation velocity of a particle is directly proportional to gravitational acceleration, the density difference between particle and fluid, and the particle diameter. Generally, the slow separation velocities of particles smaller than 50 microns limit the use of gravity settling to very large chips and free oils.

Skimming

Free oils readily separate from the MWF and will rise to the surface of an MWF sump. The oil can be removed from the surface by taking advantage of the natural affinity of hydrophobic materials such as oil for other hydrophobic materials such as polypropylene. As polypropylene ropes, belts, or disks pass over the oil surface, the oil binds to the polypropylene and can be removed from the MWF. Figure 6 illustrates a disk-skimming device attached to a process tank. In this geometry, the disk spins beneath the surface of the fluid and attracts free oils. A blade removes the oil from the disk for separate collection.

Coalescers

Coalescers also utilize the natural affinity of oil to plastic. Byers et al.[5] describes the use of the coalescing principle in a vertical geometry. As an oil-contaminated

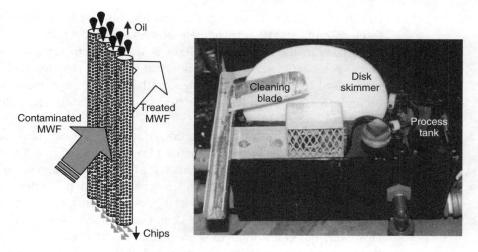

Figure 6 Vertical coalescer and disk skimmer.

feed-stream flows through perforated plastic material, the oil concentrates within the tubes and rises out of the tank. Figure 6 shows the basic operation of a vertical coalescer. Several geometrical advances with increased surface area of coalescence have been developed and are widely available. Often, skimming devices are used to remove oil separated in a coalescing system. Removal efficiencies of free oil can exceed 99 percent under some conditions.

Flotation

Flotation uses the natural affinity of metal chips and oils to bubbles of air rising through a MWF. The bubbles attach to the contaminants, and the buoyant force of the bubble causes the particle to rise to the surface. Particles that have a higher density than the MWF can thus be made to rise to the surface if their mass is not too large. The rising of oil particles with a lower density than the MWF can also be facilitated by the injection of air bubbles.[34]

Sköld describes the use of flotation to enhance separation in a novel device resembling a hydrocyclone.[35] In small-scale pilot operation, the device was able to remove over 70 percent of the particulate and a significant fraction of the contaminant oil. The full performance characteristics of this technology have not yet been tested. Flotation has also been used in conjunction with settling and pasteurization in a system described by Schenach.[36] Small-scale experimentation with this system demonstrated particulate and tramp oil removal as well as biological inactivation.

Centrifugation

Replacing the gravitational force in gravity settling with centrifugal force can significantly increase separation velocities for contaminants both heavier and lighter than water. Centrifuges can utilize up to 10,000 times the gravitational force to remove MWF particulate down to 1 μm.[37] One key advantage of centrifugation in the MWF application is its ability to separate out both oils and chips. Although some bacterial removal has been reported with centrifuges typically used for MWF recycling, generally centrifugation must be coupled with a biological inactivation technique such as pasteurization technology and biocides to control bacterial populations.[13,38]

The operation of a disk-stack centrifuge for oil and solids separation is diagrammed in Figure 7. Contaminated MWF is injected into a spinning bowl with a stack of disks. Given the geometry of the feed channel, solids will tend to settle beneath the disk stack and migrate toward the outer wall of the bowl. The remaining oil-contaminated MWF, with a density lower than the solids, can begin migration back toward the center of the bowl, through the stacks of disks. However, the centrifugal force will also separate the lighter oil contamination from the heavier MWF to be recycled. An interface between the two phases forms at some diameter from the center of the centrifuge. A properly designed disk centrifuge will have holes at that precise diameter to allow upward migration of the heavier MWF toward an outlet port for recycling. The lighter oil is discharged through a second outlet closer to the center of the centrifuge and can generally be recycled, burned, or disposed based on its oil content.

Besides three-phase separation capability, centrifuge advantages include the absence of media that need to be replaced, lower space requirements compared

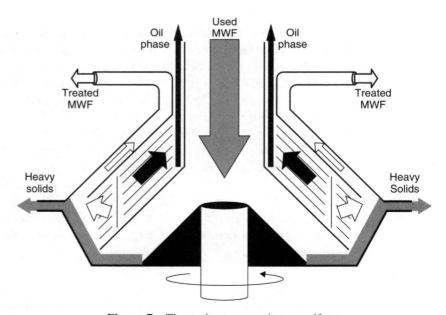

Figure 7 Three-phase separation centrifuge.

to settling tanks, high recycling flow rates, and constant, predictable output. Centrifuges, by contrast, also have significant capital costs associated with them and can require frequent bowl and disk cleaning in MWF applications. The centrifuge also tends to have more frequent downtime in the long-term owing to the high process rates and abrasive nature of MWF particles.[5]

Hydrocyclone

The hydrocyclone also exploits density separation due to the centrifugal force. Unlike the centrifuge, hydrocyclones have no moving parts. In the hydrocyclone, MWF is impinged upon a vortex-encouraging conical chamber. A pressure gradient formed between the center of the vortex and the fluid at the outer wall of the hydrocyclone provides the accelerated separation of high-density particulate and low-density MWF. The particulates concentrate toward the outer wall and fall under the force of gravity out the bottom of the hydrocyclone. Less-dense constituents of the MWF migrate to the center of the vortex and return back to the process, as shown in Figure 8. This process is generally effective for removing particles greater than 10 μm.

8.2 Magnetic Separation

Magnetic separation employs magnetic force to remove ferromagnetic solids (e.g., cast iron chips) from MWFs. A typical unit consists of a nonmagnetic drum lined on the inside with four to six stationary magnets. As MWF makes contact with the rotating drum, magnetic particles are attracted to the drum and are removed from the MWF by a blade. Apart from being limited to ferrous materials, magnetic separators generally have low throughput. Usually, they serve as

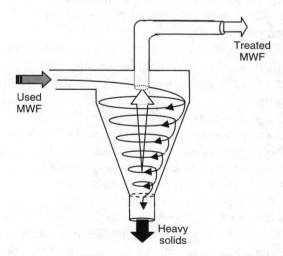

Figure 8 Hydrocyclone concept. (Adapted from Ref. 39.)

pretreatments to remove particles that will easily plug filters used in subsequent recycling steps. They are generally effective for removing magnetic particles greater than 10 µm.

8.3 Filtration

Filtration consists of the separation of two phases using a physical barrier. In the case of MWFs, this usually is the separation of chips. Large oil droplets can also be removed by filtration and, depending on the affinity of oil for the filter media, smaller oil droplets can also be removed. Bacteria can also be separated in the case of membrane filtration. Traditional filters used in MWF recycling operate with the MWF entering the filter in the same direction it exits. This is called *dead-end filtration*. In contrast, membrane filtration typically operates with the MWF entrance direction perpendicular to the direction of filtration. This is called *cross-flow filtration*.

Dead-End Filtration

Dead-end filters can be disposable or permanent. From the standpoint of the environment, permanent filters are preferred because disposable filters present a solid waste problem of their own. However, to the extent that in-plant hazards and extension of MWF life can be achieved using filtration, disposable filters can be warranted from the environmental perspective where permanent filters are ineffective or uneconomical.

A wide assortment of filtration materials (ranging from woven textiles to paper to diatomaceous earth) is commonly used. Material selection depends most on filtration rate and retention characteristics of the material in a given filter configuration. Material strength, resistance to creep, resistance to abrasion, chemical and biological stability, disposability, cost, and many other factors also influence material selection.[39]

Filtration can occur either at the surface or in the depth of the filter. Depth filters must be disposed due to an irreversible clogging of chips and oil within the filter matrix. Disposable depth filters used in MWF recycling may be in the form of bags, cartridges, or paper rolls. In contrast, surface filters are reused. They consist of a metal mesh, fabric belt, or wedge wire screen that serves to reject metal particulate found in MWFs. As rejected particulate accumulates on the surface of the filter, a cake layer is formed that serves to produce a finer filtration. As this secondary layer gets clogged, the cake layer is rinsed from the surface and the process begins again.

Dead-end filtration operations utilize either gravity, pressure, or vacuum as a driving force. As filters clog, the driving force required to maintain a given filtration rate increases, or the filtration rate decreases for a constant driving force. After some time, demands on driving force or declining filtration rates become unacceptable. Disposable media must be changed, and permanent media must be

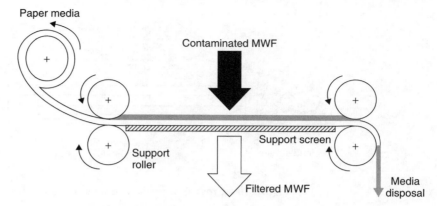

Figure 9 Gravity-driven paper-indexing filter.

cleaned. Permanent media can be cleaned with high-pressure air or metalworking fluid.

A common dead-end filtration operation used for MWF recycling is the gravity-driven paper-indexing filter illustrated in Figure 9. The contaminated MWF is fed onto the surface of a cellulose filter paper supported by a metal screen. When the process begins, the filtration rate is a maximum, but over time filter plugging causes filtration rate decline to uneconomical levels. At this point, the paper roll is indexed such that the MWF is exposed to fresh filter media. The used paper media is then disposed.

Cross-Flow Membrane Filtration
Membrane filtration utilizes a semi-permeable barrier or *membrane* capable of separating feed stream constituents according to their particle size relative to the pore sizes of the membrane.[40–42] Membranes can separate MWF ingredients from bacteria, tramp oils, and chips. The cross-flow membrane filtration concept is illustrated in Figure 10.

Membrane filtration is a surface filtration similar to that described for the permanent filters. However, membrane filters do not depend on a cake layer for fine separation. Instead a porous membrane support is lined with a thin skin where the surface filtration takes place. Pore sizes that have been used in MWF recycling range from 0.01 to 0.1 μm (ultrafiltration) and 0.1 to 1.5 μm (microfiltration). Also in contrast to dead-end filters, cross-flow membrane filtration is performed with filtration tangential to the channels of bulk fluid flow, as shown in Figure 10. This discourages the accumulation of particles on the membrane surface and allows for simple flushing or cleaning to regenerate the filter.

Membrane filtration has been well established as a means to remove not only contaminants, but also the bulk of oil- and oxygen-demanding ingredients of MWFs prior to municipal disposal.[43] As a pretreatment to disposal, the MWF

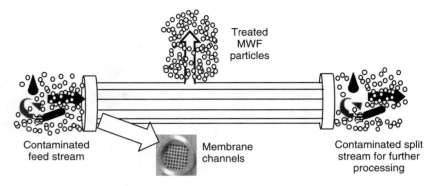

Figure 10 Cross-flow membrane filter. (Adapted from Ref. 4.)

ingredients are removed to the largest degree possible. When recycling an MWF back to a manufacturing process, however, the goal is to selectively remove particulate, oil, and bacteria from the MWF without removing any of the MWF ingredients. Achieving the separation of contaminants without disruption of the MWF chemistry is the major challenge for membrane filtration.

Figure 11 compares the particle sizes of MWF ingredients and contaminants to the pore sizes of membrane filters. On average, MWF particle sizes increase from synthetic to semi-synthetic to soluble oil formulations.[5] As seen in Figure 11, semi-synthetic and soluble oil formulations have effective particle sizes on the same order of magnitude or even larger than MWF contaminants. Therefore, membrane-filtration applications to semi-synthetic and soluble oils have been more complicated, although there has been some success in limited applications.[11]

Further complicating matters for membrane filtration recycling of MWFs is that ingredients common to synthetic, semi-synthetic, and soluble oil formulations can adsorb to membrane filters, causing loss of MWF functionality and decline of filtration rate.[4] In addition, as surfactants present in MWFs emulsify tramp oils, they are separated from the MWF and removed with the tramp oil. This can

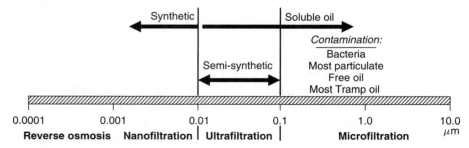

Figure 11 Size of MWF particles and contaminants compared to pore sizes of membrane filters.

destabilize the MWF and cause a loss of functionality. Due to the complex nature of the technology and the large sensitivity of membrane filtration performance to MWF chemistry, most membrane-filtration research has remained at the academic level, although limited industrial research and implementation has also occurred.

Consideration of membrane filtration for MWF recycling especially highlights the relationship between environmental improvement and process knowledge. Selection of one MWF formulation over another in a given operation can sometimes be arbitrary, but that selection can also determine whether membrane filtration can be used for recycling. As was seen with dry machining and minimal lubrication technologies, the informed implementation of membrane filtration for MWF recycling in-process will require fundamental knowledge regarding the role MWF chemistry plays in specific operations.

8.4 Biological Inhibition

Biological inhibition is not a recycling technology on its own. However, it is a required part of any MWF recycling system that cannot specifically remove microorganisms at a high enough rate to keep microbial populations low. Biological inhibition technologies that have been extensively tested include biocides, pasteurization, and ultraviolet irradiation.

Biocides

In the absence of biological inhibiting chemicals, water-soluble MWFs are essentially nutrients for bacterial and fungal growth. Thus, water-soluble formulations require biocides to keep bacterial populations from metabolizing corrosion inhibitors and surfactants, reducing product lubricity, and creating rancid odors. A wide variety of biocides are available because different microorganisms respond to different biocides and because biocide performance is sensitive to different MWF formulations.[5] Biocides lose activity over time and must be added regularly (and sometimes on an emergency basis) to maintain microbial populations at sufficiently low levels.

Heat Pasteurization

Heat pasteurization has been traditionally used to improve the storage life of products such as wine and milk. Bulk pasteurization of milk involves heating milk to approximately 65°C for 30 minutes. More typically in modern operations, a flash pasteurization of milk is conducted at 71°C for 15 seconds.[44] Flash pasteurization is relatively ineffective for reducing MWF bacterial numbers to sufficiently low levels.[45] Longer pasteurization times have been more effective in reducing bacterial populations in MWFs.[36,46]

There are practical limits to heat pasteurization imposed by the amount and frequency of heating that can be conducted with MWFs that serve as coolants. In addition, heat pasteurization does not kill all bacteria and cannot offer residual biological inhibition after heating. Researchers have also found that resistance to heat pasteurization can develop in MWF bacteria, such that bacterial regeneration times increase after successive heat treatments.[47]

For these reasons, pasteurization should be used in conjunction with biocides. It has been observed that biocides and pasteurization are more effective when conducted together.[5] Such a scheme could also lower biocide consumption and hence reduce occupational risks posed to workers by biocides (see Section 5).

UV Radiation

Ultraviolet radiation has sufficient energy to cause breaks in DNA leading to microbial death.[43] Generally, the UV radiation technique has been limited in its application to MWFs due to economics and the fact that the UV radiation cannot penetrate deep into MWFs. Research is currently being conducted to improve the penetrability of UV radiation into MWFs.[48]

8.5 Conclusions: In-Process Recycling

The MWF recycling technologies discussed in this section will never create a truly closed-loop process. Evaporation and loss of active ingredients such as surfactants, corrosion inhibitors, and lubricants will always occur by the nature of metalworking and by ever-present microorganisms. However, these in-process

Figure 12 Technologies applicable to microorganism, oil, and chip control.

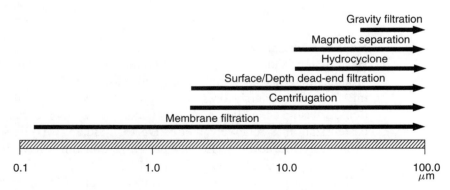

Figure 13 Separation performance of various MWF contaminant removal technologies. (Adapted from Ref. 38.)

recycling technologies can greatly extend MWF life, reduce costs and disposals, reduce health risks to workers, and improve the quality of manufacturing. The ability of in-process recycling to achieve these benefits depends on its ability to control levels of bacteria, oils, and particulate in MWF without disrupting the MWF chemistry. Figure 12 reviews the contaminants that can be removed by each technology discussed in this section. Approximate separation levels achievable by these technologies is provided in Figure 13.[38]

9 CONCLUSION

Metalworking fluids (MWFs) perform a variety of functions that make them essential to the metal products manufacturing industry. However, MWFs also have inherent environmental and health liabilities associated with them that are of concern to the industry. These liabilities can be viewed as an opportunity to reevaluate the functionality of MWFs and improve profitability of operations.

Improving environmental performance and profitability simultaneously will require informed process planning, novel process modifications, and more aggressive efforts at in-process recycling. Specific examples discussed in this chapter include comprehensive systems modeling of machining operations, dry machining, minimal MWF application strategies, mist reduction, environmentally conscious MWF formulation, and contaminant control strategies. To a large degree, industrial movements in these directions are inhibited by a paucity of models to account for MWF functionality in metalworking operations. Such models will be required to improve the environmental performance of the industry in the future. Parallel research into improving contaminant control and chemical maintenance can extend the life of MWFs currently in use.

Environmentally conscious MWF usage in the future will likely be a mixture of the strategies discussed in this chapter. Where dry machining is inapplicable or uneconomical, minimal MWF consumption strategies—in concert with comprehensive contaminant control strategies to control particulate, oils, and bacteria—will likely be implemented where possible. Such a combined reduce/reuse strategy would have benefits of improved MWF quality, reduced costs, reduced worker exposure to MWF mists, and reduced disposal. Once again, however, the requisite process models to guide this strategy remain at the research stage of development.

In some machining operations, current research has suggested that higher productivity gains will only be achieved with higher volumes of MWF consumption applied at higher pressure. In such cases, environmentally conscious process planning directed toward a closed-loop recycling strategy with minimal mist production will be required to reduce health risks, costs, and disposal volumes. The use of formulations that are more bio-resistant and compatible with recycling technologies will greatly assist achieving these goals.

10 ACKNOWLEDGMENTS

This work was made possible by the NSF/DARPA Machine Tool Agile Manufacturing Research Institute (award number DMI-9320944) and the Illinois Waste Management and Research Center. The author would also like to thank Professor Richard E. DeVor, Professor Shiv G. Kapoor, Dr. Kishore Rajagopalan, Dr. Timothy C. Lindsey, Chris Harris, and Laura Skerlos for their assistance and contributions to this effort.

REFERENCES

1. United States Environmental Protection Agency: Office of Pollution Prevention and Toxics, *Pollution Prevention Options in Metal Fabricated Products Industries*. EPA Contract No. 68-W0-0027, 1992.
2. ILMA (Independent Lubricants Manufacturing Association), "*ILMA Report*," *Lubricants World* (November 1998), 10.
3. S. Oberwalleney and P. Sheng, *Framework for an Environmental-Based Cutting Fluid Planning in Machining Facilities*. University of California at Berkeley: Department of Mechanical Engineering, 1996.
4. S. J. Skerlos, N. Rajagopalan, R. E. DeVor, and S. G. Kapoor, "Ingredient-Wise Study of Flux Characteristics in the Ceramic Membrane Filtration of Uncontaminated Synthetic Metalworking Fluids," *Journal of Manufacturing Science and Engineering, Transactions of ASME* (February 1999).
5. J. P. Byers (ed.), *Metalworking Fluids*, Marcel Dekker, New York, 1994.
6. R. S. Marano, G. S. Cole, and K. R. Carduner, "Particulate in Cutting Fluids: Analysis and Implications in Machining Performance," *Lubrication Engineering*, **47**, 376–382 (1991).
7. F. Klocke and T. Beck, "Zufuhrsysteme für den anforderungsgerechten Kühlschmierstoffeinsatz," Laboratorium für Werkzeugmaschinen und Betriebslehre. Aachen, Germany, 1996.
8. N. Rajagopalan et al., "Pollution Prevention in an Aluminum Grinding Facility," *Metal Finishing*, **96**, 18–24 (1998).
9. S. J. Skerlos, R. E. DeVor, and S. G. Kapoor, "Environmentally Conscious Disposal Considerations in Cutting Fluid Selection," ASME International Mechanical Engineering Congress and Exposition, in J. Lee, (ed.) *Proceedings of the ASME: Manufacturing Science and Engineering Division*, Anaheim, CA, Vol. 8 (November 1998).
10. NIOSH, *Criteria for a Recommended Standard: Occupational Exposure to Metalworking Fluids,* National Institute of Occupational Safety and Health, Cincinnati, OH, 1998.
11. S. J. Skerlos, and N. Rajagopalan, "Putting the Squeeze on Metalworking Fluids," *EPA Region 5 Waste Minimization Conference*, Chicago, December 1998.
12. G. Masters, *Introduction to Environmental Engineering and Science*, Prentice-Hall, Englewood Cliffs, NJ, 1991.
13. United States Environmental Protection Agency, *Development Document for the Proposed Effluent Litigations Guidelines and Standards for the Metal Products and Machinery Phase I Point Source Category*. EPA 921-R-95-021, 1995.

14. J. Kulowiec, "Techniques for Removing Oil and Grease from Industrial Wastewater," *Pollution Engineering*, **11**(49) (1979).
15. B. Savonius, H. Keskinen, M. Tuppurainen, and L. Kanerva, "Occupational Asthma Caused by Ethanolamines," *Allergy* **49**(10), 877–881 (1994).
16. H. W. Rossmore, "Antimicrobial Agents for Water-Based Metalworking Fluids," *Journal of Occupational Medicine* **23**(4) (1981).
17. E. O. Bennett, "The Biology of Metalworking Fluids," *Lubrication Engineering*, **28**, pp. 237 (1972).
18. M. Lee and A. C. Chandler, "A Study of the Nature, Growth, and Control of Bacteria in Cutting Compounds," *Journal of Bacteriology*, **41**, 373–386 (1941).
19. L. A. Herwaldt, G. W. Gorman, T. McGrath, S. Toma, B. Brake, A. W. Hightower, et al, "A New *Legionella* Species, *Legionella feeleii* Species, Nova Causes Pontiac Fever in an Automobile Plant," *Annals of Internal Medicine*, **100**, 333 (1984).
20. C. Zugerman, (1986). "Cutting Fluids: Their Use and Effects on the Skin," *Occupational Medicine: State of the Art Reviews* (1986).
21. C. R. Mackerer, "Health Effects of Oil Mists: A Brief Review," *Toxicology and Industrial Health*, **5**(3), 429–440 (1989).
22. S. G. Kapoor, R. E. DeVor, R. Zhu, R. Gajjela, G. Parakkal, and D. Smithey, "Development of Mechanistic Models for the Prediction of Machining Performance: Model building Methodology," *Machining Science & Technology*, **2**(2), 213–238 (1998).
23. J. W. Sutherland, T. Cao, C. Daniel, Y. Yue, Y. Zheng, P. Sheng, D. Bauer, M. Srinivasan, R. E. DeVor, S. G. Kapoor, and S. J. Skerlos, "CFEST: An Internet-Based Cutting Fluid Evaluation Software Testbed," *Transaction of NAMRI/SME*, **25** 243–248 (May 1997).
24. F. Klocke and G. Eisenblatter, "Dry Cutting," *Annals of the CIRP*, **46**, 1–8 (1997).
25. T. Grasson, "Cloudy Forecast for Dry Machining," *American Machinist* (March, 1997), 47–50.
26. P. Young, G. Byrne, and M. Cotterell, "Manufacturing and the Environment," *International Journal of Advanced Manufacturing Technology*, **13**(7), 488–493 (1997).
27. A. P. Malshe, W. F. Schmidt, M. A. Taher, H. Mohammed, A. Muyshondt, and H. Mohammed, "Comparative Study of Dry Machining of A390 Alloy Using PCD and CVD Diamond Tools," Technical Paper—Society of Manufacturing Engineers, pp. 1–6, MR 1998.
28. T. Wakabayashi, H. Sato, and I. Inasaki, "Turning Using Extremely Small Amounts of Cutting Fluids," *JSME International Journal Series C-Mechanical Systems Machine Elements & Manufacturing*, **41**(1), 143–148 (1998).
29. Yan Yue, J. W. Sutherland, and W. W. Olson, "Cutting Fluid Mist Formation in Machining via Atomization Mechanisms," *Design for Manufacturing and Assembly*, **89**, 37–46 (1996)
30. D. Bell, J. Chou, L. Nowag, and S. Liang, "Modeling of the Environmental Effect of Cutting Fluid," *Lubrication Engineering*, **42**(1), 168–173 (1999).
31. Sanjay Kalhan, Steve Twining, Richard Denis, Richard Marano, and Rebecca Messick, "Polymer Additives as Mist Suppressants in Metalworking Fluids Part IIa: Preliminary Laboratory and Plant Studies—Water-soluble Fluids," *Design and Manufacture for the Environment SAE Special Publications*. vol. 1342 SAE, Warrendale, PA, 1998, p. 47–51.
32. R. Marano, R. Messick, J. Smolinski, and L. Toth, "Polymer Additives as Mist Supressants in Metalworking Fluids Part I: Laboratory and Plant Studies—Straight Mineral Oil

Fluids," SAE International Congress and Exposition, Detroit, MI, February 27–March, 2, 1995.
33. J. Childers, S. J. Huang, and M. Romba, "Metalworking Fluid Additives for Waste Minimization," *Lubrication Engineering* (June 1990).
34. L. Metcalf, and H. Eddy, *Wastewater Engineering*. McGraw-Hill, New York, 1991.
35. Rolf O. Sköld, "Three Novel Physical Methods of Contaminant Control in Aqueous Metalworking Fluids," *Lubrication Engineering* **52**(5), 393–400 (1996).
36. T. Schenach, "Simple Process for the Recycling of Spent Water-Based Metalworking Fluids," *Lubrication Engineering* **55**(2), 15–22 (1999).
37. W. W-F. Leung, *Industrial Centrifugation Technology*, McGraw-Hill, New York, 1998.
38. P. S. Sheng and S. Oberwalleney, "Life-cycle Planning of Cutting Fluids—A Review," *Journal of Manufacturing Science & Engineering, Transactions of the ASME*, **119**(4B), 791–800 (1997).
39. R. J. Wakeman and E. S. Tarleton New York, *Filtration: Equipment Selection, Modeling, and Process Simulation*, Elsevier Advanced Technology, New York, 1999.
40. R. R. Bhave, *Inorganic Membranes: Synthesis Characteristics and Applications*, Van Nostrand Reinhold, New York, 1991.
41. M. Cheryan, *Ultrafiltration Handbook*. Technomic, Pennsylvania, 1986.
42. W. S. W. Ho and K. K. Sirkar, *Membrane Handbook*, Van Nostrand Reinhold, New York, 1992.
43. ILMA (Independent Lubricants Manufacturing Association), *Waste Minimization and Wastewater Treatment of Metalworking Fluids*, 1990.
44. M. Madigan, J. Martinko, and J. Parker, *Brock Biology of Microorganisms*, 8th ed., Prentice Hall, Englewood Cliffs, NJ, 1997.
45. J. Wright, "Coolant Pasteurization—A Promising Answer?" *Manufacturing Engineering* **93**(5), 83–84 (1984).
46. J. Howie, T. Schenach, and S. Shurter, "Zero Coolant Discharge—It Can Be Done," *Society of Manufacturing Engineers*. Dearborn, MI. MR 92-376 (1992).
47. H. W. Rossmoore, L. A. Rossmoore, and A. L. Kaiser, *Biodeterioration 7*. D. Houghton, R. Smith, and O. Eggins, (eds.). Elsevier, New York, 1988, pp. 517–522.
48. M. Valenti, "Lighting the Way to Improved Disinfection," *Mechanical Engineering* (July 1997), 83–86.

CHAPTER 6
METAL FINISHING AND ELECTROPLATING

Timothy C. Lindsey, Ph.D
Illinois Waste Management and Research Center, Champaign, Illinois

1 METAL-FINISHING PROCESSES 124
   1.1 Electroplating 124
   1.2 Electroless Plating and Immersion Plating 125
   1.3 Chemical and Electrochemical Conversion 125

2 OTHER SURFACE FINISHING TECHNOLOGIES 126
   2.1 Cladding 126
   2.2 Case Hardening 126
   2.3 Dip/Galvanized 126
   2.4 Electropolishing 127
   2.5 Vapor Deposition of Metallic Coatings 127

3 THE FINISHING PROCESS 127

4 ENVIRONMENTAL AND REGULATORY ISSUES ASSOCIATED WITH METAL FINISHING 128
   4.1 Water Issues in Metal-finishing Operations 129
   4.2 Wastewater Issues in Metal-finishing Operations 131
   4.3 Solid and Hazardous Waste 132
   4.4 Air Emissions 133

5 FEDERAL REGULATIONS AFFECTING METAL FINISHING 133
   5.1 Clean Air Act 134
   5.2 Clean Water Act 134
   5.3 Resource Conservation and Recovery Act 135
   5.4 Toxics Release Inventory Reporting 137

6 ENVIRONMENTALLY CONSCIOUS MANUFACTURING PRACTICES 137
   6.1 Incremental ECM Practices 138
   6.2 Conductivity Control System in an Electroplating Shop 140

7 INNOVATIVE ECM TECHNIQUES 140
   7.1 Rinsewater Purification and Recovery Techniques 140
   7.2 Vendor Roles in ECM 142

Metal-finishing processes are utilized by many industrial sectors to improve the appearance and/or performance of their products. Through the prevention of corrosion and wear, metal-finishing processes can increase the lifespan of a wide range of products many times over. Additionally, finishing processes can be used to enhance electrical properties, form and shape components, and enhance the bonding of adhesives or organic coatings. Decorative appearances can also be enhanced through finishing processes.

Metal-finishing processes alter product surfaces to enhance the following:
- Appearance and reflectivity (color, brightness)
- Corrosion resistance
- Wear resistance
- Chemical resistance
- Hardness
- Electrical conductivity
- Electrical resistance
- Torque tolerance
- Solder performance
- Tarnish resistance
- Ability to bond to other materials

Based on the wide range of performance improvements that can be achieved through finishing processes, it is not surprising that these processes are utilized in industrial sectors as diverse as automotive, aerospace, electronics, heavy equipment, hardware, appliances, telecommunications, and jewelry.

1 METAL-FINISHING PROCESSES

Metallic coatings change the surface properties of the workpiece from those of the substrate to that of the metal being applied. Most industries that manufacture metal parts utilize metal-finishing processes that alter the surface of a workpiece to achieve a certain property. Common metal finishes include paint, lacquer, ceramic coatings, and other surface treatments. This chapter focuses mainly on the plating and surface treatment processes.

The metal-finishing industry generally categorizes plating operations as electroplating and electroless plating. Surface treatments consist of chemical and electrochemical conversion, case hardening, metallic coating, and chemical coating. The following sections briefly describe the major plating and surface treatment processes.

1.1 Electroplating

The *electroplating* process is performed by passing an electric current through a solution containing dissolved metal ions and the metal object to be plated. The metal object serves as the cathode in an electrochemical cell, attracting ions from the solution. Ferrous and nonferrous metal objects are plated with a variety of metals including aluminum, brass, bronze, cadmium, copper, chromium, gold, iron, lead, nickel, platinum, silver, tin, and zinc. The process is regulated by controlling a variety of parameters, including voltage and amperage, temperature, residence times, and purity of bath solutions. Plating baths are almost always aqueous solutions. Therefore, only those metals that can be reduced in aqueous solutions of their salts can be electrodeposited. The only major exception to this principle is aluminum, which can be plated from organic electrolytes.[1]

Plating systems can be operated in either batch or continuous mode. Typically, in batch operations metal objects are dipped into a series of baths containing various reagents for achieving the required surface characteristics. Operators can place the workpiece on racks or in barrels to carry the part from bath to bath. Barrels rotate in the plating solution and hold smaller parts.[2] In continuous plating operations, the plating lines are fully automated such that transfer of workpieces from tank to tank is accomplished with PLC-controlled systems.

The sequence of unit operations in an electroplating process is similar in both rack and barrel-plating operations. A typical plating sequence involves various phases of cleaning, rinsing, stripping, and plating. Electroless plating uses similar steps but involves the deposition of metal on metallic or nonmetallic surfaces without the use of external electrical energy.[1]

1.2 Electroless Plating and Immersion Plating

Electroless plating is accomplished through the chemical deposition of a metal coating onto an object using chemical reactions rather than electricity. The basic ingredients in an electroless plating solution are a source metal (usually a salt), a reducer, a complexing agent to hold the metal in solution, and various buffers and other chemicals designed to maintain bath stability and increase bath life. Copper and nickel electroless plating are commonly used for printed circuit boards.[3]

Immersion plating is a similar process in that it uses a chemical reaction to apply the coating. However, the immersion plating differs in that the reaction is caused by the metal substrate rather than by mixing two chemicals into the plating bath. This process produces a thin metal deposit by chemical displacement, commonly zinc or silver. Immersion-plating baths are usually formulations of metal salts, alkalis, and complexing agents (e.g., lactic, glycolic, or malic acids salts). Electroless plating and immersion plating commonly generate more waste than other plating techniques, but individual facilities vary significantly in efficiency.[3]

1.3 Chemical and Electrochemical Conversion

Chemical and electrochemical conversion treatments deposit a protective and/or decorative coating on a metal surface. Chemical and electrochemical conversion processes include phosphating, chromating, anodizing, passivation, and metal coloring. Phosphating prepares the surface for further treatment, especially painting. Chromating uses hexavalent chromium in a certain pH range to deposit a protective film on metal surfaces. Anodizing is an immersion process in which the workpiece is placed in a solution (usually containing metal salts or acids) where a reaction occurs to form an insoluble metal oxide. The reaction continues and forms a thin, nonporous layer that provides good corrosion resistance and pretreatment for painting. Passivating also involves the immersion of the workpiece into an acid solution, usually nitric acid or nitric acid with sodium dichromate. The passivating process is used to prevent corrosion and extend the life of the product. Metal coloring involves chemically treating the workpiece to impart a decorative finish.[1]

2 OTHER SURFACE FINISHING TECHNOLOGIES

In addition to the plating or chemical and electrochemical conversion processes, other commonly used finishing technologies include cladding, case hardening, dip/galvanizing, electropolishing, and vapor deposition. The following sections provide brief overviews of these processes.

2.1 Cladding

Cladding is a mechanical process in which the metal coating is metallurgically bonded to the work-piece surface by combining heat and pressure. An example of cladding is a quarter. The copper inside is heated and pressed between two sheets of molten nickel alloy, bonding the materials. Cladding is used to deposit a thicker coating than electroplating, and requires less preparation and emits less waste. However, equipment costs are higher than electroplating.[3]

2.2 Case Hardening

Case hardening is a metallurgical process that modifies the surface of a metal. The process produces a hard surface (case) over a metal core that remains relatively soft. The case is wear-resistant and durable, while the core is left strong and pliable. In case hardening, a metal is heated and molded and then the temperature is quickly dropped to quench the workpiece. An example of a material made with case hardening is the Samurai sword. The hardened surface can be easily shaped; however, the sword remains pliable. This method has low waste generation and requires a low degree of preparation. Operating difficulty and equipment cost are approximately the same as for anodizing, although case hardening imparts improved toughness and wear.[3]

2.3 Dip/Galvanized

Dip/galvanized coatings are applied primarily to iron and steel to protect the base metal from corroding. During the dipping process, the operator immerses the part in a molten bath commonly composed of zinc compounds. The metal part must be free of grease, oil, and other surface contaminants prior to the coating process. Operating difficulty and equipment costs are low, which makes dipping an attractive coating process for most industrial applications. However, dipping does not always provide a high-quality finish.[3]

2.4 Electropolishing

In electropolishing, the metal surface is anodically smoothed in a concentrated acid or alkaline solution. For this process, the parts are made anodic (reverse current), causing a film to form around the part that conforms to the macro-contours of the part. Because the film does not conform to the micro-roughness, the film is thinner over the micro-projections and thicker over the micro-depressions. Resistance to the current flow is lower at the micro-projections, causing a more rapid dissolution. Many different solutions are available for electropolishing, depending on the substrate.[2]

2.5 Vapor Deposition of Metallic Coatings

Vapor deposition allows the workpiece to become a composite material with properties that generally cannot be achieved by either material alone. The coating's function is usually as a durable, corrosion-resistant protective layer, while the core material provides a load-bearing function. Common coating materials include aluminum, coated lead, tin, zinc, and combinations of these metals. The facility applies the coating metal in a powdered form at high temperatures (800 to 1,100°C) in a mixture with inert particles such as alumina or sand, and a halide activator. Vapor deposition produces high-quality pure metallic layers, and can sometimes be used in place of plating processes.[4]

3 THE FINISHING PROCESS

Most objects undergo three stages of processing associated with metal finishing, each of which involves moving the workpiece through a series of baths containing chemicals designed to complete certain steps in the process. Each of the basic finishing stages and the steps typically associated with them are described briefly as follows:

- *Surface preparation:* The surface of the workpiece is cleaned to remove greases, soils, oxides, and other materials in preparation for application of the surface treatment. Detergents, solvents, caustics, and/or other media are used first in this stage, followed by rinsing of the workpiece. Frequently, the cleaning is accomplished in two separate baths to ensure maximum cleaning. Next, an acid dip is often used to remove metal oxides from the workpiece, followed by another rinse. The part is now ready to have the treatment applied. Figure 1 shows the steps in the process of preparing a work-piece for electroplating.
- *Surface treatment:* This stage involves the actual modification of the workpiece surface. The finishing process includes a series of baths and rinses to achieve the desired finish. For example, a common three-step plating system is copper–nickel–chrome. The copper is plated first to improve the adhesion of the nickel to the steel substrate and the final layer, chrome,

Figure 1 Surface preparation for metal finishing.

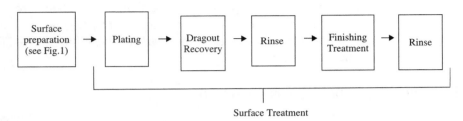

Figure 2 Metal-finishing processes.

provides additional corrosion and tarnish protection. Following the application of each of the plate layers, workpieces are rinsed to remove the process solution. Figure 2 presents an overview of the metal-finishing process.

- *Post treatment:* The workpiece, having been treated, is rinsed. Further finishing operations can follow. These processes are used to enhance the appearance or add to the properties of the workpiece. A common example of a post-treatment process is heat-treating to relieve hydrogen embrittlement or stress. Chromate conversion is another post-treatment process that often follows zinc or cadmium plating to increase corrosion resistance.[4]

Each of these stages can result in significant impacts to environmental resources. Likewise, opportunities for reducing waste and potential impacts to the environment exist at each process step. Summaries of environmental issues and regulatory constraints are provided in the next section, along with methods for making metal-finishing operations more environmentally conscious.

4 ENVIRONMENTAL AND REGULATORY ISSUES ASSOCIATED WITH METAL FINISHING

To varying degrees, all metal-finishing processes tend to create pollution problems and to generate hazardous and solid wastes. Unlike other manufacturing operations, the vast majority of chemicals that metal finishers use end up as waste. Of particular concern are those processes that use highly toxic or carcinogenic ingredients that are difficult to destroy or stabilize. Some of these processes are:

- Cadmium plating
- Cyanide-based plating including zinc, cadmium, copper, brass, bronze, and silver plating

- Chromium plating and conversion coatings using hexavalent chromium compounds
- Lead and lead-tin plating[5]

The metal-finishing process often produces undesirable byproducts or wastes including air emissions, wastewater, and hazardous and solid wastes.

Metals (dissolved salts in the plating baths) used during the surface treatment stage are also of environmental concern if emitted to the environment because of their tendency to bio-accumulate. Cyanide, used in many plating baths, is also a pollutant of concern. Table 1 provides a summary of these pollutants and their sources.

4.1 Water Issues in Metal-finishing Operations

Misunderstanding the true cost of utilizing raw materials such as water can frequently result in poor management choices regarding how the raw materials are used. Metal-finishing operations are notorious for using large quantities of water in their processes. The common perception is that water is cheap so, it can be used liberally to ensure that workpieces are adequately cleaned, rinsed, and coated. Although it may be true that the actual purchase of the water itself is relatively inexpensive, the cost to use the water within the processes may be considerably more expensive. A 2004 assessment performed by the Illinois Waste Management and Research Center on a major automotive assembly plant's phosphating processes revealed that it perceived its water costs to be only $2.20 per 1,000 gallons (the cost to purchase water from the city). At this low cost, water was used liberally throughout the plant to ensure adequate quality of cleaning and coating processes. Consequently, conservation measures were difficult to justify from an economic standpoint. However, when the process was broken down on a step-by-step basis and all costs associated with using the water were considered, it was concluded that the true cost of using the water was much higher. Including the value of process chemicals, energy, water purification measures, and wastewater treatment, the total average cost of using water increased to $80 per 1,000 gallons (a 36-fold increase).

Figure 3 shows the breakdown of water use costs on a stage-by-stage basis. As shown, some stages were considerably more costly than others. In general, rinsing stages that used municipal water (Stages 3, 4, and 7) were the cheapest, while the most expensive step was the degreasing stage (Stage 2), costing $200 per 1,000 gallons. An analysis of the total costs associated with Stage 2 is provided in Figure 4. As shown, process chemicals make up the vast majority of process inputs (82%) in this stage, followed by wastewater treatment (9%), de-ionized water (5%), and heat energy (4%).

Prior to performing this analysis, the plant had been using about 90 million gallons of water annually because it perceived that water was cheap and conservation measures were not warranted. The company estimated that the water

130 Metal Finishing and Electroplating

Table 1 Process Inputs and Pollution Generated (EPA 1995b)

Process	Material Input	Air Emission	Process Wastewater	Solid Waste
Surface Preparation				
Solvent Degreasing and Emulsion Alkaline and Acid Cleaning	• Solvents • Emulsifying agents • Alkalis • Acids	• Solvents (associated with solvent degreasing and emulsion cleaning only) • Caustic mists	• Solvent • Alkaline • Acid wastes	• Ignitable Wastes • Solvent wastes • Still bottoms
Surface Finishing				
Anodizing	• Acids	• Metal ion-bearing mists • Acid mists	• Acid wastes	• Spent solutions • Wastewater treatment sludges • Base metals
Chemical Conversion Coatings	• Dilute metals • Dilute acids	• Metal ion-bearing mists • Acid mists	• Metal salts • Acid • Base wastes	• Spent solutions • Wastewater treatment sludges • Base metals
Electroplating	• Acid/alkaline solutions • Heavy metal-bearing solutions • Cyanide-bearing solutions	• Metal ion-bearing mists • Acid mists	• Acid/alkaline • Cyanide • Metal wastes	• Metal • Reactive wastes
Plating	• Metals (e.g., salts) • Complexing agents • Alkalis	• Metal ion-bearing mists • Acid mists	• Cyanide • Metal wastes	• Cyanide • Metal wastes
Other Metal Finishing Techniques (including polishing, hot dip coating, and etching)		• Metal fumes • Acid fumes • Particulates	• Metal • Acid wastes	• Polishing sludges • Hot dip tank dross • Etching sludges • Scrubber residues

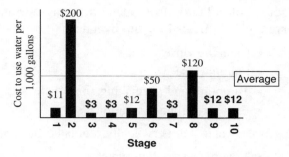

Figure 3 Cost to use water in an automotive plant's phosphate line.

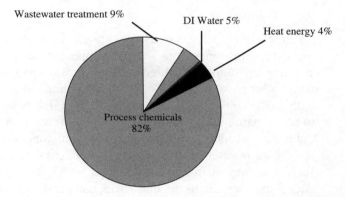

Figure 4 Cost breakdown of Stage 2 degreasing process in an automotive plant's phosphate line.

cost it about $200,000 per year. The results of the assessment showed that using this quantity of water was actually costing the company more than $7 million per year. Within one year of the assessment, numerous conservation measures had been implemented such that water usage was reduced by nearly one-third, resulting in cost savings of over $2 million per year. Understanding the full cost of using raw materials such as water—not just the cost of purchasing the raw materials—can provide the necessary incentives and justification for making changes that improve both economic and environmental performance.

4.2 Wastewater Issues in Metal-finishing Operations

Rinsing processes are the primary source of wastewater generated in metal-finishing operations. Rinsing removes plating solutions or cleaners from the workpiece. Metal-finishing chemicals are wasted as they are carried into the rinse water with a workpiece as it is moved from the processing solution into the rinse. This phenomenon is defined as *drag-out*, and is the largest volume source of pollution from metal-finishing operations. Sources of wastewater that are typically treated on site include:

- Cleaning rinse water
- Plating rinse water
- Tumbling and burnishing rinse water
- Exhaust scrubber solution

Wastewater that is typically regulated but not treated includes:

- Noncontact cooling water
- Steam condensate
- Boiler blowdown
- Stormwater

Spent cleaning baths contribute greatly to wastewater issues associated with metal-finishing operations. Many companies have switched from solvent-based cleaning operations to aqueous-based systems to reduce air pollution associated with solvent evaporation. The resulting volume of spent cleaning chemicals laden with surfactants, chelates, oil, grease, and dirt has increased dramatically. This trend is particularly burdensome on wastewater treatment operations because industrial cleaning and degreasing solutions function in the exact opposite direction of wastewater treatment systems. Industrial cleaning solutions are formulated such that they strip soils from parts and hold them in solution and/or emulsion using ingredients such as surfactants and chelates. This feature is important because it keeps the soils from recoating the parts once they have been removed. However, this same phenomenon creates demanding wastewater issues because wastewater treatment systems are designed to separate the water from its chemical contaminants. Further compounding the issue is the fact that when the cleaning baths are dumped, they usually become mixed with other wastewater sources that contain other contaminants such as heavy metals that also mix well with the spent cleaning solution.

4.3 Solid and Hazardous Waste

All process tanks utilized in the metal-finishing process can potentially become waste if impurities build up or critical ingredients are depleted to a point that the chemistry no longer performs to acceptable standards. Accidental discharges of these chemicals also can occur (e.g., when a tank is overfilled). These concentrated wastes are either treated on site or hauled to an off-site treatment or recovery facility. In general, the sources of hazardous and solid wastes at a metal finishing shop include the following:

- Spent plating baths
- Spent etchants and cleaners

- Strip and pickle baths
- Exhaust scrubber solutions
- Industrial wastewater treatment sludge, which can contain materials such as cadmium, copper, chromium, nickel, tin, and zinc
- Miscellaneous solid wastes such as absorbants, filters, empty containers, and abrasive blasting residue
- Solvents used for degreasing

Outdated chemicals are another example of wastes that metal finishers typically do not attribute to the production process. These obsolete or expired materials often accumulate and can violate waste storage requirements and must either be returned to the supplier or disposed of appropriately.[4]

4.4 Air Emissions

Metal-finishing facilities can be the source of several air-emission issues. Those of greatest environmental concern are chrome plating and anodizing processes that use hexavalent chromium and solvents used in degreasing operations. Chromium emissions frequently are controlled by wet scrubbers. The discharge of these systems is sent to wastewater treatment and combined with other waste waters for processing.

Historically, organic-based air emissions from metal finishing operations have resulted from the use of organic halogenated solvents, ketones, aromatic hydrocarbons, and acids used during the surface preparation stage. However, in the past decade, most metal finishers have switched to aqueous cleaning processes that minimize air emissions but tend to increase water-based waste issues. Solvents can evaporate substantially during degreasing operations. This is especially true in vapor degreasing applications that heat the solvent to intentionally create vapors that clean workpieces. Contaminated liquid solvents are recovered either by distillation (on site or off site) or sent for disposal (incineration). Most shops do not have controls for organics. However, some larger plants use carbon adsorption units to remove hydrocarbons.[4] Metal finishing results in a variety of hazardous compounds that are released to the land, air, and water. As a result, facilities are required to comply with numerous regulations. Regulations for metal finishers are promulgated at the federal, state, and local level. The requirements are complex and can vary, not only from state to state, but also from municipality to municipality.

5 FEDERAL REGULATIONS AFFECTING METAL FINISHING

The metal-finishing industry is regulated under numerous federal, state, and local environmental statutes. Three major federal laws regulate releases and transfers from the metal-finishing industry: the Clean Air Act as amended in

1990 (CAAA), the Clean Water Act (CWA), and the Resource Conservation and Recovery Act (RCRA). Also, the emissions reporting requirements under EPA's Toxics Release Inventory (TRI) cover many of the chemicals used in metal finishing. Summaries of these regulatory requirements are provided in the following sections.

5.1 Clean Air Act

The Clean Air Act required EPA to regulate 189 air toxics and also gave it authority to require pollution-prevention measures (installation of control equipment, process changes, substitution of materials, changes in work practices, and operator certification training). It also required the phase-out of the production of chlorofluorocarbons (CFCs) and several other ozone-depleting chemicals, which led to replacing solvent-based cleaning systems with water-based systems. New air pollution sources located in nonattainment areas must use more stringent controls and emissions offsets that compensate for residual emissions.[5]

5.2 Clean Water Act

The Clean Water Act regulates the amount of chemicals/toxics released via direct and indirect wastewater/effluent discharges. EPA has promulgated effluent guidelines and standards for different industries under the CWA provisions. These standards usually set concentration-based limits for the discharge of a given chemical. EPA defines two types of discharges: direct and indirect. Both types have different requirements.

A facility that is discharging directly into a body of water is regulated under the National Pollution Discharge Elimination System (NPDES) and must apply for an NPDES permit. The permit specifies what type of pollutants can be discharged and includes a schedule for compliance, reporting, and monitoring. The NPDES regulation limits the amount of metals, cyanides, and total toxic organics that a facility can discharge.

Most metal-finishing facilities discharge their wastewater to publicly owned treatment works (POTWs). These indirect dischargers must adhere to specific pretreatment standards because the POTWs are designed to deal mainly with domestic sewage, not industrial discharges. Often, specific state or local water regulations require more stringent treatment or pretreatment requirements than those in the federal effluent guidelines because of local water-quality issues. All facilities discharging to a POTW are governed by the General Pretreatment Standards. These standards state that discharges must have a pH greater than 5.0 and cannot do any of the following:

- Create fire or explosion
- Obstruct the flow of wastewater through the system
- Interfere with sewage plant operations

- Contain excessive heat
- Contain excessive petroleum, mineral, or nonbiodegradable oils

In addition, two CWA regulations affect the metal-finishing industry: the Effluent Guidelines and Standards for Metal Finishing (40 CFR Part 433) and the Effluent Guidelines and Standards for Electroplating (40 CFR Part 413).

The Effluent Guidelines and Standards for Metal Finishing are applicable to wastewater generated by any of the following processes:

- Electroplating
- Electroless plating
- Anodizing
- Coating
- Chemical etching and milling
- Printed circuit board manufacturing

If a facility performs any of the processes just listed, it is subject to this standard. In addition, discharges from 40 additional processes, including cleaning, polishing, shearing, hot dip coating, and solvent degreasing could be subject to this standard.

The metal-finishing and electroplating standards include daily maximum and monthly maximum average concentration limitations. The standards, which are based on milligrams per square meter of operation, determine the amount of wastewater pollutants from various operations that can be discharged.

The Effluent Guidelines and Standards for Electroplating cover wastewater discharges from electroplating and related metal finishing operations. This standard was developed prior to the metal finishing standard and has less stringent requirements than the metal-finishing standard. Facilities that are currently regulated by the electroplating standard can become subject to the more stringent metal-finishing standard if they make modifications to their facility's operations. EPA has made some exceptions to this rule—for example, printed circuit board manufacturers (primarily to minimize the economic impact of regulation on these relatively small firms). EPA defines independent printed-circuit-board manufacturers as facilities that manufacture printed circuit boards primarily for sale to other companies. Also excluded from the metal-finishing standard are facilities that perform metallic plate-making and gravure cylinder preparation within printing and publishing facilities.[5]

5.3 Resource Conservation and Recovery Act

The Resource Conservation and Recovery Act (RCRA) directly regulates several metal-finishing wastes as hazardous wastes. RCRA requires all hazardous waste generators, including metal finishers, to certify that they have a program in place to reduce the volume or quantity and toxicity of the waste they generate.

It addresses solid (Subtitle D) and hazardous (Subtitle C) waste management activities. Regulations promulgated under Subtitle C establish a *cradle-to-grave* system that governs these wastes from point of generation to disposal. A material is classified under RCRA as a hazardous waste if the material meets the definition of solid waste and exhibits one of the characteristics of a hazardous waste (i.e., corrosiveness, flammability, toxicity, or reactivity, designated with the code D) or if it is specifically listed by EPA as a hazardous waste (designated with the code F). Metal finishers generate a variety of hazardous wastes during the plating process. Within RCRA subtitle C, EPA includes hazardous waste from nonspecific sources in a series of F listings. The universe of RCRA listed wastes is constantly changing. In some states, the list of specific wastes might be different because they have added to EPA's list of hazardous chemicals.

To determine what a metal finisher must do to comply with RCRA requirements, the facility first must determine its generator status. Generator status is based on the amount of waste generated on a monthly basis. The following criteria determine the quantity of waste that is regulated by RCRA:

- Material remaining in a production process is not counted as waste until it is no longer being used in that process.
- Waste discharged directly and legally to a POTW in compliance with CWA pretreatment standards is not counted toward RCRA generation total.
- Any material that is characteristic or listed as a hazardous waste, and is accumulated after its removal from the process before being sent off site for treatment, storage, or disposal, is counted toward RCRA Subtitle C generation total.

In general, there are three classes of generators. Although individual states might have different names for them, EPA classifies them as follows:

1. *Large quantity generators:* Facilities generate more than 1,000 kilograms (2,200 pounds) of hazardous waste per month, or generate or accumulate more than 1 kilogram (2.2 pounds) of acute hazardous waste at one time.
2. *Small quantity generators:* Facilities generate between 100 kilograms (220 pounds) and 1,000 kilograms (2,200 pounds) of hazardous waste in any calendar month.
3. *Conditionally exempt small quantity generators:* Facilities generate less than 100 kilograms (220 pounds) of hazardous waste per month, or less than 1 kilogram (2.2 pounds) of acute hazardous waste in any calendar month.

Each state has varying degrees of regulation for the three generator classes. Individual operators should consult with state and local regulatory authorities to identify specific requirements that might be more stringent than those promulgated under federal law.

5.4 Toxics Release Inventory Reporting

Metal finishers must publicly report many of the chemicals they use in plating under the federal Toxic Release Inventory (TRI) reporting requirement. Facilities report information on a TRI data form (Form R) for each toxic chemical that is used over the threshold amount. Form R contains this basic information:

- Facility identification
- Parent company information
- Certification by corporate official
- SIC code
- Chemical activity and use information
- Chemical release and transfers
- Off-site transfer information
- On-site waste treatment
- Source reduction and recycling activities

The releases and transfers reported on a Form R include the following:

- Emissions of gases or particulates to the air
- Wastewater discharges into rivers, streams, and other bodies of water
- Releases to land on site, including landfill, surface impoundment, land treatment, or other mode of land disposal
- Disposal of wastes in underground injection wells
- Transfers of wastewater to POTWs
- Transfers of wastes to other off-site facilities for treatment, storage, and disposal

A facility must fill out Form R if these conditions apply:

- It is included in Standard Industrial Classification codes 20 to 39.
- The company has 10 or more full-time employees.
- It manufactures, processes, or "otherwise uses" any listed material in quantities equal to or greater than the established threshold for the calendar year.

6 ENVIRONMENTALLY CONSCIOUS MANUFACTURING PRACTICES

Environmentally conscious manufacturing (ECM) practices include any practices that reduce the environmental footprint of the operation. Specifically, practices that reduce waste, pollution, and energy usage are of most interest. ECM practices can be either incremental or innovative in nature. Incremental practices can usually be implemented, with little capital expenditure and minimal disruption to existing methods. Frequently, incremental changes involve basic modifications in the way materials are handled, stored, or used, and can be implemented at the lowest levels of the organization. Activities such as material substitution, waste stream segregation, and preventative maintenance are examples of incremental activities that can result in significantly reduced environmental impacts.

More innovative ECM practices often require significant investment and lead to more disruptive changes within the operation. Frequently, new technologies that improve the efficiency of materials usage are involved. Innovative ECM practices are usually more difficult to implement and often require downtime for conversion to the technology, along with workforce training. Innovative ECM practices require buy-in and commitment from the highest levels of the organization to ensure success. Frequently, pilot testing of the new technology is required prior to making the commitment to implementing innovative ECM practices.[6]

6.1 Incremental ECM Practices

Many ECM practices are low-cost, low-risk alternatives to hazardous waste disposal. Most of the approaches do not require a great deal of sophisticated technology and can be relatively inexpensive. Significant ECM gains can be made by simply improving standard operating procedures:

- Implement an effective inventory control system to prevent waste generation due to unnecessary or excessive purchases and through expiration of a product's shelf life
- Segregate waste streams to allow for certain wastes to be recycled or reused and to keep nonhazardous materials from becoming contaminated.
- Prevent and contain spills and leaks by installing drip trays and splash guards around processing equipment.
- Ensure that product and waste containers are kept closed except when material is added or withdrawn.
- Track wastes and include careful labeling to ensure safe handling of materials and identifying wastes that have the potential for recycling or reuse.

As previously noted, *drag-out* is process chemistry that is wasted by being carried over into the rinse water as a processed part is moved from the processing solution into the rinse. The lifetime of many solutions used in metal finishing is limited by the accumulation of impurities and/or by depletion of constituents due to drag-out. Numerous techniques have been developed to control drag-out. The efficacy of any technique depends on the geometry of the workpiece processed, operator technique used, racking methods, transfer times, dwell times, and numerous other variables.

Reducing the drag-out reduces the amount of rinse water needed. Also, less of the process solution ingredients (e.g., metals) leave the process, which ultimately produces savings in raw materials, treatment, and disposal costs. Drag-out reduction techniques include the following:

- Lower the concentration of process tank ingredients (many operators use concentrations in excess of specifications). This will save in raw material costs as well as drag-out.
- Withdraw workpieces at a slower rate to allow maximum drainage back into process tanks.
- Install drainage boards between tanks to route drag-out into the correct process tank while the workpiece is being transferred.
- Orient workpieces onto racks to improve drainage.
- Train personnel or program automatic machinery to ensure optimum drain times.
- On hand-operated lines, install drain bars to allow workers to rest the parts and drain them before moving to the next process tank.
- Increase solution temperature to reduce viscosity and surface tension of the solution.
- Use nonionic wetting agents to reduce solution surface tension.
- Install fog sprays (on hot tanks with high evaporation losses) or air knives to push drag-out back into process tank.

Water tends to be the chemical most used by metal finishers. And, as described previously, the indiscriminate use of water can be very costly. Consequently, it is important for metal finishers to implement measures that minimize water usage while ensuring that quality finishes are maintained. Some examples of measures that metal finishers can employ to minimize water usage are described as follows:

- *Install multiple rinse tanks in a countercurrent configuration.* In countercurrent rinsing, water flows through tanks in opposite direction of workpieces. Rinsing should begin in the dirtiest tank and finish in the cleanest tank.
- *Agitate rinsing baths mechanically or with air.* This increases rinsing efficiency. (*Note:* Blown air is far more energy efficient than compressed air)
- *Use sprays or mist to rinse off excess process solutions.* This can often be accomplished above process tanks, such that the process solution chemistry can be removed from the workpiece and returned to the tank it was purchased for. At the same time, the added rinse water replaces evaporation losses.
- *Install flow restrictors to limit the volume of rinse water flowing through a rinse system.* This method will maintain a constant flow of fresh water to the rinse process and prevent operators from manipulating flow rates such that the flow of water exceeds what is actually needed.
- *Install automatic flow controls that add water to the system only when it is needed.* The next section describes how an automatic flow controller can reduce rinse water usage.

6.2 Conductivity Control System in an Electroplating Shop

The quality of rinse water can be correlated to the conductivity level of the rinse tank, which rises due to the addition of ionic contaminants dragged in by workpieces. A conductivity control system reduces water consumption by allowing fresh water to flow into a rinse tank only when the conductivity level within the rinse tank exceeds a previously determined value. The conductivity control system thus consists of: (1) a sensor that detects the conductivity level of a solution; (2) an analyzer that monitors the conductivity level relative to a preset maximum conductivity value; and (3) a solenoid valve that receives a signal to open or close from the analyzer, allowing or restricting the flow of fresh water into the rinse tank as determined by the conductivity level. Early versions of this technology required frequent maintenance and were not very reliable. However, innovations in electroless conductivity control systems have resulted in much better reliability and performance.

One Chicago-area electroplater had run its lines for many years at flow rates that maintained conductivity levels around 1,000 micro-siemens per centimeter (uS/cm). Through testing and installation of the electroless conductivity control system, they found that they could operate the system between 3,400 and 3,600 uS/cm without affecting quality. They were able to reduce rinse water flow rates by 92 percent, achieving a payback of four and a half months on a $2,000 investment for the system.

7 INNOVATIVE ECM TECHNIQUES

7.1 Rinsewater Purification and Recovery Techniques

Many techniques are available to recover water, metals, or acids from the metal-finishing process. In some cases, contaminants are removed from within the actual metal-finishing process (purification techniques), while in other cases, the materials are removed from the system and processed with recovery/recycling techniques that remove contaminants or replace depleted ingredients to facilitate reuse. This can be accomplished either on site or at a vendor facility.

Rinse water can be recovered and reused at the point in the process where it is used for rinsing or further downstream after it has commingled with other streams. In general, the further upstream in the process the water is recovered, the easier it will be to reuse. In some instances, it can be reused several times with no treatment or possibly minimal treatment, such as coarse filtering. In other cases, it may require sophisticated technology to remove contaminants. One of the most important considerations for metal finishers is to understand precisely what quality level of water is required to ensure a quality finish on their products. This information can often be obtained with assistance from their chemical supplier.

In some cases rinse solutions that have become too contaminated for their original purpose can be used in other rinse applications. For example, effluent from a rinse system following an acid-cleaning bath sometimes can be reused as influent to a rinse system, following an alkaline cleaning bath (reactive rinsing). If both rinse systems require the same flow rate, 50 percent less rinse water could be used to operate the system.

Chemical Recovery

Facilities can manage the captured drag-out solution from rinse-water recovery in three ways: (1) recycling solution back into the process; (2) on-site recovery; and (3) shipment off site for disposal or recovery. The appropriate choice depends on the type of process bath, composition of the drag-out, and the cost of the technique.

Metal finishers must understand the chemical properties of a waste stream to assess the potential for reusing the waste as a raw material. The operator must ensure that the spent solutions are compatible with the application where they intend to use it. For example, because spent cleaners often contain high concentrations of metals, they should not be used for final pH adjustments. Facilities should check with chemical suppliers to determine whether they have reclamation services for plating baths. Some states classify reclamation as waste treatment under the RCRA program, requiring compliance with additional regulatory requirements.

Metal Recovery

The metal-finishing industry pours millions of dollars down the drain each year in valuable metals. In-process recovery systems such as evaporation, ion exchange, reverse osmosis, electrolysis, and electrodialysis, can reduce rinse-water volumes and facilitate the recovery of metal salts for reuse in plating baths. According to Gallerani, it would be economically feasible to recover 80 to 90 percent of copper, 30 to 40 percent of zinc, 90 to 95 percent of nickel, and 70 to 75 percent of chromium presently disposed of as sludge.[7]

The savings achieved through metal recovery are site-specific. These factors determine whether metal recovery is economically justified:

- The volume of waste that contains metals
- The concentration of the metals
- The potential to reuse some of the metal salts
- Treatment and disposal costs

Table 2 outlines technologies that can be used for chemical recovery, metal recovery, and chemical solution maintenance.[8] Each of the technologies described in Table 2 requires significant resources for implementation. Capital investment, combined with changes in procedures and training for employees, are required to achieve improvements in process efficiency with these technologies. Many vendors that supply these technologies will provide them on a trial basis to allow potential adopters to test them in their application and reduce uncertainty associated with on-site performance prior to purchase.

Table 2 Overview of Applications for Recycling and Recovery Equipment

Recycling/Recovery Method	Chemical Recovery	Chemical Solution Maintenance	Metal Recovery
Acid Sorption		X	
Diffusion Dialysis		X	
Evaporators	X		X
Electrolytic Metal Recovery	X		X
Electrodialysis	X		X
Ion Exchange	X	X	X
Ion Transfer		X	
Microfiltration		X	
Membrane Electrolysis		X	

Note: From Ref. 8.

7.2 Vendor Roles in ECM

In many metal-finishing operations, the management of activities associated with chemicals, wastes, and environmental compliance are not considered to be part of the operation's core business. These operations frequently depend heavily on certain suppliers for support and may specifically structure vendor contracts to ensure that these needs are met. Large facilities associated with automotive and aerospace manufacturing operations often issue a single contract to a *chemical manager* who is responsible for all chemical management activities within the plant.

In traditional supplier–customer relationships, the chemical supplier's profitability is a function of the volume sold. The more chemicals sold, the higher the revenue for the supplier. Meanwhile, the buyer has an opposite incentive—to reduce costs or the amount of chemicals purchased. In the chemical management model, suppliers become chemical management providers and are paid for successfully delivering and managing chemicals. Thus, the supplier's profitability is based on better performance, not on selling more chemicals. This is often achieved by paying the supplier on a price per unit (e.g., price per car produced) basis or a flat monthly fee whereby the supplier provides all chemicals needed for one fixed fee. In these scenarios, the supplier can actually make more money by providing less chemicals.[9]

By aligning the incentives of suppliers and customers, both can achieve bottom line benefits via reduced chemical use, costs, and waste. Fewer chemicals in the overall equation can also result in significant environmental benefits. Reductions in chemical use also result in reduced cost, emissions, and exposure to liabilities. Beyond the tangible benefits of reduced costs and waste, there are not-so-quantifiable benefits to well-managed chemical programs, such as reducing accidents; maintaining a good reputation in the community; and staying clear of environmental agencies' spotlights. These are all positive aspects of a sound chemical management program.

REFERENCES

1. U.S. EPA, *Profile of the Fabricated Metal Products Industry*. Office of Enforcement and Compliance Assurance, Washington, DC, 1995.
2. Christopher J. Ford and Sean Delaney. *Metal Finishing Industry Module*. Toxics Use Reduction Institute, Lowell, MA, 1994.
3. Harry M. Freeman, *Industrial Pollution Prevention Handbook*. McGraw-Hill, New York, 1995.
4. U.S. EPA, *Metal Plating Waste Minimization*. Waste Management Office, Office of Solid Waste, Arlington, VA, 1995.
5. The Northeast Waste Management Officials' Association (NEWMOA), *Pollution Prevention for the Metal Finishing Industry: A Manual for Pollution Prevention Technical Assistance Providers*, 183 pp., 1997.
6. Timothy C. Lindsey, *Key Factors for Promoting P2 Technology Adoption: "How-To" Knowledge is the Key*. Pollution Prevention Review. John Wiley and Sons. Winter 2000.
7. Peter A. Gallerani, *Good Operating Practices in Electroplating Rinsewater and Waste Reduction*. Massachusetts Department of Environmental Protection, Boston, 1990.
8. Gary E. Hunt, "Waste Reduction in the Metal Finishing Industry," *The International Journal of Air Pollution Control and Waste Management* (May 1988), 672–680.
9. Thomas Bierma and Frank Waterstraat, "Chemical Management Services—Focused Studies: 1. CMS in Small and Medium Enterprises 2. A CMS Standard," Illinois Waste Management and Research Center, 2004.

CHAPTER 7

AIR QUALITY IN MANUFACTURING

Julio L. Rivera, Donna J. Michalek, John W. Sutherland
Michigan Technological University Houghton, Michigan

1	INTRODUCTION	145
2	PARTICULATE CHARACTERIZATION	149
	2.1 Particulate Classification	149
	2.2 Particle Size	150
	2.3 Composition	152
	2.4 Instrumentation	153
3	HEALTH EFFECTS AND REGULATIONS	153
	3.1 Health Effects	154
	3.2 Workplace Air-quality Regulations	155
	3.3 Trends in Standards and Regulations (PM_{10} to $PM_{2.5}$)	156
4	ORIGIN OF AIRBORNE PARTICLES	156
	4.1 Welding	156
	4.2 Wet Machining	157
	4.3 Dry Machining	158
	4.4 Grinding	158
	4.5 Casting	159
	4.6 Powder Manufacture—Carbon Black	160
5	TRADITIONAL CONTROL TECHNOLOGIES	161
	5.1 Ventilation Systems	161
	5.2 Enclosures	163
	5.3 Chemical Treatment	163
	5.4 Personal Protection Systems	164
6	ENVIRONMENTALLY RESPONSIBLE PARTICULATE MITIGATION/ELIMINATION	164
	6.1 Alternative Process Plan	164
	6.2 Alternative Process Sequence	166
	6.3 Process Change	166
7	MANUFACTURING TRENDS AND POTENTIAL PARTICULATE CONCERNS	167
	7.1 Nanoparticles	168
	7.2 Bioaerosols	170
8	SUMMARY AND CONCLUSIONS	171
9	ACKNOWLEDGMENTS	172

1 INTRODUCTION

In recent years, increased attention has been devoted to the environmental performance of products and their associated manufacturing processes. Toward this end, tools such as life-cycle analysis (LCA) are being used more frequently to quantify the environmental effects of decisions made during the engineering design process. In many cases, for the first time, decision makers have begun to emphasize material and energy flows throughout the life cycle of a product, such as that depicted in Figure 1.[1] In this context, it is worth noting that the manufacturing design community has historically largely ignored the waste streams produced

146 Air Quality in Manufacturing

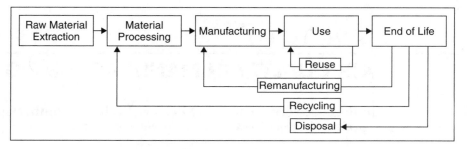

Figure 1 Product life cycle. (Adapted from Ref. 1.)

by manufacturing operations. One of these waste streams, process-generated airborne particulate, has received much emphasis within the scientific literature of the industrial hygiene and safety engineering communities. All too often, however, the individuals who have the power to do something about these emissions, manufacturing decision makers, have never effectively dealt with these issues. With this in mind, this chapter is aimed at the manufacturing R & D community, and endeavors to address a variety of topics related to airborne particulate arising from manufacturing processes. It will emphasize the origins of the particulate and the potential hazards that it represents, and will examine the role of traditional control technologies in addressing particulate and more environmentally conscious approaches that focus on dealing with particulate at its source.

In considering the product life cycle depicted in Figure 1, engineers often focus on the product use stage, often to the exclusion of all the other stages. Yet, manufacturing remains an important life-cycle stage and accounts for a significant portion of environmental burdens, employment and community presence, and economic impacts of the product life cycle. The U.S. Census Bureau reports that "manufacturing" is the second largest industry sector, trailing only "wholesale trade."[2] In considering the analysis of manufacturing processes, it is often convenient to think of them in terms of their inputs and outputs. Figure 2 depicts the inputs and outputs associated with a typical manufacturing operation.

Airborne particulate is produced by a variety of manufacturing processes via such mechanisms as atomization, vaporization/condensation, and combustion. For

Figure 2 Input–output description of a manufacturing process.

example, the use of casting processes can create a number of combustion byproducts; the application of metalworking fluids (MWF) in machining may result in metalworking fluid mist, and welding processes can produce metal fumes. This airborne particulate can be in either liquid or solid form, and since it is exists within an air medium, it is termed an aerosol—a multiphase substance in which solid or liquid particles are suspended in air.[3] Workers that are exposed to this particulate-laden air may suffer a variety of health maladies. Worker exposure can happen via three different pathways: ingestion, skin absorption, and inhalation. In the United States, an employee may inhale from 4 to 10 m^3 of air during a typical 8-hour workday, and this air may contain a variety of particles of differing material types.[4]

The human respiratory system takes in air from the atmosphere. As shown in Figure 3, the air passes through the nose, pharynx, larynx, trachea, bronchi, and bronchioles, and enters the alveoli.[5] Gas exchange takes place in the alveoli, with oxygen passing into the bloodstream and carbon dioxide entering the alveoli; upon exhalation, CO_2 is removed from the body. Particulate matter (or PM) differs in terms of chemical composition and size. As indicated in Table 1, particle size influences the depth to which the PM penetrates into the respiratory system. The human lung has evolved over many millennia to function in the particle-filled air of the natural environment.[6] Particle clearance mechanisms in the tracheobronchial and alveoli are effective at removing most deposited particles, but excessive particulate exposure can overwhelm the natural clearance mechanisms.

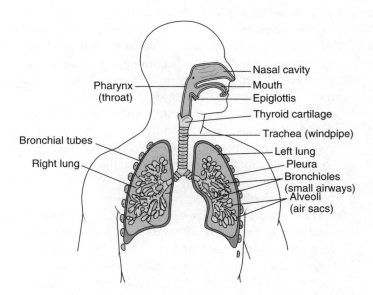

Copyright © 2005 Mckesson health Solutions, LLC. All Rights Reserved.

Figure 3 Parts of the respiratory system. (From Ref. 5.)

Table 1 Size and Fate of Particles

Size (μm)	Fate of Particles
>10	Deposited in the nose
5–10	Trapped in the nasal pharynx
2–5	Removed by mucociliary escalator in the tracheobronchial area
0.01–2	Alveolar deposition in the lung
<0.01	Exhaled

Note: From Ref. 7.

The U.S. Environmental Protection Agency (EPA) is required by law to set National Ambient Air Quality Standards (NAAQS) for pollutants (including PM) considered harmful to public health and the environment. While the EPA focuses on air pollution in the world around us, the mission of the U.S. Occupational Safety & Health Administration (OSHA) is to assure the safety and health of America's workers by setting and enforcing standards—including those for workplace air quality. Some publications suggest that OSHA's existing air-quality standards might not be protective of human health.[8–10] One recent trend with respect to PM is the attention given to particles smaller than 2.5 μm, usually called fine particles, which epidemiology studies have linked to work-related diseases. Research findings over the last few years have advanced our knowledge of fine particles effects and work is progressing to achieve a comprehensive understanding of these effects at low exposure levels.[9–11]

The responsibilities of a manufacturing enterprise extend far beyond that of designing and producing manufactured products and generating a profit for owners/shareholders. It is becoming ever clearer that to ensure long-term competitiveness, the enterprise must also provide a safe and healthy work environment, and a company must responsibly consider social and environmental issues. The authors believe that corporate responsibility and profitability are not necessarily competing goals; product and process innovations focused on being more environmentally responsible can simultaneously create more profit. This approach to environmentally responsible manufacturing (ERM) calls for manufacturers to seek out win–win opportunities, and while this approach can be challenging, the resulting benefits can be significant.

This chapter focuses on manufacturing air-quality issues, with an emphasis on airborne particulate. It starts with a discussion on particle characterization to help the reader understand the vocabulary and basic concepts used by the air-quality community, and the instrumentation used to characterize particles. Then it presents a discussion of the health issues and regulations that relate to PM. This chapter also covers airborne particle emissions, traditional engineering controls used to manage airborne particulate, innovations that seek to address particulate at its source, and manufacturing trends and potential concerns related to air quality.

2 PARTICULATE CHARACTERIZATION

Although it is of paramount interest to understand the health impacts associated with aerosol emissions from manufacturing processes, medical researchers are still developing knowledge regarding this linkage. It is understood, however, that the health impacts depend on the characteristics of the particles (e.g., the particle size, mass, number, surface area, and chemical composition). It has been suggested that the health effects of exposure depend on the chemical makeup of the particles. For example, the OSHA asbestos standard considers the particle number concentration rather than the particle mass concentration. Some have reported that the particle surface area is just as important as the particle chemistry.[9,12–14] Hypotheses addressing particle number concentrations are largely independent of aerosol mass concentrations since fine particles dominate the number distribution.[9] Historically, most air quality standards have focused on PM_{10} particulate (particulate matter less than 10 μm). However, since the fine particles penetrate to the gas-exchange region of the lungs, more attention has begun to be devoted to this size range.[3,15] In the absence of specific knowledge regarding particle health effects, emphasis focuses on quantifying various types of particulate characteristics.

2.1 Particulate Classification

A number of different words and phrases are used to describe aerosols. The American National Standards Institute (ANSI) has published definitions for dusts, fumes, and mists;[16] Hinds has described other aerosol types.[3] Table 2 describes the origin and typical size of these aerosols.[17] These aerosols, which are frequently found in an industrial environment, are defined as follows:

- *Dust* is a group of solid particles generated by the mechanical breakdown (e.g., handling, crushing, and grinding) of organic or inorganic materials such as rock, metal, and grain. Dust particles do not tend to adhere to one another except under electrostatic forces, and they do not diffuse in air but settle under the influence of gravity.
- *Fumes* are solid particles generated by the condensation of a gas, often after volatilization from a molten metal, and are often accompanied by a chemical reaction such as oxidation. Fume particles often clump together (flocculate) and sometimes combine or coalesce.
- *Mist* is a set of suspended liquid droplets generated by condensation from the gaseous to the liquid state or by liquid breakup via splashing, foaming, or atomizing.
- *Smoke* is an aerosol that results from incomplete combustion. Particles may be either solid or liquid.
- *Spray* is a droplet aerosol formed by the mechanical breakup of a liquid.
- *Homogeneous aerosol* is an aerosol in which the particles have the same chemical composition.

Table 2 Description of Airborne Particles

Name	Origin	Usual Size Range (μm)
Dust	Mechanical dispersion	>2
Fumes	Condensation	<1
Smoke	Condensation	<1
Mists	Condensation or atomization	5–1,000
Sprays	Breakup of a droplet	>1

Note: Adapted from Ref. 17.

- *Monodisperse aerosol* is an aerosol that has particles of the same size.
- *Polydisperse aerosol* is an aerosol with particles of different size and whose characterization requires statistical methods.

2.2 Particle Size

Particle size is one of the important attributes in terms of aerosol behavior and toxicological effects. The size of a particle dictates its behavior in air.[3] Usually, particle size refers to the particle diameter. In some circumstances, a particle might be spherical (or nearly so), as in the case of liquid particles. A solid particle, by contrast, might have an irregular shape. For such a case, the aerodynamic diameter, d_a, is often used to describe the effective diameter of an irregular particle. The aerodynamic diameter is defined as the diameter of a spherical particle with a density of a water droplet that has the same settling velocity as the particle of interest.[3] The settling velocity is the velocity at which the drag (resistance) force of the air affecting the particle equals the force of gravity. Settling times for different particle aerodynamic diameters are shown in Table 3.[18] In some circumstances, a *Stokes diameter*, defined as the diameter of a sphere that has the same density and settling velocity as the particles of interest, is used. As the particle size gets smaller, the force of gravity becomes less significant and the behavior of the particle is determined by diffusion. At this point, a mobility equivalent diameter is commonly used.[19] The U.S. Environmental Protection Agency has classified particles based on their size, as seen in Table 4.[20]

Table 3 Time Required for a Particle to Settle 5 Feet in Still Air

Particle Size (d_a in μm)	Settling Time
0.5	41 hours
1	12 hours
3	1.5 hours
10	8.2 seconds
100	5.8 seconds

Note: From Ref. 18.

Table 4 Particle Size Categories

Name	Particle Size (aerodynamic diameter, d_a) (μm)
Super coarse	>10
Coarse	$2.5 < d_a < 1.0$
Fine	$0.1 < d_a < 2.5$
Ultrafine	<0.1

Note: Adapted from Ref. 20.

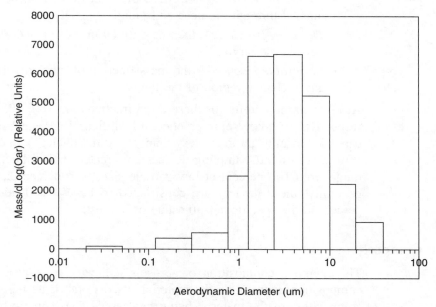

Figure 4 Sample mass distribution of particles. (From Ref. 21.)

The particulate generated by a process has a variety of different shapes and sizes, and as a result, statistical methods are needed to characterize size distributions. For example, if a particulate sample is collected from the workspace air surrounding a process, the number of particles within each size range of interest may be recorded, and then a histogram may be prepared displaying the particle count as a function of diameter. The same approach could be followed to characterize the mass distribution as a function of particle size (see Figure 4).[21] Several statistics may then be calculated from the data, including the mean, median, and mode, for further analysis.[3] The *mean* is defined as the sum of all the particle sizes divided by the number of particles. The *median* is the middle value if the data are organized in ascending order, and the *mode* is the size that is found most frequently in the data.

Other values are calculated when analyzing aerosols. Two of the most common are the *mass concentration* that is expressed in terms of mass per unit volume and *number concentration* that is expressed in terms of number of particles per unit volume. The mass concentration is most often used to evaluate aerosols in working environments to ensure regulatory compliance. More details about regulations, exposure limits, and standards will be provided shortly.

Particles will deposit in different parts of the respiratory system, with the particle size serving as a principal factor in determining the deposition location. Relative to deposition concerns, ISO uses the following particle size classifications:[22]

- *Inhalable fraction*— the mass fraction of total airborne particles that is inhaled through the nose and mouth
- *Thoracic fraction*— the mass fraction of inhaled particles penetrating beyond the larynx
- *Respirable fraction*— the mass fraction of inhaled particles that reach the gas exchange region of the lungs

At present, the cutoffs for these three fractions are 100, 10, and 4 micrometers, respectively. Of course, in practice, it is difficult to create a sampling device that can take in only particles less than 100 μm (inhalable fraction), for example. With this in mind, sampling devices are created with collection efficiencies that range from 100 percent at low particle sizes to 0 percent at high particle sizes; generally, these devices are constructed to have a collection efficiency of 50 percent at the specific cut-off value of interest.

2.3 Composition

The chemical composition of aerosols will vary from process to process. The composition depends on the materials used to fabricate the desired product and composition of the material being processed. The scenario becomes more complex if there are chemical reactions within the aerosol (secondary aerosols). One of the primary reasons for performing chemical characterization is to identify the contaminants that are present in the working environment to allow industrial hygiene personnel to determine whether applicable standards relating to specific chemical constituents are being met. Several devices can be used for identifying particle composition:

- X-ray fluorescence spectrometry (XRF)
- Proton-induced X-ray emission spectrometry (PIXE)
- X-ray diffraction (XRD)
- Optical emission spectrometry and mass spectrometry

Jenkins and Eagar provide an excellent summary of techniques and devices to characterize particle chemical composition.[23] As a reminder, in the case of low-toxicity fine particles it has been shown that inflammatory activity is linked

Table 5 Common Sampling Instrumentation

Air Sampling Device	Average Size Range (μm)
Scanning Mobility Particle Sizers (SMPS)	0.015–0.7
Aerodynamic Particle Sizer (APS)	0.2–20
TSI DustTrak	0.3–1.0
Micro-Orifice Uniform Deposit Impactors (MOUDI)	0.06–20
Electric aerosol analyzer (EAA)	0.01–1
Microscopy Device	
Scanning electron microscope (SEM)	0.5–50
High-resolution SEM (HRSEM)	0.002–1
Electron probe microanalysis (EPMA)	0.5–50
Transmission electron microscope (TEM)	0.001–1
Light microscopy	1–400

to the high surface area of the particles.[24] This suggests that the particle size is the characteristic that "promotes" toxic reactions and that chemical composition is of secondary significance.[25]

2.4 Instrumentation

A variety of instruments are available for sampling and analyzing aerosols. Brandt defined four reasons for the collection of particles: (1) to assess the workers' exposure, (2) to determine the average exposure of a worker that seems to have become ill, (3) to determine the type of controls that could be incorporated, and (4) to study the efficiency of particle control equipment.[16] Another reason for sampling is to quantify the effect of process variables on the resulting particle size distribution. To learn more about sampling instruments, refer to Cohen and McCammon.[26] Table 5 lists some common aerosol measuring instruments and the size range in which they may be utilized.

3 HEALTH EFFECTS AND REGULATIONS

Air quality in manufacturing environments has been a social concern since the 1700s, when Bernardino Ramazzini, in his work called *De Morbis Artificum Diatriba* (*Diseases of Workers*), reported diseases that resulted from exposure to dust and fumes.[27] The U.S. manufacturing industry experienced a boom increase after World Wars I and II due to higher demands for products. Even though this increase marked an important stage in strengthening the U.S. economy, it gave birth to a new problem. The increase in production resulted in an unhealthy environment for workers—poor air quality—that immediately had a negative impact on the workers' health, representing a new challenge for employers and regulatory agencies. Epidemiological studies have linked air pollution to adverse health effects, including respiratory diseases and increased mortality and morbidity.[28–30]

This section focuses on reviewing some of the most common occupational diseases caused by the inhalation of airborne particles.

3.1 Health Effects

In 2004 the manufacturing industry represented less than 14 percent of the private employment sector and accounted for 42 percent of the nonfatal workplace illnesses, as published by the U.S. Bureau of Labor Statistics.[31] A portion of work-related illnesses are lung diseases, which are associated with poor air-quality conditions. These statistics were compiled in 2004: (1) inhalation and accumulation of dust were responsible for 2,860 deaths in 2000; (2) about 20 to 30 percent of asthma cases in adults are caused by work-related exposure; and (3) the fourth national leading cause of death, chronic obstructive pulmonary disease, was estimated to be 15 percent related to workplace exposure.[32] It is evident that the manufacturing sector still has a significant impact on the health of workers. Inhaled particles have also been linked to increased lung cancer cases.[33]

Once particles are inhaled they can promote adverse biological effects in the human body, leading to local or systemic reactions (e.g., impact the area of first contact or other organ(s) as the particles are transported through the body). The biological impacts include allergenic and irritative reactions, as well as carcinogenic effects; these may lead to lung inflammation, respiratory diseases, and lung tumors.[30,33,34,35] Health effects will depend on different factors, including length and frequency of exposure, susceptibility of the workers, and type of chemicals present in the inhaled aerosol. Symptoms may be noticed in a short-term or in a long-term period. Short-term exposures might lead to *acute* poisoning, while long-term or repeated doses might lead to *chronic* poisoning, depending on the toxicity of the inhaled material.

In the case of welding, short-term health effects might include irritation of the nose, chest, and eyes, coughing, bronchitis, edema, pneumonitis, nausea, and vomiting. Long-term health effects include diseases such as asthma, emphysema, silicosis, and cancer in the lungs, larynx, and urinary tract.[36] A common disease found in welding workers is the *metal fume fever*. Welding has also been linked to Parkinsonism.[37]

Silicosis is thought to be the oldest occupational disease.[4] It is caused by the inhalation and accumulation of silica particles in the workplace. Symptoms include shortness of breath, fever, coughing, fatigue, and chest pain. *Fibrosis* is a disease caused by the accumulation of fibrous material in human tissue (e.g., the lungs), with symptoms including shortness of breath and cough. Other occupational diseases include chronic beryllium disease (CBD), asbestosis, and pneumoconiosis.

Government agencies set and enforce occupational standards to avoid or prevent adverse health effects. The following section provides a description of workplace standards and regulations in the United States.

3.2 Workplace Air-quality Regulations

Workers' health has been affected by work-related activities since the Industrial Revolution. However, it was not until the 1900s that the U.S. government took major actions (see Table 6) to address occupational health and safety issues.[40] The 1936 Walsh–Healy Public Contract Act provided for basic labor standards and established maximum working hours, as well as health and safety standards.

In 1970, the U.S. Congress recognized that employees should be provided a healthy working environment with the enactment of the *Occupational Safety and Health Act of 1970* [Public Law 91–596], considered the most important legislation dealing with occupational issues. It created the Occupational Safety and Health Administration (OSHA)—under the Department of Labor—for the enforcement of safety and health regulation, and also established the National Institute of Occupational Safety and Health (NIOSH)—under the U.S. Department of Health and Human Services.[39] The act authorizes NIOSH to "develop and establish recommended occupational safety and health standards" to prevent work-related illness, disability, and death, and to conduct research to make recommendations for current and new standards. This agency is also responsible for educating and training in the occupational safety and health field. Under the act, the secretary of labor is authorized to establish health and safety standards. The first standards were adopted from the Walsh–Healy Public Contract Act and from the American National Standards Institute.[40] In addition, the Institute has authority for coal and mine research under the *Federal Mine Safety and Health Amendments Act of 1977*.[39] The states are given jurisdiction under section 18 of the act to decide if they want to be in charge of the enforcement of health and safety standards.

For airborne particles, OSHA publishes standards called permissible exposure limits (PELs), aimed at protecting the workers against health effects that may

Table 6 Occupational Related Action in the United States

Date	Activity
1852	First safety law in the state of Massachusetts
1908	Congress passed a law to compensate work-related injured workers
1910	First national conference in occupational diseases
1913	Creation of the Department of Labor
1934	Labor standards established by the Department of Labor
1936	Walsh–Healy Public Contract Act
1941	Federal Mine Inspection Act
1946	American Conference of Governmental Industrial Hygienists (ACGIH)
1966	The Metal and Nonmetallic Mine Safety Act
1969	The Federal Coal Mine Health and Safety Act
1970	Occupational Health and Safety Act
1977	Federal Mine Safety and Health Amendments Act

Note: Adapted from Ref. 40.

arise from exposure to hazardous substances.[41] The PELs are described as the amount or concentration of a substance in air, based on an 8-hour time-weighted average (TWA) exposure. For particles, the PELs are given in mass per unit volume (mg/m^3). Up to 2003, 500 PELs had been established by OSHA. The airborne particulate standard (OSHA) is 5 mg/m^3 for particulates not otherwise regulated (PNOR).

In 1946, the American Conference of Governmental Industrial Hygienists (ACGIH) was created with the purpose of advancing the field of occupational health by educating, developing and disseminating technical and scientific knowledge. The ACGIH publishes recommended air-quality standards called the threshold limit values (TLV) and biological exposure indices (BEI). These standards may be used by a company in addition to those required by law.

3.3 Trends in Standards and Regulations (PM$_{10}$ to PM$_{2.5}$)

Since its creation, the EPA has established standards for particulate matter, which can be composed of small particles and liquid droplets suspended in air.[42] The EPA established that one of the characteristics of particulate associated with negative health impacts is the presence of particles smaller than 2.5 µm. Particles this small can be inhaled and may reach the gas exchange region of the lungs and even translocate to other organs, such as the olfactory bulb.[43] In 1987, the EPA set a standard for particulate matter smaller than 10 µm (PM$_{10}$), and in 1997 the agency published a standard to include particles smaller than 2.5 µm (PM$_{2.5}$). This decision was not well received by industry and other organizations, and establishment of the new standard was contested before the U.S. Court of Appeals for the D.C. Circuit. The EPA's 1997 standard was rejected by the D.C. Circuit in 1999. The EPA, in conjunction with the U.S. Department of Justice (DOJ), appealed the decision, and in the year 2002 the D.C. Circuit allowed the EPA to follow with implementation of the PM$_{2.5}$ standard.

4 ORIGIN OF AIRBORNE PARTICLES

Airborne particulate in the workplace environment originates from a variety of sources. Particles can be generated directly by the manufacturing process as the product is being created or indirectly as a result of part cleaning, product packaging, part handling, and other processes. The airborne particulate that is generated varies with the type of manufacturing process and the associated particle formation mechanism. This section is aimed at describing the various mechanisms and modes that lead to the formation of aerosols and examines particle generation/formation for a variety of manufacturing processes.

4.1 Welding

There are a variety of different types of welding operations that serve to join two or more metal components. In welding operations, the workpieces are melted in the desired contact area to form a strong permanent joint as the melted material fusions and gets cooler. Common welding processes include metal inert gas (MIG) arc welding, tungsten inert gas (TIG) arc welding, shielded metal arc welding (SMAW), gas metal arc welding (GMAW), plasma-arc welding (PAW), carbon-arc welding (CAW), submerged arc welding (SAW), and electroslag welding (ESW).[44] During many welding operations, fumes are produced.[45,46,47] Fume constituents produced by welding operations include aluminum, beryllium, cadmium, chromium, copper, iron, lead, manganese, molybdenum, nickel, vanadium, zinc, and oxides of various types. Welding fumes contain particles of various sizes, with a significant number of the particles lying within the nanoscale regime (<0.1 μm). The number of particles, particle size, and specific chemical components of the fume depend on the process type, process conditions, and the materials being used.

Welding particles can take on an assortment of morphologies, depending on the forces acting among them, which may lead to particle agglomeration and the final size of the airborne particulate. When analyzing the welding fumes, individual particles may be seen, as well as particle chains and agglomerates as a result of such forces. Spiegel-Ciobanu[46] reports that for manually operated arc welding on Cr-Ni steel, particle sizes may be as small as 20 nm, with agglomerates on the order of 500 nm. For an automated version of the process, particle sizes were observed as small as 10 nm and agglomerates were approximately 100 nm in size.

The constituents within a welding fume depend on the (1) base and filler metals, (2) electrode coating, (3) contaminants on the base metal, and (4) the reactions that occur during the welding process.[48] Chan et al. found that for a shielded metal arc welding (SMAW) operation, the particle size, composition, and generation rate depend on electrode type and diameter, welding current, welding speed, welding angle, and current type (AC or DC).[45] For this case, the electrode was found to be the dominant factor in the chemical composition of the fume, and the fume generation rate was linked to the current type, with an AC current decreasing the fume generation rate. Similar studies for the characterization of SMAW emissions have also been reported.[49,50]

4.2 Wet Machining

Some manufacturing processes (e.g., tapping, turning, drilling, and milling), incorporate metalworking fluids (MWF) in an effort to improve machined surface quality, enhance tool life, provide lubrication, and remove heat from the working zone.[51,52,53] The interaction of the cutting fluid with the machine tool and the workpiece provides several mechanisms for the formation of mist. Figure 5 displays two potential mechanisms that may lead to the formation of fluid mist.[54] When fluid is exposed to hot elements within the machining system (cutting tool, workpiece, etc.), some of the fluid may vaporize and subsequently condense to form mist. In the case of processes with high-speed rotating elements,

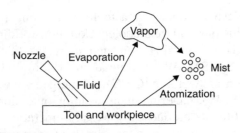

Figure 5 Mist formation from MWF in machining process. (Adapted from Ref. 54.)

mechanical energy may lead to fluid breakup via atomization. Gunter and Sutherland found that for a rotating cylindrical workpiece, the dominant process variables for atomization are the spindle speed and the workpiece diameter.[55] Other factors that affect the mist formation are the type of nozzle, distance of nozzle from point of application, and fluid application mode.[56]

4.3 Dry Machining

In recent years, dry machining has been investigated as an option for the elimination of metalworking fluids.[57–61] Thus, avoiding the environmental, health, and safety challenges associated with wet machining. In the absence of metalworking fluids, some workpiece materials have the propensity to form dusts. Therefore, dry machining should also be given attention from an air-quality standpoint. Sutherland et al. demonstrated that the speed, feed, and depth of cut were key variables in the formation of dust during the machining of cast iron.[62] Research findings reported that aerosol concentration during dry machining were 12 to 80 times less than those produced by wet-machining operations. Particles sizes were found to be between 1 and 4 µm. In a similar study, the particle distribution was found to have a MMAD of 10 µm.[21]

4.4 Grinding

Several studies have been performed to characterize aerosols produced during grinding processes.[21,63–65] Zimmer and Maynard[65] proposed three mechanisms/modes, or sources, for the formation/generation of aerosols from grinding: (1) emission of particles from the motor of the machine tool itself; (2) combustion of the workpiece material; and (3) volatilization and subsequent condensation of grinding materials at the wheel/workpiece interface.

Dasch et al. considered a wet grinding operation performed on gray cast iron with a straight cutting fluid from Castrol.[21] Temporal sampling plots revealed that particle concentration increased when the machine tool enclosures were periodically opened. The concentration of PM_{10} was reported to be 2.5 times the concentration of $PM_{1.0}$. In terms of the particle distribution, the MMAD was reported to be 2.5 µm.

Zimmer and Maynard examined surface grinding of various substrates (granite, clay, ceramic, steel, aluminum, polytetrafluoroethylene-PTFE, and oak hardwood) under dry conditions and characterized the particulate as having sizes ranging from 4.2 nm to 20.5 μm.[65] In some instances, it was possible to identify the particle mechanism of formation though the morphology of the particles (e.g., in the case of granite grinding, particles larger than 1 μm were attributed to the pullout of wheel abrasives). When PTFE was ground, a large number of nanoparticles (<0.1 μm) formed; grinding steel produced similar results. The mechanism for aerosol generation for PTFE and steel were said to be through nucleation, since the temperature in the working zone was high enough for vaporization or combustion to occur. Aerosol from wood grinding was attributed to frictional heating. Transmission electron microscopy (TEM) was used to conclude that the source of nanoparticles in the grinding of PTFE was associated the workpiece material. The source of nanoparticles for the other workpiece materials was not identified due to technical limitations of chemically characterizing particles at the nanoscale.

4.5 Casting

Many components of products are created through casting processes. It is estimated that 13 million tons of castings are produced each year in the United States.[66] There are a number of different types of casting processes, including sand casting, investment casting, permanent mold casting, continuous casting, and lost foam casting. Each process has unique stages, but most of them share four basic steps: (1) pattern making; (2) mold and core preparation and pouring; (3) furnace charge preparation and metal handling; and (4) shakeout, cooling, and sand handling.[66] Although not technically associated with casting, other processes are often linked with it: quenching, finishing, cleaning, and coating. Stages producing particulate for a typical green sand foundry include sand and binder mixing, core forming, mold making, sand preparation and treatment, metal melting, tapping, treatment, slag and dross removal, mold pouring and cooling, casting shakeout, and riser cutoff and gate removal.[66]

Several studies have been performed to characterize the aerosol emissions from a casting process. Chang et al. characterized the particle ($PM_{2.5}$) number and mass distribution, and chemical composition of gray iron for single-pour/no-baked molds.[67] They performed sampling in several stages of the process including pouring, cooling, and shakeout. During the study, they observed that particle number and mass increased during the pouring process and a rapid decrease occurred during the cooling period. Overall, the maximum particulate emissions were seen in the shakeout process, with particle sizes ranging between 1.5 to 2.6 μm.

Chang et al. observed a large number of ultrafine particles (<0.1 μm) and attributed these to condensation and coagulation mechanisms.[67] The melting of gray iron takes place around 1,500°C. They hypothesized that the interaction of the hot metal with the cold mold (21°C) led to chemical reactions, vaporization, and combustion of some organic compounds that form particles through nucleation or condensation, depending on the changes in the medium (e.g., vapor pressure). It was also found that process changes such as different pouring and shakeout variables influence both the particle size and the chemical presence of elements. The sampled particulate mostly consisted of organic and elemental carbon, and oxides of aluminum, silicon, calcium, and iron.

Lost foam casting (LFC) is an alternative method with advantages over the traditional casting process, such as the reuse of casting sand. However, as with other casting processes, LFC can produce significant amounts of airborne particulate, and there is also the concern about the decomposition of the foam pattern (often made of expanded polystyrene, EPS). Behm et al. investigated the generation of particulate matter from LFC using an EPS pattern with the variation of three process variables: pouring temperature, pattern coating, and pattern thickness using a 2^3 full factorial experiment. Many of the emissions appeared to be associated with the decomposition of the EPS pattern. They concluded that process variables play a significant role in the generation of airborne particles.[68]

An initiative to improve the air quality associated with the foundry industry started in 1994 with the establishment of the Casting Emissions Reduction Program (CERP). CERP is a collaborative effort between the U.S. government, the auto industry, and academia aimed at helping industry to meet air-quality standards through R&D in a real-world foundry environment. CERP also pursues research for emission measurements and energy technologies that will advance the casting industry in compliance with environmental regulations.[69]

4.6 Powder Manufacture—Carbon Black

Many industries (e.g., pharmaceutical and chemical industries) produce large quantities of powders that are used for medical products, cosmetics, paint pigments, etc. In many of these industries, there is a high probability that workers will be exposed to suspended powders. A powder that is produced in great quantities is carbon black. In fact, a million metric tons of carbon black was produced globally in 1999.[70]

Carbon black (CB) has long been used by industry for many purposes. It is used as an additive in the manufacturing of tires, toners, reinforcement filler, and paint pigment. While carbon black can result from incomplete combustion, this section only focuses on exposure to manufactured carbon black.[71,72]

Carbon black manufacturing usually consists of three steps: (1) reactor processing; (2) pelletizing; and (3) packaging.[71,72] The reactor processing stage is used to fabricate the primary particles with sizes in the 1 to 500 nm range, but due to the high concentration of particles, the individual particles aggregate to form larger particles that are the primary entity of CB. The pelletizing stage is used to make very large clumps of particles. Once the particles are dried following pelletizing, particles are packaged in different bag sizes and shipped.

Kuhlbusch et al. conducted an investigation to characterize aerosols during the packaging of carbon black in three European plants, concentrating on ultrafine particles.[72] They detected ultrafine particles (<0.10 μm) only in plants 2 and 3, which they attributed to the use of diesel or propane forklift machines and to the heaters used in plant 3. Emissions from heaters have also been reported by Stephenson et al. in a welding facility.[50] Elevated mass concentrations found in the filling area were linked to carbon black. CB particles in the bagging area had a lower size limit of 0.4 μm.

The propensity and mechanisms of powders (in the nanoscale) to become airborne are not clearly understood. Attempts have been undertaken to determine the propensity of powders to become airborne, known as *dustiness*.[3] Lyons and Mark investigated the dustiness of CB.[73]

5 TRADITIONAL CONTROL TECHNOLOGIES

A majority of the air-quality control technologies that are classified as traditional were initially implemented upon identification of the existence of an airborne hazard. The sole objective of these technologies is to control the identified airborne particulate to the level specified in the relevant regulations. The type of traditional control technology employed depends primarily on three factors: (1) the type of particulate to be controlled; (2) the process by which the particulate is produced; and (3) the interaction of the worker with the manufacturing process. The latter consideration can include the possibility of worker exposure, as well as the level of worker intervention in the process. Brandt described a hierarchy of principles that are used for particle control (Table 7).[16] Several broad categories of traditional control technologies are discussed in this section.

5.1 Ventilation Systems

Ventilation systems are used to control particulate that can be easily transported by the movement of air in the workplace. This includes smoke, dust, fumes,

Table 7 Traditional Approach for Particle Control

Eliminating the source of aerosols	Selection of "less toxic" materials Adjusting the process parameters Housekeeping
Prevention of contaminant dispersion	Incorporation of Enclosures Local exhaust ventilation Maintenance of equipment Education of employees
Protection of employees	Use of general ventilation Personal protection equipment

and fine mists but not larger droplets or metal particles. Ventilation systems range in configuration from typical building HVAC units to more specialized ducted configurations dedicated to a single piece of equipment. Regardless of the configuration, ventilation systems employ a collection mechanism specific to the particulate being removed from the workers' environment. Collections systems may or may not include filters. The adaptability of ventilation systems makes them suitable for a wide range of manufacturing operations and worker interaction levels.

Ventilation systems remove particulate from the workplace air to a location where the particulate is trapped or collected. Such equipment often functions by drawing particulate-laden air through HEPA (high-efficiency particulate air) filters and/or by using centrifugal air flow to force particles to side walls to be collected and possibly recycled (collected MWF mist can be returned to the sump). The location of the ventilation system within the workspace is critical. The system must remove particulate from the workspace without drawing it closer to the worker. Therefore, systems that create a downdraft are considered more effective than those that create an updraft, which can draw particulate into the workers' breathing space.

While such methods can be effective in substantially reducing particulate concentrations, they do not eliminate all the related health risks. Strong air flow induced by ventilation fans can force fine liquid particles to break away from liquid pools and saturated filter elements, thereby increasing the particulate concentration in the work environment. Evaporation or vaporization of a collected liquid can lead to increased amounts of fluid vapor, which can easily pass through collectors and air filters. This vapor may condense to form mist if it is recirculated from a warm duct to a cooler workroom.[74] Therefore, even in the presence of a 100% efficient particulate collector, hazardous particulates in the form of vapor may still pass through removal systems and recondense to form fluid mist.[75]

Another important issue regarding ventilation systems is the implementation of a routine maintenance program. A study conducted to analyze the performance of industrial mist collectors indicated that performance of individual collections stages, as well as assembled collectors, varied substantially. A protocol was developed to evaluate the systems that filter the mist produced by MWFs that suggested overall mist collector efficiency may be somewhat suspect.[76] Although filter efficiency increased with solid loading, efficiency decreased with liquid loading. In some cases, it was found that mist-loaded filters developed negative efficiencies and became generators of submicrometer droplets. In addition, the inability of liquid saturated filters was found to fully drain under normal continuous operation.[77] The recommendation is to enforce a rigorous maintenance program with implementation of air cleaner or air filter rotation. In closing, while they can be effective, collectors within ventilation systems are expensive to purchase, install, operate, and maintain.

5.2 Enclosures

Maintaining the appropriate air quality in the workplace can be accomplished by containing the particulate with the use of an enclosure. A wide variety of enclosures, including OEM (original equipment manufacture) and retrofit types, are used in manufacturing processes that require minimal worker interaction. It is not uncommon for a ventilation system to be used in combination with an enclosure. Enclosures can control a wide variety of airborne particulate, provided they are properly designed.

The purpose of an enclosure is to contain the particulate, which may be released into the workers' environment if the enclosure is opened. Hands et al. investigated the efficacy of machining enclosures, in which exposure from three different control methods used for MWF aerosols were compared.[78] It indicated that OEM enclosures equipped with exhaust ventilation provided significantly lower exposures, as compared to retrofit enclosures or no enclosure. It was also found that operations with retrofit enclosures have exposures that are not significantly different from those with no enclosure at all. Furthermore, retrofit enclosures were twice as costly to fabricate and install as OEM enclosures, as they were inherently difficult to design due to operation and maintenance requirements and their efficacy may degrade over time.

Enclosures are most effective in manufacturing environments in which all processes can be enclosed. This is not always the case, as illustrated by a study conducted to assess the exposure to MWFs in a typical automotive manufacturing facility. Of the 295 machines studied, 35 percent had no enclosures, 45 percent were partially enclosures, and 20 percent were completely enclosed.[79] Furthermore, 88 percent of the machines had no local exhaust in place. Heitbrink et al. found that nonventilated machine operations appeared to disperse mist concentrations throughout the particular machining plant they investigated.[80] Half of the worker exposure was attributed to the machine operation, while the other half was attributed to this background concentration of mist.

5.3 Chemical Treatment

The use of chemicals as a control technology for airborne hazards is limited to manufacturing processes that produce MWF mists. These can be used in any process that utilizes MWF without consideration of the level of worker input to the process.

A method developed for reducing the amount of suspended mist in machining operations involves the use of chemical additives called mist suppressants. It was found that the addition of high molecular weight polymer polyisobutylene (PIB) increased the elongational viscosity of the MWF, thus increasing the size and settling rate of the mist drops that are created and reducing the quantity of suspended mist. This particular mist suppressant has not been found to be as effective in reducing mists generated from water-based MWFs.[81,82] Therefore, studies have been conducted to investigate the use of high molecular weight polymer polyethylene oxide (PEO) as a mist suppressant for water-based MWFs.[82,83] The physical theory behind mist suppressants is based on the atomization mechanism of mist formation, and, as a result, such methods may have little effect on reducing mist practices generated via vaporization/condensation.

5.4 Personal Protection Systems

Under the Occupational Health and Safety Act of 1970 employers are required to provide a healthy working environment. That is, it must be free of substances that might impact workers' health. Personal protection equipment is required after efforts have been undertaken to control the sources of particulate and the elimination of particulate generation has not been accomplished. *Respirators* serve two purposes: (1) filter airborne particles; and (2) supply clean air to the worker.[84]

6 ENVIRONMENTALLY RESPONSIBLE PARTICULATE MITIGATION/ELIMINATION

The airborne emissions from several manufacturing processes have been examined, along with traditional control methods. Attention now turns to identifying environmentally responsible actions that may be employed to improve the performance of these operations. As opposed to the control actions presented in the previous section, these efforts are not focused on simply containing or capturing the emissions, but on reducing or eliminating the emissions at their source. Of course, as Figure 6 suggests, it is desirable to establish environmentally responsible mitigation/eradication technologies early in product/process design.[85] The earlier these technologies can be considered, the more options will be available and the more effective the resulting technology development effort. In addition, developments early in the design activity typically cost much less to implement than developments that are introduced later. These facts have been reported elsewhere for characteristics such as quality and ease of manufacture.[86]

From a manufacturing standpoint, there are at least three general actions that can be taken to address emissions in an environmentally responsible manner:

1. Alternative process plan
2. Alternative process sequence
3. Process change

6.1 Alternative Process Plan

The set of operations to be employed in producing a manufactured product can be identified once sufficient product specification detail is available. There is generally significant flexibility in deciding the operations to employ (and their

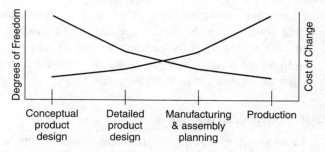

Figure 6 Degrees of freedom and change costs at various points in product and process design. (Adapted from Ref. 85.)

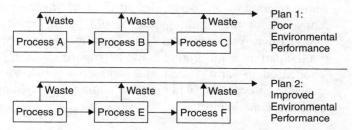

Figure 7 Alternative process plans producing different performance. (Adapted from Ref. 85.)

order) in creating a product. This process-planning task has historically been based primarily on the economics and cycle times associated with the various operations. Product designers have traditionally developed product designs based on product use considerations, but nowadays must also consider the downstream impacts on manufacturing. The objective of environmentally responsible manufacturing requires that all product and process decision makers consider new metrics in addition to the traditional ones.

As noted, process planners generally have some flexibility in deciding the manner by which a product is to be created. It is quite possible that one candidate process sequence could be preferable to another in terms of airborne emissions. Figure 7 illustrates this point with two candidate process plans.[85] Assuming that both plans are comparable in terms of cost, cycle time, quality, and so on, the figure indicates that Plan 2 is preferable to Plan 1 because it is more environmentally responsible (i.e., it produces less harmful aerosols). In general, process plans can involve completely different sets of operations, or they may be the same except for a single operation. Eliminating an operation from a process plan would be considered equivalent to creating a new process plan (this could involve changing the settings of the remaining operations).

Examples that illustrate this approach to ERM include: (1) avoiding the emissions from a painting operation by employing a grinding operation that produces an aesthetically pleasing product surface appearance; (2) avoiding cutting fluid mist by creating the desired feature via a casting operation; or (3) using hard turning to eliminate a grinding operation and its concomitant dust. It is worth noting that a more environmentally responsible process plan may also produce parts at lower cost or reduced cycle times.

6.2 Alternative Process Sequence

In addition to specifying the operations within a plan, process planners may have some discretion in varying the order of said operations. In the context of a plan, such as that of Plan 2 in Figure 7, this may mean that the process order is switched from D-E-F to D-F-E. Although such a switch would mean that the same operations are performed, it does not necessarily follow that the aerosol emissions of the resulting plan would be the same. For example, consider a product that is laser welded following a machining operation. MWFs that remain on the part after machining could be vaporized during welding and subsequently condensed to form a mist that represents a risk to worker health. It may be possible to consider switching the cutting and welding operations to avoid this problem completely.

6.3 Process Change

The previous approaches have focused on the set of operations (and their sequence) that are used to manufacture a given product. With the process plan specified, decision-making degrees of freedom are reduced to the point that environmental improvement must be sought within an operation. The problem at hand is the reduction/elimination of process aerosols given the degrees of freedom available only at the process level. For that, we must critically examine such operation characteristics as process settings, inputs, and process procedure.

Process Method and Inputs

Even though a given process has been specified, this does not mean that there is no latitude in the performance of the operation. Certainly, those methods and equipment types that provide for reduced aerosol emission should be favored, assuming they still affect the same desired change to the product being operated upon. For example, Hands et al. reported that machine tools with original equipment manufacture (OEM) installed enclosures are better than retro-fitted structures at containing fluid mist.[78] As another example, researchers have reported some success with an innovative dry drilling approach.[38,87–89] In the absence of a cutting fluid, fluid mist will not be produced. With this approach, the drill is subjected to axial modulations to produce chips that can be easily channeled out of the drill flute in the absence of a cutting fluid. Another innovative approach to address cutting fluid mist employs kinematic coagulation, the capture of smaller particles by larger collector droplets.[90]

Manufacturing operations take some input component and modify it in a particular way to accomplish an objective. Often, other materials (apart from the product material itself) are used to achieve this goal. The mission of environmentally responsible manufacturing requires that these materials be selected and used wisely.

An example of a well-known material substitution is of chlorofluorocarbons (CFC). CFCs were banned as refrigerants in the 1970s due to their role in depleting the ozone layer. Nowadays, solvents are still used in the auto, electronic, chemical, and pulps industries as process inputs, cleaning agents, and dispersants.[91] Additional concerns of solvents include the emission of volatile organic compounds (VOCs) and flammability. To avoid these, some efforts have been undertaken to mitigate or reduce the environmental impacts of solvents.[91] An example taken from the auto industry is the use of water-based paint, which has proven to reduce the use of solvent, therefore minimizing their impacts.

Processing Conditions

This refers to such characteristics as machine settings and the ambient conditions under which an operation is performed. Again, conditions should be selected with consideration for aerosol emissions, along with attention to other factors such as cost, cycle time, and quality. As already suggested, dry machining may require a significant process change to be successful. However, for some situations, few substantive changes may be required, and dry machining can be implemented by simply turning off the fluid. When compared to wet machining, the aerosol concentration generated by dry machining in turning was reported to be 12 to 80 times less than that generated from wet machining.[62] However, dry machining is still a research topic, as many challenges need to be overcome (e.g., product quality). If some fluid is needed for machining, minimum quantity lubrication (MQL) represents a strategy that can offer technological and economic advantages over traditional fluid applications in machining.[92,93] As the name implies, MQL seeks to reduce the amount of cutting fluid used in an operation, and with less fluid, less MWF mist may be generated.

7 MANUFACTURING TRENDS AND POTENTIAL PARTICULATE CONCERNS

Thus far, this chapter has focused on air-quality issues associated with traditional manufacturing processes. Recently, there has been a growth in interest in such emerging manufacturing trends as nanotechnology and biotechnology within industry and academia. In the case of nanotechnology, the focus has been to exploit the improved properties that materials show at the nanoscale, in contrast to their bulk counterparts. Manufacturing-related biotech applications include drug development, biomaterials, and tissue engineering. A review of the literature reveals that air-quality research related to these emerging technologies has been limited. With this in mind, this section focuses on exposure to carbon nanotubes and microorganisms.

7.1 Nanoparticles

In the last decade, we have experienced a boom in nanotechnology research. Nanotechnology has been defined "as the science and technology that will enable the understanding, measurement, manipulation, and manufacturing at the atomic, molecular, and supra-molecular levels, with the aim of creating materials, devices, and systems with fundamentally new molecular organization, properties, and functions."[94] The emerging field of nanotechnology is leading to unprecedented understanding and control over the composition of all physical things. Nanotechnology is expected to change the way in which products are designed and manufactured.[95]

When reviewing the literature associated with nanotechnology, the case is often made that this technology will positively affect society. However, it is unclear at this point, the effects that the *creation* of these nanoscale products will have on the environment and society. Of course, the manufacturing of nanomaterials such as carbon nanotubes, nanowires, and quantum dots should raise concerns about worker exposure. Currently, there are no standards regarding exposure to nanomaterials in the United States, and the literature related to the toxic effects of nanomaterials is limited.

Manufacturing of nanomaterials may give rise to new sources of nanoscale airborne particles. During the manufacturing of nanomaterials, possible stages during which workers could be exposed to nanomaterials include synthesis, material handling, equipment cleaning (e.g., reactor), and facility maintenance.[96] The level of exposure will vary by process as the synthesis methods change. Common commercial processes include: (1) vapor-phase, (2) colloidal, (3) gas-phase, and (4) attrition. Table 8 provides a comprehensive description of the exposure that might result from each of these processes.

As can be seen in Table 8, it is unlikely that there will be a concentration of nanoparticles during the manufacturing of the nanomaterials. During synthesis, exposure might result if there are leaks from the reactor.[96] Therefore, reactor design will play an important role in ensuring that the synthesis process limits the exposure of the worker to nanoparticles. It is important to mention that if the product becomes airborne, it may start to agglomerate and the exposure could be less significant, since the particles may no longer be in the nano-regime. At this moment, it is unclear if agglomerated particles would have the same toxicity as individual particles. However, it is believed that the particles will be disaggregated following inhalation, thus causing the same possible effects on worker health.

Exposure to Carbon Nanotubes

Nanoparticles could become airborne, and therefore may threaten worker health if inhaled. Figure 8 depicts confirmed and potential exposure routes for

Table 8 Possible Route of Exposure to Nanoparticles by Process and Type

Synthesis Process	Particle Formation	Inhalation Risk	Dermal/Ingestion Risk
Gas phase	In air	Leakage from reactor Product recovery Packaging	Air contamination Handling of product Cleaning/maintenance
Vapor phase	On substrate	Product recovery Packaging	Dry contamination Handling Cleaning/maintenance
Colloidal	Liquid suspension	Drying of product	Spillage of workplace Handling of product Cleaning/maintenance
Attrition	Liquid suspension	Drying of product	Spillage Handling of product Cleaning/maintenance

Note: From Ref. 96.

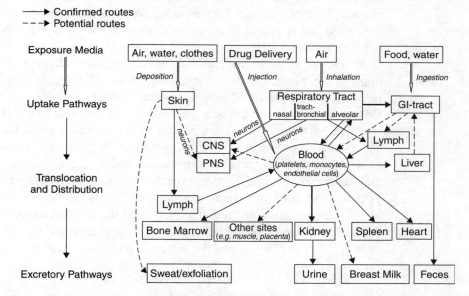

Figure 8 Exposure routes of nanoparticles. (Reproduced with permission from Ref. 97.)

nanomaterials.[97] Maynard et al. conducted the first assessment of exposure to carbon nanotubes (CNTs) by two different production processes: laser ablation and high pressure carbon monoxide (HiPCO®).[98] This project investigated the propensity of carbon nanotubes to become airborne when agitated in a laboratory environment. CNT material from laser ablation did not lead to a significant amount of airborne CNT, and detectable particles were bigger than 100 nm. For the case of HiPCO® there were a number of nanoscale particles (<100 nm). In both cases, the concentration of particles in air decreased very quickly with time and measured concentrations never exceeded the OSHA-mandated personal exposure limit currently set at 5 mg/m^3 (for particulates not otherwise regulated, PNOR). Maynard et al. also reported on experiments to investigate the propensity of carbon nanotubes to become airborne during material removal/handling at four facilities, but the results from these experiments were not conclusive.[99]

7.2 Bioaerosols

Bioaerosols are aerosols of biological origin. Different types of aerosols can be found in the occupational environment, including viruses, bacteria, fungal spores, and pollen. Common settings for exposure to viruses and bacteria are hospitals and health clinics. Tuberculosis is an example of an occupational related disease cause by the bacterium *mycobacterium tuberculosis*. Bioaerosols can contain very small components; for example, viruses exist at the nanometer scale. In manufacturing environments, bioaerosols might originate from HVAC systems, machining operations that use metalworking fluids (MWF), or manufacturing biotech applications. Table 9 provides the size range for several bioaerosols. A comprehensive list of the origin/source of bioaerosols in addition to bioaerosols related diseases is given in Table 10.

It has long been known that MWFs contain diverse microbial populations.[100–102] Some of these bacteria are believed to cause respiratory problems in humans, including MWF-associated nontuberculosis mycobacteria (NTM). Endotoxin levels can also be significant due to the relatively high levels of gram-negative bacteria.[103,104] A recent review has concluded that bacterial endotoxin is a primary component of MWF toxicity. There are no regulations or recommendations regarding worker exposure to microorganisms, and it is not apparent that the current air-quality recommendations adequately protect worker health, as evidenced in 2001 at a TRW automotive plant in Mount Vernon, Ohio. Despite an OSHA inspection of the plant that found the level of exposure to MWF to be within "acceptable" limits, 107 workers were placed on medical restrictions due to respiratory problems, and 37 suffered long-term disability.

Table 9 Type and Size of Bioaerosols

Type of bioaerosol	Size (μm)
Viruses	0.02–0.3
Bacteria	0.3–10
Fungal Spores	0.5–30
Pollen	10–100

Note: From Ref. 3.

Table 10 Common Bioaerosols, Related Diseases, and Typical Sources

Bioaerosol	Examples of Diseases	Common Sources
Pollen, spores, plant parts	Hay fever, allergic rhinoconjunctivitis, asthma, upper airway irritation	Plants, trees, grasses, ferns—harvesting, cutting, shiploading
Fungi	Asthma, allergic diseases, infection, toxicosis, tumors	Plant material, skin, leather, oils; bird, bat and animal droppings; feathers, soil nutrients, glues, wool
Bacteria	Endotoxicosis, tuberculosis, pneumonia, respiratory and wound, infections, legionellosis, Q and Pontiac fever	Humans, birds, and animals (e.g., saliva, blood, dental secretions, skin, vomit, urine, feces); water sprays and surf, humidifiers, hot tubs, pools, drinking water, cooling towers
Other allergen sources, arthropods, Vertebrates	Esthma, dermatitis, hypersensitivity, pneumonitic	Mite excreta, insect parts (cockroach, spiders, moths, midge), dander and saliva from cats, dogs, rabbits mice and rats, bird serum, farm animal dander
Viruses	Respiratory infections, colds, measles, mumps, hepatitis A, influenza, chicken pox, Hanta virus	Infected humans, animal excreta, insect vectors, protozoa

Note: Adapted from Ref. 101.

In the last decade, the biotechnology industry has experienced an increase in production as biotechnology finds its way into the development of innovative new products such as pharmaceuticals, nonchemical pesticides, and useful microorganisms (e.g., for waste biodegradation). Fungal enzymes play an important role in the production of these products.[99] With the growing use of these microorganisms within industrial applications, it is likely that they (or some component or byproduct of the microorganism) will become aerosolized within the working environment.

8 SUMMARY AND CONCLUSIONS

This chapter has focused on air quality (aerosols) issues in manufacturing environments. It has covered a variety of topics, such as the importance of occupational health, regulations, and aerosol characterization. The chapter has also reviewed common industrial processes in terms of relevant aerosol generation mechanisms, particle mass concentration, and particle size. Finally, particulate control technologies have been examined, and efforts to mitigate/eliminate particulate in a more environmentally responsible manner have been studied. The chapter has concluded with an examination of air-quality concerns for innovative new manufacturing technologies. A large number of references have been cited in the chapter, and the reader is encouraged to consult these works for further details.

The emphasis of this chapter on manufacturing process aerosols is driven by concerns for human health as a result of exposure to airborne particles. Given this statement, it must be noted that all the manufacturing studies that have been cited in the chapter reported compliance with applicable U.S. health/safety standards. However, as has been noted, there is some concern that current standards might not be protective of human health, especially from exposure to ultrafine particles. The authors believe that as more knowledge is accumulated regarding the potential health effects of workplace aerosols, in particular for novel processing methods that are of growing importance, emphasis on air quality will increase. New/improved standards will be developed, and many manufacturers will be forced to scramble to react to these changes; such reactionary approaches are usually not cost effective, and may be counterproductive. Traditional particulate control approaches often seek only to contain or mask process generated aerosols, but more and more emphasis will need to be placed on innovative techniques to avoid/eliminate airborne particulate from manufacturing processes. The development of such techniques will likely require an understanding of the relevant mechanisms associated with the creation of process aerosols that is vastly improved over the knowledge that exists today.

9 ACKNOWLEDGMENTS

This work was funded, in part, by the Sustainable Futures IGERT project sponsored by the U.S. National Science Foundation (under Grant No. DGE 0333401).

REFERENCES

1. K. Haapala, J. L. Rivera, and J. W. Sutherland, "Environmentally Responsible Process Selection Via Life Cycle Analysis," Proceedings of 2006 ISFA 2006 International Symposium on Flexible Automation Osaka, Japan, July 10–12, 2006.
2. U.S. Census, "1997 Economic Census: Company Statistics Series," U.S. Census Bureau, September 2001.
3. W. C. Hinds, 2nd ed., *Aerosol Technology*, John Wiley, New York, 1999.
4. J. E. Peterson, *Industrial Health*, Prentice-Hall, Englewood Cliffs, NJ, 1977.
5. University of Michigan Health System, "Illustration of the Respiratory System," http://www.med.umich.edu/1libr/aha/aha_respirat_art.htm. Accessed in June 2006.
6. G. Wright, "The Pulmonary Effects of Inhaled Inorganic Dust," *Patty's Industrial Hygiene and Toxicology*, G. Clayton and F. Clayton, ed., John Wiley, New York, 1991.
7. F. C. Lu, and S. Kacew, 4th ed. *Basic Toxicology*, Taylor & Francis, New York, 2002.
8. D. Dockery, C. Pope III, X. Xu, J. Spengler, J. Ware, M. Fay, B. Ferris, and F. Speizer, "Mortality Risks of Air Pollution: a Prospective Cohort Study," *New England Journal of Medicine*, **329**, 1753–1759 (1993).

9. G. Oberdörster, "Pulmonary Effects of Inhaled Ultrafine Particles," *Int. Arch. Occup. Environ. Health*, **74**(1) 1–8 (2001).
10. M. Utell and M. Frampton, "Acute Health Effects of Ambient Air Pollution: The Ultrafine Particle Hypothesis," *J. Aerosol Med.* **13**(4), 355–359 (2000).
11. R. Vincent, P. Kumarathasan, P. Geogan, S. G. Bjarnason, J. Guenette, D. Berube, I. Y. Adamson, S. Desjardins, R. T. Burnett, R. J. Miller, and B. Battistini, *Research Report 104: Inhalation Toxicology of Urban Ambient Particulate Matter: Acute Cardiovascular Effects in Rats*, Health Effects Institute, Cambridge, MA, 2001.
12. D. Lison, C. Lardot, F. Huaux, G. Zanetti, and B. Fubini, "Influence of Particle Surface Area on the Toxicity of Insoluble Manganese Dioxide Dust," *Arch. Toxicology*, **71**, 725–729, (1997).
13. T. Osunsanya, G. Prescott, and A. Seaton, "Acute Respiratory Effects of Particles: Mass or number?" *Occup. Envion. Med*, 58 (3), 154–159 (2001).
14. K. R. Spurny, "On the Physics, Chemistry and Toxicology of Ultrafine Anthropogenic, Atmospheric Aerosols (UAAA): New Advances," *Toxicol. Lett.*, **96–97**, 253–261 (1998).
15. J. H. Vincent and C. F. Clements, "Ultrafine Particles in Workplace Atmospheres," *Phil. Trans. R. Soc. Lond. A.*, **358**, 2673–2682 (2000).
16. A. D. Brandt, *Industrial Health Engineering*, John Wiley, New York, 1947.
17. NAS, "Airborne Particles," National Academy of Science, National Research Council. Subcommittee on Airborne Particulate, 1979.
18. P. Baron, "Generation and Behavior of Airborne Particles (Aerosols)," National Institute of Occupational Health and Safety, http://www.cdc.gov/niosh/topics/aerosols/pdfs/Aerosol_101.pdf. Retrieved on June 2006.
19. A. D. Maynard, and E. D. Kuempel, "Airborne Nanostructured Particles and Occupational Health," *Journal of Nanoparticle Research*, **7**, 587–614 (2005).
20. U.S. Environmental Protection Agency, "Basic Concepts in Environmental Sciences," http://www.epa.gov/eogapti1/module3/index.htm. Accessed on June 2006.
21. J. Dasch, J. D'Arcy, A. Gundrum, J. W. Sutherland, J. Johnson, and D. Carlson, "Characterization of Fine Particles from Machining in Automotive Plants," *Journal of Occupational and Environmental Hygiene*, **2**, 1–14 (2005).
22. ISO, International Organization for Standarization, "Air Quality—Particle Size Fraction Definitions for Health-related Sampling," ISO 7708, 1995.
23. N. T. Jenkins and T. W. Eagar, "Chemical Analysis of Welding Fumes," supplement to the *Welding Journal* (June 2005).
24. D. M. Brown, M. R. Wilson, W. MacNee, V. Stone, and K. Donaldson, "Size-Dependent Proinflammatory Effects of Ultrafine Polystyrene Particles: A Role for Surface Area and Oxidative Stress in the Enhanced Activity of Ultrafines," *Toxicology and Applied Pharmacology*, **175**, 191–199 (2001).
25. K. Donaldson, X. Y. Li, and W. MacNee, "Ultrafine (nanometer) Particle Mediated Lung Injury," *Journal of Aerosol Science*, **29**, 553–560 (1998).
26. B. S. Cohen and C. S. McCammon, 9th ed., *Air Sampling Instruments*, ACGIH. ISBN: 1-882417-39-9, 2001.
27. F. A. Patty, "Industrial Hygiene: Restrospect and Prospect," *Patty's Industrial Hygiene and Toxicology*, G. Clayton and F. Clayton, ed., John Wiley, New York, 1978.

28. D. W. Dockery and C. A. Pope III, "Acute Respiratory Effects of Particulate Air Pollution," *Annu. Rev. Public Health*, **15**, 107–132 (1994).
29. HEI, "Airborne Particles and Health: HEI Epidemiologic Evidence," *Health Effects Institute (HEI) Perspectives* (June, 2001).
30. G. Oberdörster, R. M. Gelein, J. Ferin, and B. Weiss, "Association of Particulate Air Pollution and Acute Mortality: Involvement of Ultrafine Particles?" *Inhal. Toxicol*, **7** 111–124 (1995).
31. Bureau of Labor Statistics, "Work Place Injuries and Illnesses in 2004," U.S. Department of Labor, http://www.bls.gov/iif/oshwc/osh/os/osnr0023.pdf. Retrieved on March 2006.
32. NIOSH, "Work-Related Lung Diseases," U.S. National Institute of Occupational Health and Safety, U.S. Department of Health, Center for Disease Control and Prevention. http://www.cdc.gov/niosh/docs/pib/pdfs/2004PIB-lungDis.pdf. 2004.
33. A. M. Knaapen, P. J. A. Borm, C. Albrecht, and R. P. F. Schins, "Inhaled Particles and Lung Cancer Part A: Mechanism," *International Journal of Cancer*, **109**, 799–809 (2004).
34. C. Dasenbrock, L. Peters, O. Creutzenberg, and U. Heinrich, "The Carcinogenic Potency of Carbon Particles with and without PAH after Repeated Intratracheal Administration in the Rat," Toxicol Lett. Vol. 88(1–3):15–21, 1996.
35. K. J. Nikula, M. B. Snipes, E. B. Barr, W. C. Griffith, R. F. Henderson, J. L. Mauderly, "Comparative Pulmonary Toxicities and Carcinogenicities of Chronically Inhaled Diesel Exhaust and Carbon Black in F344 Rats," *Fundam. Appl. Toxicol*, **25**, 80–94 (1995).
36. UCLA, "Welding Fumes: What You Need to Know," UCLA Labor Occupational Safety & Health Program (LOSH), 2003.
37. B. A. Racette, L. McGee-Minnich, S. M. Moerlein, J. W. Mink, T. O. Videen, and J. S. Perlmutter, "Welding-related Parkinsonism: Clinical Features,Treatment, and Pathophysiocology," *Neurology*, **56**, 8–13 (2001).
38. B. Ackroyd, W. D. Compton, and S. Chandrasekhar, "Reducing the Need for Cutting Fluids in Drilling by Means of the Modulation Assisted Drilling," Proceedings of the Symposium on Manufacturing Science and Engineering, ASME Bound Vol.-MED-Vol. 8: pp. 405–411, 1998.
39. NIOSH, "NIOSH Origins and Mission," U.S. National Institute of Occupational Health and Safety. http://www.cdc.gov/niosh/about.html. Accessed in June 2006b.
40. W. E. MacCormik, "Legislation and Legislative Trends," *Patty's Industrial Hygiene and Toxicology*, G. Clayton and F. Clayton, ed., John Wiley, New York, 1978.
41. OSHA, U.S. Occupational Safety and Health Administration. http://www.dol.gov/. Accessed in April 2006.
42. U.S. Environmental Protection Agency, "Particulate Matter," http://www.epa.gov/oar/particlepollution/index.html. Retrieved on March 2006.
43. G. Oberdörster, Z. Sharp, V. Atudorei, A. Elder, R. Gelein, W. Kreyling, and C. Cox., "Translocation of Inhaled Ultrafine Particles to the Brain," *Inhalation Toxicology*, **16** (6–7), 437–45 (June 2004).
44. J. A. Schey, 3rd ed., *Introduction to Manufacturing Processes*, McGraw-Hill, New York, 2000.
45. W. Chan, K. L. Gunter, and J. W. Sutherland, "An Experimental Study into the Fume Particulate Produced by the Shielded Metal Arc Welding Process," *Transactions of the*

NAMRI/SME, **30**, 581–588 (2002). Also appeared as SME Technical Paper, No. MS02-2001.

46. V. E. Spiegel–Ciobanu, "Ultrafine Particles Created by Welding and Allied Processes," Ultrafine Aerosols at Workplaces, Workshop held on August 21–22 at the BG Institute for Occupational Safety and Health—BA, Sankt Augustin, Germany, 2003.
47. A.T. Zimmer and P. Biswas, "Characterization of Aerosols Resulting from Arc Welding Processes," *Journal of Aerosol Science*, **32**, 993–1008 (2001).
48. Battelle-Columbus Laboratories, "The Welding Environment, Parts IIA, IIB, and III;" American Welding Society, Miami, 1973.
49. P. Hewett, "The Particle Size Distribution, Density and Specific Surface Area of Welding Fumes from SMAW and GMAW Mild and Stainless Steel Consumables," *American Industrial Hygiene Association Journal*, **56**(2), 128–135 (1995).
50. D. Stephenson, G. Seshadri, and J. M. Veranth, "Workplace Exposure to Submicron Particle Mass and Number Concentrations from Manual Arc Welding of Carbon Steel," *AIHA Journal*, **64**, 516–521 (2003).
51. L. De Chiffre, "Function of Cutting Fluids in Machining," *Lubrication Engineering*, **44**, 514–518 (1999).
52. M. A. El Baradie, "Cutting Fluids: Part 1. Characterization," *Journal of Materials Processing Technology*, **56**, 786–797 (1996).
53. P. S. Sheng and S. Oberwalleney, "Life Cycle Planning of Cutting Fluids—A Review," *Journal of Manufacturing Science and Technology*, **119**(11), 791–800 (1997).
54. Y. Yue, J. W. Sutherland, and W. W. Olson, "Cutting Fluid Mist Formation in Machining via Atomization," *ASME Design for Manufacturing and Assembly*, **89**, 37–46 (1996).
55. K. L. Gunter, and J. W. Sutherland, "An Experimental Investigation into the Effect of Process Conditions on the Mass Concentration of Cutting Fluid Mist in Turning," *Journal of Cleaner Production*, **7**(5), 341–350 (1999).
56. J. P. Byers, *Metalworking Fluids*, Marcel Dekker, Inc, 1994.
57. D. P. Adler, W. W-S. Hii, D. J. Michalek, and J. W. Sutherland, "Examining the Role of Cutting Fluids in Machining and Efforts to Address Associated Environmental/Health Concerns," *Machining Science and Technology*, **10**(1) (2006).
58. R. B. Aronson, "Why Dry Machining?" *Manufacturing Engineering*, **114**(1), 33–36 (1995).
59. C. Granger, "Dry Machining's Double Benefit," *Machinery*, **152**(3873), 14–15, 17, 19–20, 1994.
60. H. Popke, T. Emmer, and J. Steffenhagen, "Environmentally Clean Metal Cutting Processes—Machining on the Way to Dry Cutting," Proceedings of the Institution of Mechanical Engineers, Part B: *Journal of Engineering Manufacture*, **213**(3), 329–332 (1999).
61. P. Young, G. Byrne, and M. Cotterell, "Manufacturing and the Environment," *The International Journal of Advanced Manufacturing Technology*, **13**(7), 488–493 (1997).
62. J. W. Sutherland, V. N. Kulur, and N. C. King, "Experimental Investigation of Air Quality in Wet and Dry Turning," *CIRP Annals—Manufacturing Technology*, **49**(1), 61–64 (2000).

63. J. W. Martyny, M. D. Hoover, M. Mroz, M. Margaret, K. Ellis, L. A. Maier, K. Sheff, L. S. Newman, "Aerosol Generated During Beryllium Machining," *Journal Occup Environ Med*, **42**(1), 8–18 (2000).
64. F. S. Rosenthal and B. L. Yeagy, "Characterization of Metalworking Fluid Aerosols in Bearing Grinding Operations," *American Industrial Hygiene Association Journal*, **62**, 379–382 (2001).
65. A. T. Zimmer and A. D. Maynard, "Investigation of the Aerosol Produced by a High-speed Hand-held Grinder Using Various Substrates," *Annals of Occupational Hygiene*, **46**(8), 663–672 (2002).
66. U.S. Environmental Protection Agency, "Office of the Compliance Sector Notebook Project: Profile of the Metal Casting Industry," EPA/310-R-97-004, 1998.
67. M.-C. O. Chang, J. C. Chow, J. G. Watson, C. Glowacki, S. A. Sheya, and A. Prabhu, "Characterization of Fine Particles Emissions from Casting Processes," *Aerosol Science and Technology*, **39**, 947–959 (2005).
68. S. U. Behm, K. L. Gunter, and J. W. Sutherland, "An Investigation into the Effects of Process Parameter Settings on Air Emission Characteristics in the Lost Foam Casting Process," AFS Transactions, 2003. Paper #03–031, appeared on CD-ROM.
69. CERP, "Casting Emission Reduction Program," http://www.cerp-us.org/. Accessed on June 2006.
70. I. Poliski, "Carbon Black: The State of Art," *Modern Paint and Coatings*, **89** 40–41 (1999).
71. K. Gardiner, I. A. Calvert, M. J. A. Tongeren van, J. M. Harrington, "Occupational Exposure to Carbon Black in Its Manufacture: Data from 1987 to 1992," *Annals of Occupational Hygiene*, **40**, 65–77 (1996).
72. T. A. J. Kuhlbusch, S. Neumann, and H. Fissan, "Number Size Distribution, Mass Concentration, and Particle Composition of PM1, PM 2.5, and PM10 in Bag Filling Areas of Carbon Black Production," *Journal of Occupational and Environmental Hygiene*, **1**, 660–671 (2004).
73. C. P. Lyons and D. Mark, "Development and Testing of a Procedure to Evaluate the Dustiness of Powders and Dusts in Industrial use," Health and Safety Executive Contract Research Report No62/1994, HSE Books, ISBN 0 7176 0727 5, 1994.
74. S. J. Cooper and D. Leith, "Evaporation of Metalworking Fluid Mist in Laboratory and Industrial Mist Collectors," *American Industrial Hygiene Association Journal*, **59**(1), 45–51 (1998).
75. P. C. Raynor, S. Copper, and D. Leith, "Evaporation of Polydisperse Multicomponent Oil Droplets," *American Industrial Hygiene Association*, **57**(12), 1128–1136 (1996).
76. D. Leith, P. C. Raynor, M. G. Boundy, and S. J. Cooper, "Performance of Industrial Equipment to Collect Coolant Mist," *American Industrial Hygiene Association Journal*, **57**(12), 1142–1148 (1996).
77. D. A. Cozzens, W. W. Olson, and J. W. Sutherland. "An Experimental Investigation into the Effect of Cutting Fluid Conditions on the Boring of Aluminium Alloys," *Concurrent Product and Process Eng.*, ASME Bound Volume-MED **1**(85): 251–257 (1985).
78. D. Hands, M. J. Sheehan, B. Wong, and H. B Lick, "Comparison of Metalworking Fluid Mist Exposures from Machining with Different Levels of Machine Enclosure," *American Industrial Hygiene Association*, **57**(12), 1173–1178 (1996).

79. S. R. Woskie, M. A. Virji, D. Kriebel, S. R. Sama, D. Eberiel, D. K. Milton, S. K. Hammond, and R. Moure-Eraso, "Exposure Assessment for a Field Investigation of Acute Respiratory Effects of Metalworking Fluids. I. Summary of Findings," *American Industrial Hygiene Association Journal*, **57**(12), 1154–1162 (1996).
80. W. A. Heitbrink, J. M. Yacher, G. J. Deye, and A. B. Spencer, *Mist Control at a Machining Center, Part 1: Mist Characterization, American Industrial Hygiene Association*, **61**(2), 275–281, 2000.
81. E. Gulari, C. Manke, J. Smolinski, R. Marano, and L. Toth, "Polymer Additives as Mist Suppressants in Metalworking Fluids: Laboratory and Plant Studies," The Industrial Metalworking Environment: Assessment and Control, Symposium Proceedings, November 13–16, pp. 294–300, 1995.
82. R. S. Marano, J. M. Smolinski, and C. W. Manke, et al., in Polymer Additives as Mist Suppressants in Metal Cutting Fluids, Technical Paper FoMoCo, presented at 52nd annual meeting, Kansas City, Missouri (May 1997), also appeared as SAE technical paper No.980097 (1997).
83. S. Kalhan, S. Twining, R. Denis, R. Marano, and R. Messick, "Polymer Additives as Mist Suppressants in Metalworking Fluids Part IIa: Preliminary Laboratory and Plant Studies—Water Soluble Fluids," SAE 980097, pp. 47–51, 1997.
84. NIOSH, "Respirators," U.S. National Institute of Occupational Health and Safety, U.S. Department of Health, Center for Disease Control and Prevention http://www.cdc.gov/niosh/npptl/topics/respirators/. Accessed in June 2006a.
85. K. L. Gunter, and J. W. Sutherland, "Environmental Attributes of Manufacturing Process," *Handbook of Environmental Conscious Manufacturing*, ed. C. N. Madu. Boston: Kluwer Academic Publishers, 2001.
86. R. E. Devor, T. H. Chang, and J. W. Sutherland, *Statistical Quality Design and Control: Contemporary Concepts and Methods*, Macmillan, New York, 1992.
87. A. J. Filipovic, and J. W. Sutherland, "Development of a Magnetostrictive-Actuated Tool Holder for Dry Deep Hole Drilling," *Transactions of NAMRI/SME*, **33**, 437–444, 2005.
88. McCabe, J., "Advances in Dry Machining of Aluminum," *Cutting Tool Engineering* (February 2002).
89. H. G. Toews, W. D. Compton, and S. Chandrasekhar, "A Study of the Influence of Superimposed Low-Frequency Modulation on the Drilling Process," *Precision Engineering*, **1**(22), 1–9 (1998).
90. S. S. Kinare, D. J. Michalek, C. Ju, J. Sun, and J. W. Sutherland, "An Experimental Investigation of Cutting Fluid Mist Removal via a Novel Atomizer System," *Environmental Sustainability in the Mobility Industry: Technology and Business Challenges*, SAE SP-1865, March 2004.
91. J. M. DeSimone, "Practical Approaches to Green Solvents," *Science*, **297** (August 2002).
92. F. Klocke, A. Schulz, and K. Gerschwiler, "Saubere Fertigungstechnolgien—Ein Wettbewerbsvoteil von morgen? Aachener Werkzeug-maschinen-Kolloquium," Conference Proc. June 13–14, pp. 4.35–4.108, 1996.
93. K. Weinert, I. Inasaki, J. W. Sutherland, and T. Wakabayashi, "Dry Machining and Minimum Quantity Lubrication," *Annals of CIRP*, **53**(2), 511–537 (2004).
94. U.S. Congress, "21st Century Nanotechnology Research and Development Act," Public Law 108–153. December 2003, Washington, D.C. 2003.

95. NSTC, National Science and Technology Council, "Nanotechnology Shaping the World Atom by Atom," http://www.wtec.org/loyola/nano/IWGN.Public.Brochure/. 1999. Accessed in May 2006.
96. HSE, 2004. Health and Safety Executive, "Nanoparticles: An Occupational Hygiene Review." http://www.hse.gov.uk/research/rrpdf/rr274.pdf. Revisited February 2006.
97. G. Oberdörster, E. Oberdörster, and J. Oberdörster, "Nanotoxicology: An Emerging Discipline from Studies of Ultrafine Particles," *Environmental Health Perspectives*, **113**(7) (2005).
98. A. D. Maynard, P. A. Baron, M. Foley, A. A. Shvedova, E. R. Kisin, and V. Castranova, "Exposure to Carbon Nanotube Material: Aerosol Release during the Handling of Unrefined Single-walled Carbon Nanotube Material," *J Toxicol Environ Health A*, **67**(1), 87–107 (2004).
99. D. L. MacIntosh and J. D. Spengler, "Human Exposure Assessment," International Programme on Chemical Safety, United Nations Program, International Labour Organization, and the World Health Organization. Geneva, 2000.
100. A. I. Awoskia-Olumo, K. L. Trangle, and L. F. Fallon, "Microorganism-induced skin disease in workers exposed to metalworking fluids," *Occup. Med.*, **53**, 35–40 (2003).
101. C. J. Van der Gast, C. J. Knowles, M. A. Wright, and I. P. Thompson, "Identification and Characterization of Bacterial Populations of an In-use Metal-working Fluid by Phenotypic Methodology," *Int. Biodet. Biodeg*, **47**, 113–123 (2001).
102. R.J. Wallace Jr., Y. Zhang, R. W. Wilson, L. Mann, and H. Rossmoore, "Presence of a Single Genotype of the Newly Described Species *Mycobacterium immunogenum* in Industrial Metalworking Fluids Associated with Hypersensitivity Penumonitis," *Appl. Environ. Microbiol*, **68**, 5580–5584 (2002).
103. D. M. Lewis, E. Janotka, M. P. Whitmer, and T. A. Bledsoe, "Detection of Microbial Antigens in Metal Working Fluids," *Int. Biodeter. Biodegrad*, **47**, 89–94 (2001).
104. A. T. Simpson, M. Stear, J. A. Groves, M. Piney, S. D. Bradley, S. Stagg, and B. Crook, "Occupational Exposure to Metalworking Fluid Mist and Sump Fluid Contaminants," *Ann. Occ. Hyg*, **47**, 17–30 (2003).

CHAPTER 8

ENVIRONMENTALLY CONSCIOUS ELECTRONIC MANUFACTURING

Richard Ciocci, PhD, PE
The Pennsylvania State University Middletown, Pennsylvania

1 INTRODUCTION	179	
1.1 Definitions of Terms	180	
1.2 Environmentally Conscious Manufacturing	182	
2 HISTORY—LEGISLATION AND REGULATION	183	
2.1 Early U.S. Legislation	183	
2.2 Japan Take-back Policy	184	
2.3 European Union RoHS and WEEE	184	
2.4 Federal Initiatives	185	
2.5 State Initiative	185	
3 TOOLS	185	
3.1 Life-cycle Assessment	185	
3.2 Material Flow Analysis	187	
3.3 ISO 14000	187	
4 CASE STUDIES	187	
4.1 Materials	187	
4.2 Processes	189	
4.3 Boards	191	
4.4 Components	191	
4.5 Solders	195	
4.6 Assemblies	197	
4.7 Products	198	
4.8 Testing and Qualification	199	
4.9 Reliability and Performance	202	
4.10 Formation of Tin Whiskers	203	
5 FUTURE PLANS	204	
5.1 Conductive Adhesives	204	
6 SUMMARY	205	

1 INTRODUCTION

Protecting the environment is a prerequisite for *sustainable development*—economic growth that meets the needs of the present and does not adversely affect the ability of future generations to meet their own needs. A goal-oriented principle of sustainability is pollution prevention, which calls for the reduction and/or elimination of pollutants and wastes. It implies the use of the proper materials to meet the goal of zero waste, as well as the practices of reducing the use of hazardous materials and conserving energy and other resources. The strategies include increasing efficiency in energy use and analyzing products on a total life-cycle basis.

Increased industry interest in the environment has resulted in a more proactive approach to environmental problems. Attention to the consumption of resources and to disposal of used materials has shifted from traditional end-of-pipe waste treatments to preproduction design considerations. A number of initiatives by designers and manufacturers have the ultimate goal of environmental stewardship. Waste treatment has been replaced by pollution prevention, with specific goals such as the reduction and/or elimination of hazardous materials. These efforts intend to improve the treatment of the environment while not affecting adversely the performance, reliability, and cost of existing and future products.

One goal of pollution prevention is to reduce the use of those materials that may be detrimental to the environment. However, the materials chosen as replacements must not adversely affect the performance of the product. Both materials and process operations are candidates for change; however, the product must meet performance requirements, regardless of material substitutions. In some cases, quality control and inspection criteria are subject to revision due to pollution prevention efforts, but these revisions must be made to ensure that the product meets all quality and reliability requirements. The pollution prevention initiatives developed during the 1990s can be incorporated into industry as long as they do not adversely affect the product or its performance.

1.1 Definitions of Terms

Protecting the environment is a prerequisite for sustainable development. Possibly the broadest term describing how design and manufacturing affect the environment is *industrial ecology*. The term has been defined as the means by which the human species can deliberately and rationally approach an optimal global carrying capacity.[1] The concept behind industrial ecology is that economic systems exist within the systems that surround them, rather than as separate entities. Viewing a manufacturing system as operating within the environmental system mandates that industrial operations follow a systems approach to optimize the materials cycle from virgin resource through usable material, component, product, waste product, and disposable material. This optimizes all resources, including energy and capital. The key to the concept is recognizing that the manufacturing system for any product exists within the larger global ecological system.[2]

A more goal-oriented concept is pollution prevention, a new emphasis rather than a new approach. Pollution prevention calls for the reduction and/or elimination of pollutants and wastes. It implies the use of the proper materials to meet the goal of zero waste, as well as the practices of reducing the use of hazardous materials and conserving energy and other resources. Prevention can be described as a set of techniques and strategies, the goal of which is to achieve sustainable resource use. These strategies include increasing efficiency in energy use and analyzing products on a total life-cycle basis.

There are several reasons to engage in pollution prevention:

- Elimination and reduction of hazardous materials is potentially the most cost-effective means of dealing with wastes.
- Reduction of waste is a measure of quality improvement.

- An approach that considers prevention in all areas provides a longer-lasting solution to the problem than traditional waste disposal methods that merely transfer pollutants from one part of the environment to another.[3]
- The primary goal is the betterment of the environment through elimination, reduction, and prevention.
- If smaller quantities of materials, hazardous or not, are used, less material needs to be discarded eventually.
- Preventing waste makes it easier to dispose of waste products.

These philosophies suggest an ideal of zero waste, but industries faced with regulation, legislation, and customer environmental awareness need more practical approaches.

In order to focus on pollution prevention, environmental concerns must be addressed in the product and process design phase. A proactive approach is design for the environment (DfE), which calls for the design of products in order to minimize waste during item production, use, and disposal.[4–6] DfE considers the environment at a product's design phase, including all materials and energy requirements and their effects over the life cycle of the product. DfE principles must also be applied to manufacturing processes. Energy usage includes the energy required to manufacture the product, to use the product, and to dispose of the product. The product life cycle goes beyond what is typically considered as the useful life, beginning at the initial stages of raw material extraction and continuing through the final product disposition.

A few models of pollution prevention and design for the environment describe the levels and hierarchies of an environmental effort. The models are helpful in visualizing the ordered steps of the effort. The correct definition of a DfE hierarchy can lead to an understanding of how to handle materials that become part of the life cycle as they enter a manufacturing stage. The levels in the hierarchy may vary with specific products, although their order should be consistent. The hierarchy should be used at the design phase for all new products, and should be implemented wherever practical for existing products. This may not always be possible; for example, it may not be practicable to break down components that were never designed for disassembly.

The U.S. Environmental Protection Agency (EPA) published a pollution prevention hierarchy in 1988.[7] The model includes DfE initiatives such as reduction, recycling, and reuse. Characteristic of the pollution prevention effort is the high level of opportunity in source reduction. A model focusing on the levels of recycling includes the recovery of parts and materials after consumer usage.[8] At the product level the post-consumer products can be repaired or reused. Two additional models present a life-cycle sequence and a supply chain. In the life-cycle model, the process from raw material extraction through final disposal is shown as a system.[9] The supply-chain model has two goals: minimizing resource use and minimizing waste.[10] The main feature of the model is a feedback loop that collects products and materials at the various stages of the chain, which include processing, manufacturing, and distribution, and returns them to the material flow. Another model includes the entire product life cycle and material flows through the cycle.[11] The materials extraction and waste generation steps identify the boundaries of a product's life cycle. Materials recycling, product reuse, and remanufacturing are return loops for post-consumer scrap. DfE considerations must cover the life cycle of a product, from supply to use.[12] The model considers that substitutions and reductions of materials and wastes need to occur at all product life phases.

The models can be combined to form a hierarchical set of actions that must be considered during product and process design. In both the product and the process hierarchies, the first step is eliminating any materials that do not add value to the product. Next comes reducing the amount and variety of the materials used on a product or in the process, to minimize later disposal volumes. Eliminating and reducing processing materials is a move toward resource conservation. The third level is reusing any assemblies, subassemblies, or process materials. In addition to environmental improvement, reusing a durable product decreases costs due to lessened demand for new products and decreased disposal costs.

The next level is repairing any assemblies or subassemblies that could subsequently be reused. The similar but distinct step of rework is next. Repair is done in the field, while rework usually takes place in the factory. For example, in the electronics industry the objective of rework and repair is "to remove/replace/reorient a defective or misaligned component without affecting adjacent components." Process inspection will generally identify opportunities for rework, typically needed because of a process-related problem or defective component. Rework is more controllable, since repair is a field operation.[13]

Following rework is disassembly. At this stage, component parts are separated for reuse and recycling. Most disassembly of components occurs during rework or repair. Careful removal of damaged components is essential to keep the rest of the product unaffected. However, if the intent is to disassemble the entire product, more extensive disassembly procedures can be used. Another level in the hierarchy is recycling the process materials from all possible operations. The last level, already common, is proper disposal of product materials that cannot be otherwise recovered. The more effort that is placed on the earlier levels of the product and process hierarchies, the less material that will need proper disposal. Table 1 summarizes the steps of the hierarchy for both product- and process-related actions.

1.2 Environmentally Conscious Manufacturing

In line with the prevention of waste and pollution is environmentally conscious manufacturing (ECM). With ECM, the manufacturer resolves to exceed regulated environmental standards, striving toward the goal of pollution prevention.

Table 1 Summary of DfE Hierarchy Actions

Product-related Actions	Process-related Actions
Eliminate	Eliminate
Reduce	Reduce
Reuse	Reuse
Repair	
Rework	
Disassemble	
Recycle	Recycle
Dispose	Dispose

This philosophy assumes that by minimizing wastes, costs can be reduced, and attempts to optimize processes and procedures by using materials more efficiently.[4] One ECM goal is source reduction, which calls for the elimination of wastes and byproducts by replacing the materials that cause them. The most efficient use of raw materials, the establishment of a long-life service philosophy, and the development of an efficient material reclamation process are all parts of a source reduction effort. As with other programs that call for a change in plant culture, ECM and source reduction require leadership from the top levels of an organization.[14]

2 HISTORY—LEGISLATION AND REGULATION

2.1 Early U.S. Legislation

The first U.S. anti-lead legislation intended to control lead in electronics was proposed in the early 1990s as the Lead Exposure Reduction Act. A successful lobbying effort by the electronics industry at the time saved lead-based solder from restrictions in the proposal. However, the act required EPA to inventory lead-containing products and list those products that would present an unreasonable risk of injury to human health. The 1993 Lead Tax Act was enacted to place a per-pound charge on all lead smelted in the United States and on the lead content of imported products. Neither has had a real effect on the electronics industry. As the industry fought restrictions on lead in the United States, the move to reduce the health risk due to lead in solder shifted to Europe. In 1994, the Improved Design Life and Environmentally Aware Manufacturing of Electronics Assemblies by Lead-free Soldering (IDEALS) program was established in Europe. That program was the precursor to current lead-free initiatives.[15]

Lead may enter the environment during its mining, ore processing, smelting, refining, use, recycling, or disposal. Between 1.5 and 2.5 percent of all lead applications are used in electrical and electronic equipment (EEE). The main application of lead in EEE includes soldering of printed circuit boards. In any case, consumer electronics constitute 40 percent of lead found in landfills. The main concern in regard to the presence of lead in landfills is the potential for the lead to leach and contaminate drinking water supplies.[16]

The U.S. EPA under two presidential administrations adjusted the lead reporting level in an effort to drive a reduction in the material's use. In an action that was proposed in the final days of the Clinton administration and affirmed in the early days of the Bush administration, EPA now requires certain manufacturers, processors, and users of 100 pounds or more of lead to report that use to the Toxic Release Inventory. The threshold for lead reporting accountability had been 25,000 pounds annually for manufacturers and processors and 10,000 pounds for users.

The proposal had a dual purpose. EPA wanted to increase the information that is available to the public about the amount of lead that is used in U.S. communities. Also, EPA noted that, from past practice, having to report toxic use was enough of a driver to get companies to decrease emissions, which led to environmental and public health benefits. Since first collecting chemical emissions data reports under the Emergency Planning and Community Right-to-Know Act of 1986, EPA has witnessed substantial decreases in toxic releases and expects the new reporting threshold to show similar results.[17]

2.2 Japan Take-back Policy

Japan began its version of take-back legislation effective in 2001 for a variety of domestic products. The Electric Household Appliance Recycling Law passed the obligation for collection and recycling of waste appliances to the producers of those appliances. The appliance law is part of "the Basic Law" for establishing the Recycling-based Society in Japan. This law is one of seven, which include the Waste Management and Public Cleansing Law and the Law on Promoting Green Purchasing. The approach is to phase-in the law by adding a specific number of products that must return to the producer each year. In 2001, color televisions were included, thus directly including the electronic industry. The other products included during the first year of the law's jurisdiction were air conditioners, refrigerators, and washing machines.[18]

2.3 European Union RoHS and WEEE

On January 27, 2003, the European Parliament and the Council of the European Union (EU) passed a pair of directives aimed at reducing the effects of the production, use, treatment, and disposal of waste electrical and electronic equipment on human health and the environment. For years, the EU considered and debated Directive 2002/95/EC on the restriction of the use of certain hazardous substances in electrical and electronic equipment (RoHS)[19] and Directive 2002/96/EC on waste electrical and electronic equipment (WEEE).[20] Together they identify lead as a material not allowed in most electrical and electronic equipment marketed after July 1, 2006. mercury, cadmium, hexavalent chromium, polybrominated biphenyls, and polybrominated diphenyl ethers face similar restrictions.

The directives place electrical and electronic equipment into categories such as household appliances, telecommunications equipment, lighting equipment, toys, tools, and other consumer products. Waste electrical and electronic equipment includes all components and subassemblies that are part of the product at the time the user discards it.[20]

2.4 Federal Initiatives

In 1990, Congress passed the Pollution Prevention Act, which stated that the national policy of the United States is that pollution should be prevented or reduced at the source. Pollution that cannot be prevented should be recycled in an environmentally safe manner. Pollution that cannot be prevented or recycled should be treated in an environmentally safe manner. Disposal or other release of hazardous wastes into the environment should be used only as a last resort and be conducted in an environmentally safe manner.

With that direction, Congress set up a hierarchy of management options in descending order: prevention, recycling, treatment, and disposal. The EPA developed its definition of waste minimization from the congressional terms *source reduction* and *environmentally safe recycling*. "Source reduction is any practice that reduces the amount of any hazardous substance, pollutant, or contaminant entering any waste stream or otherwise released into the environment prior to recycling, treatment, or disposal, and reduces the hazards to public health and environment associated with the release of such substances, pollutants, or contaminants."[21] Environmentally sound recycling is the next preferred alternative for managing those pollutants that cannot be reduced at the source. A recycled material, by the EPA's definition, is used, reused, or reclaimed, where used and reused materials are employed as ingredients to make a product or as substitutes for a commercial product, respectively. Reclaimed material is recovered as a usable product or is regenerated. However, the EPA does not consider those recycling procedures that closely resemble conventional waste management activities as waste minimization. For example, treatment for the purposes of destruction or disposal is not considered waste minimization.

2.5 State Initiative

In 2003, California was the first of the United States to pass legislation to eliminate stockpiles of electronic waste. Recycling whenever possible formed a significant portion of the legislative initiative.[22]

3 TOOLS

3.1 Life-cycle Assessment

Life-cycle assessment (LCA) is a tool that can be used to evaluate the environmental consequences of a product or process over its entire life. There are four distinct, but complementary, components of the assessment: scope setting, inventory, impact, and improvement analyses. Life-cycle inventories generated through the analyses can be used, both internally and externally. The only difference is a higher degree of accountability for inventories used for external purposes. Definition of the scope is critical to its success. Boundary conditions must be established so the analysis is all inclusive, yet practical to complete. The inventory analysis can be used for process analysis, material selection, product evaluation, and decision and policy making. Since LCA does not deal with discrete components, environmental benefits can be generated during any one of the three analysis steps. One result of the analysis as a whole is the definition of goals and initial steps toward the development of an environmental management system.

LCA originated in the 1960s, with concerns about limited raw materials and energy resources. More recently, the Society of Environmental Toxicology and Chemistry (SETAC) has been the focus for technical developments in LCA. Researchers in an EPA project used SETAC work to define the stages of a life cycle, as shown in Table 2.

An alternate set of life-cycle stages is defined and shown in Table 3. The primary difference in stages is that the second set separates product delivery from manufacturing for the purpose of highlighting the life-cycle impact of transportation processes.

Other representations include a sixth life-cycle stage, which is largely packaging and transportation. Energy consumption, transportation requirements, and waste management practices are issues considered at all stages of the life cycle.[23]

Table 2 Life Cycle Stages

Number	Stage
1	Raw materials acquisition
2	Manufacturing
3	Use/Reuse/Maintenance
4	Recycle/Waste management

Note: From Ref. 23.

Table 3 Life-cycle Stages

Number	Stage
I	Premanufacture
II	Product manufacture
III	Product delivery
IV	Product use
V	Recycling, disposal

Note: From Ref. 6.

3.2 Material Flow Analysis

Beyond product and process design, material flow analysis (MFA) has two primary purposes on a global scale. MFA shows how different economies have different material flows suggesting that the uniqueness of a national economy is largely based on its material flows. Also, when applied to substances, MFA shows that the various receiving locales (measured by nation or other geographic unit) have differing material flows, which have associated environmental and human health impacts.[24] Understanding of environmental and economic impacts is, therefore, based on recognizing the material flows in various locations.

In addition to environmental and economic measures, MFA and the broader life-cycle analysis must expand to include a social dimension. While environmental impact is the primary focus of typical LCA, analyzing material flows can enable quantification of economic impact. As national economies vary because of their materials usage, the economies can be characterized based on that usage. The social dimension is more closely tied to a human-health-oriented LCA. Impacts on both the environmental and human health measures are identified and quantified by considering materials flows and usages.[25]

3.3 ISO 14000

ISO 14000 is a series of voluntary international standard documents for environmental management systems. The primary standard is 14001, "Environmental Management Systems—Specification with Guidance for Use." ISO 14001 is the standard by which organizations can be audited objectively to see how they meet standard practices. Included in the standard are guidelines for the establishment of an environmental policy, identification of goals and objectives, and specification of internal and external reporting systems.[26] The five initial standard documents were released in September 1996 and international organizations developed methods to register companies using the standards.

4 CASE STUDIES

4.1 Materials

For electronic products and processes to be more environmentally compatible, new materials must be developed and substituted. These developments have generally come as new formulations of materials already common to industry.

Chromium
Often used as a plating material because of its hardness and wear resistance, chromium is another material that is considered a hazardous waste, as either airborne mist or wastewater. During the plating process, a hexavalent chromium mist develops, on the order of four pounds of mist for each pound of chromium

plated. Studies have been conducted to find alternative materials and processes to traditional chromium plating. Nickel-tungsten-silicon carbide, as an electroplating alloy, has many of the desirable properties of chromium; the coating has a high as-plated hardness that can be increased with heat treatment. Its abrasive wear resistance is comparable to that of chromium. The alloy has a higher plating efficiency than chromium, plating between two and three times faster than chromium.[27]

Cleaning Chemicals

Replacing hazardous materials with less hazardous ones can result in better waste disposal alternatives and reduced operating costs. In printed wiring board production, traditional processes use materials that are corrosive to products and equipment and toxic to operating personnel. Examples of a more-informed selection of processing materials can be found in the etching and photoresist operations of board production. In the copper-etching operation, chrome trioxide/sulfuric acid has been replaced with an ammonia-based mixture. Although the ammonia etchant has an unpleasant odor, the replacement material removes the copper without affecting the plated pattern on the board. Carcinogenic methylene chloride has been replaced as a photoresist stripper with a solution of ethanolamine, which does present the hazards of slight respiratory and skin irritation. Replacing the photo-imagable resist, which is xylene-based, with a high-resolution dry film resist produces the same process capability but greatly reduces the hazards to minor skin irritation.[28] In addition to these materials substitutions, the stripping process has been mechanized with an enclosed conveyor operation that limits exposure to replenishing chemicals and system maintenance.

Plating Chemicals

Tin and tin-lead alloys have been used for electroplating where resulting solderability is necessary. In traditional plating and stripping baths, highly corrosive fluoride chemicals, such as fluoboric acid and fluoride salts are used. These hazardous substances have been replaced with methanesulfonic acid (MSA), which is safer to use and less detrimental to the environment. MSA has been shown to be around 75 percent biodegradable after 20 days and 83.3 percent after 55 to 68 days. MSA, an organic sulfonic acid, has a chemical structure of CH_3SO_3H. The acid is slightly corrosive, with an acid strength greater than hydrochloric and nitric but slightly less than sulfuric and fluoboric acids, and is an irritant to the skin and the eyes. Tests of the effects of MSA on the bone marrow of mice and the life of fish have shown it not to be a carcinogen. The acid is controlled in the process by the stripping rate, bath temperature, and stripper concentration. The substitution of MSA for fluoride mixtures is better environmentally; however, additional attention must be paid to the composition of the chemical bath. Use of MSA requires a bath purification procedure that can be performed as part of a preventive maintenance schedule.[29]

Fluxes

Aides to soldering processes, fluxes are typically chosen based on product specifications and design requirements. Flux materials are natural, in the case of rosin, or manufactured, in the case of resin. In the past, flux has had to be removed from the assembly after the soldering process to prevent electrical degradation.[30] However, the environmentally conscious movement in industry has spurred changes in the flux used in electronics assembly processes. Hughes Aircraft made a change when an engineer developed a new natural flux. Solvents had been used to remove the rosin flux on its products, but when directives were issued to eliminate some solvents, including CFCs, the cleaning processes had to be reevaluated. Hughes determined that a change in the flux material would also make cleaning easier. Mild acids like vinegar were tried as fluxes, but the best results were achieved using a lemon juice-based flux. The final mixture took two years of development; the new flux was then used with a simplified soldering operation, and could be removed with water.[31]

Adhesives

Critical electronic package assembly materials include adhesives and coatings that often come into contact with the silicon. The most useful adhesives and coatings must be easy to apply and cure and must also closely match silicon's rate of thermal expansion. In an effort to find a material that also is environmentally friendly, General Electric developed a wide assortment of silicones for such operations as coating semiconductor junctions and encapsulating integrated circuit modules. The silicones can be produced in both hard molding compounds and soft gels, depending on the application. The coefficient of thermal expansion is similar to that of silicon, as required for adhesion or encapsulation. From an environmental viewpoint, the silicones are solvent-free and need no primer for adhesion, nor do they risk ionic contamination.[32]

4.2 Processes

Plating Processes

Alloy deposits are electroplated in printed-wiring-board (PWB) manufacturing for two uses. One is as an etch resist, that protects the underlying electroplated copper and copper foil while the desired circuitry and ground planes are created by the etching process. The other use is as a final solder-able coating over the copper on the board. Copper, another metal that is often plated as part of PWB manufacturing, enters the waste stream in the solution used in plating and in etching to form the circuits. Environmental guidelines established by government agencies, both federal and local, have set limits on the discharge of neutralized waste streams that are low in heavy metals. At this stage, the limits vary from federal to state to local agencies. They have also been dynamic—the allowable parts per million (ppm) of different metals have been steadily reduced over the last 15 years. Tightening the limits has led to increased interest in treating waste streams to reduce trace metals before the streams are directed to sewage-treatment facilities. At the same time, efforts have been made to make all plating and etching processes as solvent-free as possible. Where solvent usage is not an issue, metal recovery has been the basis of improved process design.[33]

Cleaning Processes

As industry has become more aware of the environmental impact of its products and processes, it has focused on the elimination of chlorofluorocarbons (CFCs) and the reduction of ozone-depleting substances in manufacturing applications. Among other uses, CFCs were popular cleaners for many product lines. Since CFCs have been regulated, industry has taken steps to develop cleaning methods without the solvents. Design engineers, who were responsible for selecting alternatives, had to response to the following questions:

1. Based on the type of product and its most critical features, what technology is needed to make the product? Do cleaning processes eliminate barriers to such characteristics as high-speed performance?
2. What are the customer's specifications and requirements? What level of cleanliness does the customer require? How can the cleaning process be made most cost-effective and still achieve the desired level of cleanliness?
3. Is it necessary to develop new processes, or is an existing in-house or off-the-shelf process adequate?
4. What will the newly installed process cost? The total cost of a process must include environmental, economic, and personnel factors. The environmental impact of waste disposal and potential pollution problems must be included in the cost calculation, as must the total life cycle of the product—its manufacture, use, and disposal.[34]

No-clean Processes

The nontraditional approach of eliminating post-soldering cleaning from both reflow and wave soldering illustrates the benefits of a true DfE effort. When the cleaning operation is eliminated, so also are the capital and maintenance costs of the equipment, the materials costs of the cleaners, and the operating costs. Other ancillary improvements are a decrease in necessary floor space, a reduction in processing time, and the simplification of the entire process. In the past, some components that cannot withstand the cleaning process have had to be added in a post-solder assembly step. When the cleaning operation is eliminated, these components can go through the reflow or wave-soldering process rather than be soldered manually afterward. Environmentally, the main reason for eliminating the post-solder cleaning operation is to eliminate the cleaning materials, such as CFCs and chlorinated solvents. Dropping the cleaning operation also reduces the need for metal-laced waste stream disposal.[35] Even though some efforts in this area have not been successful, the motivation to develop a reliable no-clean process is well established.

4.3 Boards

With the increased reflow soldering temperature for lead-free alloys comes the need for printed wiring boards (PWB) to withstand the higher heating levels. A standard measure used to be the time a board would take to delaminate at 260°C. However, with the lead-free processing temperature having a designed peak at 260°C, boards will have to meet higher temperatures for extended times, and it is likely that boards will need to be tested up to 300°C for several minutes for repeated cycles. Board manufacturers are responding to the call for more environmentally considerate products with development of halogen-free PWBs. Flame-retardant materials, which are used to enhance safety of electrical and electronic products against fires, are pollutants. The movement is to replace traditional materials with those that have reduced environmental impact.

Within an extensive study to test various halogen-free materials and their applicability to existing PWB design, Toshiba Chemical Corporation tested boards at lead-free reflow temperatures.[36] Table 4 shows the three reflow temperatures used and which solder alloy each peak temperature was chosen to represent. In all cases, the bond strengths were rated "good," the boards passed 1,000 cycle testing, and there was no delamination or nonconformance on any sample for any temperature. The results indicate that the increase in reflow temperature does not necessarily translate into problems with the printed wiring boards.

4.4 Components

Component companies, in general, have been slow to react to the change to lead-free technologies. A low level of customer demand has kept further developments from occurring. The companies recognize moisture sensitivity of plastic components as a prime issue, and some large components suppliers have made the commitment to furnish lead-free compatible components long before impending regulation dates.[37] As market differentiation becomes a primary driver for electronics manufacturers, customer demand will force more active pursuit of lead-free products and processes.

Table 4 PWB Lead-free Reflow Temperature Test

Alloy Composition	Melting Temperature (°C)	Reflow Peak Temperature (°C)
Sn9Zn	199	230
Sn3Ag0.5Cu	217–219	240
SnPb	183	225

Note: From Ref. 36.

The switch to lead-free technologies raises a few issues that have been addressed to differing degrees and likely will continue to be addressed until the changeover is complete. The primary issue regarding components and the change to lead-free solder is in the ability of those components to withstand the higher lead-free soldering temperature. The most promising candidates to replace tin-lead solder are binary and ternary alloys of tin. Alloying elements include silver and copper primarily, but vary depending on processing parameters such as temperature and wettability. With only a couple of exceptions, the alternative alloys require considerably higher processing temperatures than tin-lead solder.

The raised temperature is expected to hasten thermal degradation of plastic components. Plastic packages are susceptible to moisture ingress and are rated based on JEDEC standard J-STD-020A Moisture Sensitivity Classification. JEDEC[38] Solid State Technology Association is the semiconductor engineering standardization body of the Electronic Industries Alliance specifications. Estimations are that existing plastic packages will have to be derated by at least two moisture sensitivity levels when raised the 40°C needed for the new lead-free solders. Manufacturer's *exposure time*, the compensation factor that accounts for the time after bake that the component manufacturer requires to process the components prior to bag seal, includes a factor for distribution and handling.

Most plastic molded parts will need to be evaluated for thermal degradation. This can be as simple as discoloration, or as severe as the complete meltdown of the packaging material. Thermoplastic materials currently in use will need to be monitored carefully for heat stress as parts can also shrink, warp, or have critical features move during the exposure to the higher lead-free thermal profiles. Popcorning and delamination of plastic packages are concerns. Time will be needed for the qualification of new components, the requalification of currently available ones, and the definition of new materials. New compounds need to meet higher temperature requirements and compounds should also meet proposed halogen-free requirements. One of the tests that packaged component suppliers must run is a moisture sensitivity test, as defined by JEDEC industry standards. A proposed reflow profile is listed in Table 5.

Interest in the impact of lead-free solder on the supply chain, raises the substantial issue of compatibility of currently available components with the new solder. Lead-free solder alloys have higher melting points than tin-lead and, therefore, have higher reflow profiles. Higher peaks and longer dwell times at elevated temperatures are part of the lead-free reflow profiles. Moisture trapped in plastic-encapsulated microelectronics can initiate delamination and cracking due to increased reflow temperature. Significant deformation of moisture-laden packages versus dry packages starts to increase around 90°C. An increased rate of deformation occurs beyond 138°C.[40]

With high-temperature soldering and longer reflow times, the survival of the component and its packaging must be evaluated. Surface-mount plastic packages

Table 5 Proposed Lead-free Reflow Profile[39]

Profile Elements	Convection or Infrared
Ramp rate	50°C to 100°C; 2°C/second minimum
Preheat temperature	125°C (±25°C); 60 seconds minimum
Average ramp-up rate	190–225°C; 20°C/second minimum
Temperature maintained above 217°C	60–150 seconds
Time within 5°C of actual peak temperature	10–20 seconds
Peak temperature range	260°C (−5°C/+0°C)
Ramp down rate	6°C/second maximum
Time 50°C to peak temperature	3.5 minutes minimum

Note: From Ref. 39.

are only rated for temperature exposure below 250°C, while through-hole packages are rated for a maximum 260°C solder temperature. If available, lower-temperature solders will permit more packaging flexibility.[41]

The electronics industry is concerned for the compatibility of existing components with lead-free materials and processes. As interactions between the surface finish on component leads and the solder can have unacceptable results, the two materials must be identified and the potential effects understood. The lead surface finish can diffuse into the solder and the combination of the two materials adheres to the underlying metal surface. Contaminants in the finish, solder, or base metal can affect the efficiency and the reliability of the interaction.[42]

Most lead termination finishes for active and passive components are tin-lead based. Elimination of lead raises the concern that use of lead-free solders before all components are made available with lead-free finishes will create assembly mismatches. Solder alloys containing bismuth used with finishes containing lead result in a low melting point (97°C). Successful use of palladium and nickel-palladium finishes on some components will continue. A remaining issue is the wettability of the underlying leadframe alloy as palladium dissolves into molten solder.[41]

At the component level, lead has been used for solder plating of terminal leads, solder dipping of terminal leads, and solder bumps in area array components. Among the most critical properties, which are required for semiconductor terminal lead finish, are solder wettability, solder joint strength, migration resistance, low cost, and mass productivity. Lead-free plating yields solder joint strength and wettability that are equivalent or superior to the current tin-lead plating with either lead-free or tin-lead pastes. No obvious correlations have been found between the fracture mode (solder joint fracture or component fracture) and material combination, or between the mode and measured strength.[43]

Studies indicated that none of the combinations of the palladium-nickel-finished components and promising lead-free solders tested provided the solderability or wire bondability that had been expected from traditional soldering. Wetting performance approaching that of conventional soldering did not occur until the lead-free soldering temperature reached 250°C.[37]

Motorola conducted a study to determine the effects of lead-free solders on existing plastic ball grid array (PBGA) packages. The four solders chosen were from the tin-silver-copper family, and the components were a PBGA with 1.27 mm pitch, two thin small outline packages (TSOP) with 0.5 mm pitch, and a ceramic chip capacitor. One TSOP had a copper leadframe and the other an Alloy42 leadframe. Two lead-free board finishes were used: an organic solderability preservative (OSP) and an electroless nickel/immersion gold finish. Four-point bend tests were done with pronounced differences based on board finish. The nickel-gold board finish produced inconsistent results as interfacial fractures occurred within the board. There was no statistical difference among the results of the bend test for the four solder pastes on the OSP board, and the radius of curvature was better. Motorola's conclusions were that the board solderable finish did not make a statistical difference, and there is no significant difference in the solder formulation within the tin-silver-copper family. Motorola supports the finding of Sn3.9Ag0.6Cu as the lead-free solder of choice.[44]

Manufacture of area array packages requires solder in the form of balls, where primarily tin-lead solder has been used. Elimination of lead also includes solder ball material, so researchers have conducted studies to determine the best materials for this application. Comparing lead-free alloys with tin-lead, researchers found the solder ball shear strength of the lead-free material to be better than that of the tin-lead material. From that study, standard package assembly equipment can be used without modification.[45] Solder bumping by depositing tin-silver solder, Sn2.5-2.5Ag, on the printed circuit board was successfully demonstrated for flip chip assembly. A nickel-gold finish on the aluminum pads provided sufficient wetting for manufacturability and subsequent reliability tests.[46]

Flip-chip interconnection was studied using a variety of solders, which included Sn3.5Ag, Sn0.7Cu, Sn3.8Ag0.7Cu, and eutectic, Sn37Pb. The processing behavior and reliability results were compared for these alloys combined with a variety of under-bump metallurgies. The Sn0.7Cu proved to have a large grain structure, while the silver-based alloys had fine two-phase structures of tin and Ag_3Sn. The intermetallic compounds formed for all lead-free alloys were similar; on nickel, Ni_3Sn_4 formed, and on copper, Cu_6Sn_5 and Cu_3Sn formed. Reliability testing comprised of shear strength and thermomechanical fatigue experiments. Although the Sn0.7Cu had lower shear strength, it had the highest thermomechanical fatigue resistance. All joints, except for those with Sn0.7Cu, failed during fatigue at the solder/intermetallic interface. The Sn0.7Cu solder joints failed by grain sliding, with failure occurring at the center of the joints. Frear et al. recommended the use of Sn0.7Cu for flip-chip applications based on its resistance to fatigue and its failure mechanism.[47]

4.5 Solders

Lead has a double impact on the health and safety of those who work with it and on the environment. As a major constituent in the most popular solder (tin-lead), lead has been used abundantly in industry. However, with the problems associated with lead, even the advantage of a solder melting temperature below 200°C is not enough to continue its use. Lead takes two primary avenues when entering the human bloodstream—ingestion and inhalation; it is most often absorbed through the digestive system and the lungs, where the blood stream takes it to the rest of the body. Operators can take a number of steps to avoid ingestion and inhalation of lead, but the most responsible approach is to continue to develop lead-free solder material.[48]

The alternatives to lead-based materials can themselves pose technical problems, such as in their mining, refining, processing, and disposal. The performance characteristics and properties of traditional electronic packaging materials and alternatives are common knowledge.[49] Manufacturers already use lead-free die attach materials and board and component surface finishes successfully, so elimination of the lead-based versions should not present a compatibility problem.[50,51] The greatest problem will be at the soldering operation; tin-lead eutectic solders have been the primary joining materials for electronics manufacture. Their replacements will need to perform properly for all combinations of board and component finishes.

The most promising lead-free alloys are tin-silver-copper and tin-copper combinations. There are others with alloying elements such as bismuth;[52] however, silver is expensive and is not abundant, and bismuth is a byproduct of mining lead. Ultimately, the measures of a tin-lead replacement's viability are whether the industry can make electronic products with lead-free solder that are at least as reliable and cost effective as their predecessors.[53] Product reliability, especially for legacy systems, requires further evaluation of acceleration factors.

The switch to lead-free technologies raises compatibility issues on various levels including material, process, component, design, equipment, quality, and reliability.[54] Included in the material level are the required higher temperatures for lead-free solder processing.[55] Part manufacturers establish maximum temperature ratings for their component parts.[56] Lead-free process reflow temperatures will exceed the ratings of many existing parts in the supply chain. The concerns for using parts beyond their ratings include possible failure during reflow processing and potential degradation of long-term reliability. With higher reflow temperatures for lead-free materials, designers must reconsider the life-cycle stresses and failure mechanisms that were applicable with traditional solders and finishes. Higher processing temperatures with lead-free solder will stress constituent parts more than before. The industry does not know completely the full effects of the changes on product reliability.

The temperature increase poses other concerns. The raised processing temperature requires equipment that can produce quality solder joints and withstand the increase. As the temperature increases, energy requirements increase, obviously affecting the processing costs.[57]

The electronics industry must complete testing of the most promising lead-free technologies, including die-attach materials, component finish materials, and solders. For legacy systems, lead-free material properties must compare to traditional lead-based materials to identify potential discrepancies that may affect long-term product performance. Standard qualification and conformance testing of lead-free process materials enables understanding of the performance of such alternatives during manufacture.[58] The industry continues to test existing components and boards for their reactions to new lead-free materials during manufacture and use. Brittle intermetallics and tin whiskers are issues with lead-based solders and plating materials and certainly appear as problems with their replacements.[59]

Examples of early success using lead-free solders and related processes are evident. Besides the obvious benefits associated with these successes, the improvements made show that changing to lead-free materials is possible. Most new product launches will include lead-free solder, since there is a large body of knowledge based on developmental work completed at the university and industry levels.[53] That knowledge base gives rise to additional technical innovation. Corporate examples include those of Nortel, Texas Instruments, and Motorola. Nortel tested two lead-free solders, tin-copper and tin-silver, in customer telephone handsets. Even with a reflow soldering temperature of 242°C, the tested handsets made with Sn0.7Cu were acceptable, and estimates made led to the conclusion that potential cost savings over the life of the product existed.[60] A Texas Instruments study evaluated a totally lead-free solder joint formed with palladium-nickel finished components, a tin-silver-copper solder paste, and organic solderability preservative (OSP) as the PWB surface finish. Results indicated that any difference in performance of three different finishes was visual only.[50] Motorola conducted a study to determine the effects of lead-free solders on existing plastic ball grid array packages. Motorola's conclusions were that the board solderable finish did not make a statistical difference, and there was no significant difference in the solder formulation within the tin-silver-copper family.[43]

Yields from lead-free processes in Japan have been higher than traditional ones, as the manufacturers have made commitments to understand their processes. By adopting rigorous process control, Japanese manufacturers have demonstrated that they can use lead-free solder with less than a 30°C-temperature increase in peak reflow temperature. In some applications, use of a 240°C maximum reflow temperature resulted in no degradation of boards and components. Compatibility with different component lead finishes was not a problem, except for bismuth alloys. Other finishes appeared acceptable for tin-lead or lead-free solders.[61]

The International Electrotechnical Commission (IEC) has formed a committee to develop environmental standardization for electrical and electronic products. IEC adopted a specification that details technical requirements that manufacturers must meet to ensure their products meet lead-free regulations. Such a global certification would be available for all companies to show that they follow regulations concerning lead and other hazardous substances.[22]

4.6 Assemblies

With a couple of exceptions, lead-free alloys require higher processing temperatures than tin-lead solder. The time at the processing temperature, as well as the preheating and heating ramp rates, are concerns for the temperature's effects. Especially a concern is the reaction of plastic encapsulated microelectronic parts that may crack or "popcorn" due to trapped moisture.[40,62]

Currently, industry uses a reflow profile for tin-lead eutectic solder with a maximum temperature of 225°C to 230°C. The profile provides thermal headroom to allow the largest thermal mass to rise above the alloy's 183°C melting point to form a proper joint. The standard rule is 40°C to 45°C headroom above the melting point of the solder. When applied to lead-free solder, the 217°C temperature, which is the melting point of a recommended reflow solder alloy, becomes 257°C. Parameters such as the reflow oven, the line cycle time, the solder paste alloy and flux used, and the circuit board and component masses, are what makes a reflow profile unique.[38] In a recent experiment, reaching 225°C for 10 seconds produced good joints that were reliable, so increasing the temperature somewhat might yield even better results. Wave soldering was found to be most effective in the range of 240°C to 260°C. Although a majority of components should withstand these process temperatures, not all will survive lead-free soldering temperatures.[63]

Wave soldering also needs to be simulated for component integrity testing. Tin-lead eutectic alloys are typically wave soldered between 225°C to 250°C. If the melting temperature delta between the tin-lead eutectic alloy and the Sn0.7Cu eutectic alloy of 44°C is added to the melting point, this gives a wave soldering temperature of 269°C to 294°C. The recommendation is for a wave soldering temperature of 275°C for component testing purposes. Preheating the bottom side of the circuit board to within 140°C of the wave solder temperature, at the wave entry point, minimizes thermal shock to the wave-side components.[39]

The Taguchi design-of-experiments method was applied to a study of wave soldering, which was based on the controllable factors including contact time, preheat temperature, and solder temperature. Solder pins without bridges were counted as the output characteristic. A quarternary alloy, Sn3.8Ag0.7Cu0.25Sb, was used for soldering, while the board finish used was organic solderability preservative. Then antimony, Sb, was added to lead-free alloys to increase thermal resistance, but its toxicity remained a concern. Spray fluxing, which used the smallest droplets possible, was the preferred application. Eighteen boards were run to obtain the data for the analysis of variance. The best settings for the process were a solder temperature of 275°C maximum, a contact time of 1.8 seconds, and a preheat temperature of 110°C.[64]

A Texas Instruments study evaluated a totally lead-free solder joint formed with nickel-palladium finished components, a tin-silver-copper solder paste, and organic solderability preservative (OSP) as the PWB surface finish. Mechanical tests included contact angle measurements and lead pull. Temperature cycle testing was performed to evaluate the package integrity after cycling through temperature extremes. The wetting balance test was used as a tool to determine the solder wetting time for individual components. Results showed that the contact angles do increase with nickel-palladium and nickel-palladium-gold finishes. However, those finishes achieve equivalent or better lead pull and temperature cycle testing results versus tin-lead. This indicates that any difference in performance of the three different lead finishes is visual only. A totally lead-free solder joint was formed and found to have positive results when measured against traditional reliability aspects.[51]

4.7 Products

Making a product green may involve significant changes to the design of its composition and structure. However, the Office of Technology Assessment suggested that "green design is likely to have its largest impact in the context of changing the overall systems in which products are manufactured, used, and disposed, rather than in changing the composition of products per se."[65] Ultimately, DfE initiatives should have some impact on the basic design of the product as well as on the overall system. Some products, by their designed nature, will be easier to disassemble, thereby allowing for more material recovery. Other products can be designed so that intermediate steps in assembly and some materials can be eliminated. Still other products can be designed to utilize those materials and processes that have already been deemed environmentally safe. A manufacturer must know customer specifications and the required level of product cleanliness. For example, the degree of cleanliness required after the soldering process has a large impact on the environment. The cleaning process should be as cost-effective as possible while still achieving the desired level of cleanliness.

Four stages of the lifetime of the product should be considered in the development of a green product:

1. At the raw material extraction and processing stage, the most relevant DfE consideration is the impact on the environment by each material as it is mined and prepared for manufacturing. The final product design

should include those materials that do the least harm during extraction, processing, use, and disposal.
2. At the manufacturing stage, the key issue is the simplification of assembly processes to reduce the materials needed and to make disassembly and recovery more feasible. The design is directly affected by the changes in required materials and the process sequence.
3. At the customer usage stage, use, reuse, and maintenance of the product affect product specifications. Product design, in terms of materials specified and manufacturing sequence, should make it easier for the customer to use the product.
4. At the product retirement stage, a product should be designed for recycling, remanufacturing, and environmentally safe disposal. Incorporating the proper materials in the proper manufacturing sequence leads to the least detrimental disposal methods.[66]

4.8 Testing and Qualification

Toshiba Chemical Corporation tested boards at three lead-free reflow temperatures. In all cases the bond strengths were rated "good," the boards passed 1000 cycle testing, and there was no delamination or nonconformance on any sample for any temperature.[36] The results indicate that the increase in reflow temperature does not necessarily translate into reliability problems.

Two studies measured the effects of lead-free technologies on reliability as indicated by the fatigue life of ball grid arrays (BGA) assembled to printed circuit boards. IBM Microelectronics performed an accelerated thermal cycling (ATC) test, using tin-silver-copper and tin-silver-bismuth alloys with tin-lead solder as the control. Reflow temperatures for the test assemblies were 215°C for the tin-lead soldering and 235°C for both the tin-silver-copper and the tin-silver-bismuth alloys. A nickel-gold board finish was used for the lead-free joints. Operating, 0°C to 100°C, and extended range, −40°C to 125°C, temperatures were combined with various times to cycle up to 240 minutes. The results showed that the BGAs soldered with tin-silver-copper had a greater fatigue resistance in the 0°C to 100°C range, but less in the −40°C to 125°C range than the tin-lead parts. Also, the tin-silver-bismuth solder joints were more fatigue resistant for all temperatures but only for cycle times below 42 minutes. Above that, and up to 240 minutes, tin-lead packages lasted longer.[67]

In the second study, Nokia Mobile Phones assembled plastic ball grid arrays (PBGA) on FR-4 boards, which had organic solderability preservative (OSP) or electroless nickel/immersion gold finish. Tin-silver-copper solder was used for all lead-free assemblies, and tin-lead samples were included as controls. The peak reflow temperature of 245°C was used for the lead-free profile. Thermal cycling tests of −40°C to 125°C were performed for cycles up to 1,600. Lead-free solder performance was shown to be equivalent for either the OSP or the nickel-gold boards. Also, the study showed the lead-free reliability was equal to or better than the tin-lead control assemblies.[68] In both the IBM and Nokia studies, lead-free soldering reliability was shown to be equal to or better than traditional tin-lead, at least within certain operating parameters. Each study also showed positive results when assemblies were soldered using reflow temperatures significantly below the 260°C value, as estimated by earlier studies.

Nokia also conducted an experiment with lead-free solder, nickel-gold printed circuit board finish, and off-the-shelf components, which were ball grid arrays (BGA), chip scale packages (CSP), and leadless ceramic chip carriers (LCCC). After soldering the same components with tin-lead eutectic solder, the lead-free materials were used with a reflow temperature of 245°C, which was again well below the 260°C expected peak. Popcorn cracks were found at the initial visual inspection after reflow. The lead-free components showed a 28 percent failure rate, while the tin-lead packages showed a 5 percent rate. A discoloration, which resembled a crack, was observed during drop testing. Although the discoloration darkened as the tests were completed, subsequent sectioning revealed the lines were not cracks. The various combinations produced a wide variety of solder joint metallurgy results. Elements detected in the solder joints included lead, iron, nickel, phosphorous, bismuth, copper, silver, gold, platinum, palladium, tungsten, and cobalt. Since the variety of metallurgies is produced, mixing the lead-free and lead-based technologies increases the difficulty in predicting the long-term reliability of the solder joints, although the final test results were promising. Drop testing produced no cracks on any assemblies up to 24 drops. The LCCC samples were the only ones to fail during thermal shock testing. Cracks in both lead-based control samples and lead-free assemblies were seen around the LCCC parts.[68]

Final analysis from the Nokia lead-free and lead-based mixed study showed a few interesting results.

- Board warpage was minimal, with no assembly problems noted.
- Moisture-sensitive plastic packages showed more damage during high-temperature reflow.
- Gold-tin intermetallics were difficult to detect with lead-free solder paste.
- Gold-tin intermetallics were easier to find in the region adjacent to the component when gold-finish was used for the board.
- Presence of minor elements in the solder joints was likely due to dispersion of termination metallurgies at the higher reflow temperature. Effects on physical properties, such as tensile strength and creep behavior, are unknown, but lead-free joints outperformed the lead-based joints.
- Lead-based control samples experienced shock failures earlier than those with lead-free solder, which is due to a lower coefficient of thermal expansion (CTE) mismatch with lead-free materials.

- Presence of cracks in the component termination layers may be due to the relatively greater strength of lead-free solders.

Nortel Networks developed a printed circuit assembly with no lead in the interconnect. A tin-copper solder was chosen over a tin-silver alloy, based on the cost due to the silver content. Nortel recognized that by using purchased components there would be lead in some of the parts, so the company concentrated its effort of finding a lead-free solution on its own products and processes. So, the tin-copper solder was used for the connections, and the board and component finishes were also kept lead free. The desired lead-free assemblies were not truly lead-free, and as such, they contained a mixture of lead-based and lead-free parts. Supply issues were raised in procuring lead-free components, so Nortel retinned those components with the tin-copper or other lead-free alloys. Studies gave Nortel assurances that existing assembly equipment could be used for the higher-temperature soldering operations. The company was able to achieve proper solder flow at 242°C, which was considerably below the expected level of 260°C.

Two hundred telephone boards were run in a controlled experiment with favorable results: 153 boards were assembled properly on the first reflow pass. Part placement problems and not solder flow were the reasons for the dropouts. Of the boards with reflow-processed components, 145 were presented to wave soldering, with 132 successfully assembled on the first run. Ten of the boards were completed with hand-soldering methods and were assembled into telephones. All of the installed circuit-board assemblies, with the tin copper and other lead-free finishes, passed the electrical and functional tests for telephone reliability.[42] Nortel demonstrated that inclusion of components with lead-based configurations could be assembled successfully with lead-free boards and other lead-free parts.

Another test added lead to a lead-free solder as a forced contamination in order to assess the effects of lead contamination on solder joints. Previous experimentation had shown that when a lead-free solder alloy was soldered on a lead-containing surface, the lead would contaminate the lead-free solder through a metallurgical reaction. That reaction was fundamentally a secondary alloying process, which occurred almost immediately under common soldering conditions. A small amount of lead, which was a dosage of 0.5 percent or less, was added to a tin-silver-copper-indium solder alloy with a composition of Sn3Ag0.5Cu8In. Tin-lead solder, which was Sn37Pb, was used as the reference material. Lead was added in levels of 0.1 percent, 0.2 percent, and 0.5 percent to the tin-silver-copper-indium alloy. Standard material tests were conducted including yield strength, tensile strength, Young's modulus, total strain, and number of cycles to failure. No noticeable increase in melting temperature was observed with the addition of the lead-contaminant. The addition of lead also had little effect on alloy strength. Overall, the strength of the lead-contaminated tin-silver-copper-indium solders surpassed that of the tin-lead solder. The increased amount of lead did have a noticeably negative effect on fatigue life, which was still better than that with tin-lead solder. The lead atoms were characterized as precipitating out as second phase particles. As the amount of lead particles was small, the effect on strength was minimal. Another suggestion was the relatively ductile tin-based matrix was the reason the lead had a lesser effect than it would have had in a more brittle configuration. The decrease in fatigue strength was likely due to the concentration of soft lead inclusions in the tin matrix. Lead mixing with the right lead-free alloy did not significantly decrease solder strength.[69]

4.9 Reliability and Performance

Reliability is defined as a measure of the time that a product can be used without experiencing degradation or failure under a prescribed set of operating conditions. Reliability testing applies various stresses, mechanical, thermal, and electrical, to components, assemblies, and entire systems to predict the product's useful life. In terms of reliability, the word *environment* has traditionally referred to the conditions that the component, assembly, or system will see in practice and in testing. This section distinguishes between this meaning of the term, here called the *use environment*, and the global environment impacted by the product and its processes, called simply *the environment*. Consideration of the use environment at the time of product design, called *design for reliability*, is extremely important in itself in the design and manufacture of electronic products. By contrast, as discussed previously, design for the environment considers the effects of electronic products and processes on the global environment. DfE and reliability are connected because, in order for electronic products to be reused or repaired for continued use, they must be durable; a product's durability over its life cycle is measured as its reliability. More complex relationships between reliability and the global environment are reflected in how the materials selected to enhance reliability affect the global environment as they are used or disposed.

In order to achieve success in the market, products must have excellent reliability. Ensuring this level of reliability begins at the earliest stages of product definition, with assessment of all materials, design, and manufacturing technologies that will affect the product's performance. Accelerated test methods and statistically designed experiments are used to verify that performance objectives are met. If they are not, the root cause of the deficiency is investigated so corrections can be made. Achieving reliability objectives in practical application challenges both design engineers and process engineers to work together to eliminate structural, materials, and process variations that contribute to yield loss and product degradability.

Reliability testing applies various stresses to products so that its useful life can be predicted. As materials substitutions are made in consideration of the environment, the materials need to undergo the battery of tests that assess product life. Reliability testing uses accelerated conditions to inflict stresses on the product. Typical conditions that are subjects of tests are physical and chemical mechanisms that the product will likely see in use. Stresses imposed include thermal, electrical, and mechanical stresses, humidity, radiation, shock, corrosive gas, and corrosive particles. Use environment testing is done by exposing the product to higher concentrations of humidity, airborne particles, and gases than it would see in its lifetime. The effects of these stresses are accelerated during tests by elevating the temperature. Test results are compared to field results of actual service performance when available, or can be extrapolated to estimate reliability and performance under field conditions[70].

4.10 Formation of Tin Whiskers

The consideration of tin and tin-alloys for plated surface finishes has again raised concern for an old problem: tin whiskers. The condition of tin whiskers arises when tin begins to grow tin dendrites, which have been named whiskers. The action can begin below 10°C. Solder joint reliability concern exists since the whiskers separate from the host material. Highly stressed areas of pure tin are susceptible to whiskers. Adding impurities to the plating bath can reduce the occurrence of whiskers, although the best impurity for that purpose is bismuth, which has workability problems with any lead-based materials in the system.[71]

Regardless of the specific alloy, the most promising lead-free solders are tin-based. Concern for tin whiskers, which is a long-standing issue, does exist in electronics manufacturing as the interaction between the molten tin-based solder and copper-based board substrates is the basis for printed circuit board assembly. The tin-copper system achieves interfacial continuity with the formation of Cu_6Sn_5 (η) and Cu_3Sn (ε) phase intermetallic compounds. The η phase grows into a scalloped formation with whiskers emanating along the top of the scallops.[72]

Experimental results of aging substrates plated with tin-lead and various lead-free materials showed tin whiskers formed for all samples, except those where the substrate was preplated with a layer of nickel. Brass and copper substrates were plated with Sn10Pb and tin-based lead-free solders. Half of the copper-alloy substrates were preplated with a 2 μm layer of nickel. The outer layers were 3 μm thick on all samples, with or without the nickel preplate. Samples were heat aged for three months at 55°C. Classic tin whiskers 3 μm to 4 μm in diameter and greater than 25 μm in length did form on almost all samples with brass or copper alloy 194 substrates without the nickel preplate. In the case of Sn10Pb on brass, the whiskers were moderate in length, between 5 μm and 25 μm long. In that case of the alloy 194 substrate, nickel plate, and the tin-lead and lead-free over-plates, no whiskers formed after aging.[73]

Traditional tin-lead solder and copper-based boards have been the bases for concern for tin whiskers for years. The unknown reactions of lead-free solders, as to whether they will be more susceptible to whiskers, suggest more study into the issue. Preventing tin whiskers is critical to the expanding use of lead-free solders.

The promising lead-free alloys have between 96 percent and 99 percent tin, so their susceptibility to whiskers is great, especially since thorough investigation of tin whiskers is not complete. The following approaches were suggested as the most important to preventing whiskers:[74]

- Substituting an alloy containing a few percent of lead, bismuth, antimony, or copper for pure tin
- Avoiding bright as-plated tin coatings
- Controlling tin-plating baths
- Using hot-dipped or reflow coatings instead of as-plated tin
- Baking to stress relieve coatings
- Avoiding inducing mechanical stresses in the coating
- Using conformal coatings where possible
- Using either thick, greater than eight microns, or thin coatings
- Avoiding the use of tin coating directly on brass or other copper alloys

5 FUTURE PLANS

5.1 Conductive Adhesives

A lead-free approach, which excludes solder, is the use of conductive adhesives for attaching surface mount components. The adhesive materials have been used traditionally for die attach bonding and for bonding integrated circuits to lead frames. Besides forming a mechanical bond, the adhesives provide an electrical interconnection between a device lead and a chip carrier or a printed circuit board pad. Conductive adhesives are thermosetting epoxies with metal filler materials, which carry the current through cured compounds. Among the most common filler materials are silver, nickel, copper, gold, and indium or tin oxides. Cost is an issue, as the adhesives are considerably more expensive per gram than solder, but the processing costs can be substantially lower. Component placement is important since the adhesives don't wick as do solders, so a proper circuit is formed only with accurate alignment, especially with fine pitch assembly. Rework and repair are more challenging, as more time at elevated temperature is needed for components to be removed.[75] Advantages to using conductive adhesives include lower processing temperatures versus soldering, 150°C compared to 230°C. Disadvantages include the coefficient of thermal expansion (CTE) being twice that of copper, so a significant mismatch does occur.[42]

The adhesives are made of flakes, plates, rods, fibers, or spheres of gold, silver, or nickel mixed in a polymer matrix. Gold is cost prohibitive, although it has the best properties; nickel has lower conductivity and corrodes; therefore, silver is the most widely used. In high-temperature applications the polymer matrix is either epoxy, or silicone. A recent comparative analysis at Boeing of the long-term reliability of alternatives to tin-lead solder indicated that the lead-free solders had superior performance to conductive adhesives as a group. Further studies are needed on the reliability of conductive adhesives before these materials can be considered as serious alternatives to solders at high temperatures.[76] Use of non–solder-surface finishes generally improves the contact resistance and the reliability of conductive adhesive assemblies. The primary issues, which are particularly important with larger components, are mechanical strength, adhesion, and reliability in mechanical shock.[41]

6 SUMMARY

With the dynamic increase in advanced technologies, ways of affecting the environment are also dynamic. The negative effects of these advancements must be considered, but they can also be beneficial to the environment. A growing list of environmental concerns will continually challenge those responsible for hazardous and nonhazardous wastes until a global level of sustainability is achieved.

As advances continue to be made in technology, the potential exists for environmental problems to increase as well. Moreover, with improved technology, previously undetected problems will be revealed. One solution is the adoption and implementation of an environmental management system that has as its goals sustainable development, pollution prevention and, ultimately, zero-waste production. Some methods to achieve these goals are design for environment, life-cycle assessment, and environmentally conscious manufacturing. No matter what the methods of approach, the best way for the electronics industry to meet the needs of a sustainable society is to be proactive. Consideration of the global environment at the design phase of the product-realization cycle will have the greatest impact on product configurations and manufacturing processes.

Industries are responding positively to environmental concerns. To date, the greatest push toward reducing adverse environmental effects has been most waste-generating processes. Many other areas of opportunity exist, and more can be explored with continued advances in technology. Prescriptive measures to initiate a corporate environmental system must be generated with management involvement, with employee participation, and by means of internal auditing and external benchmarking. Approaches must be based on the specific product and processes involved. Improvement must be continuous and guided by the regulations and standards of the specific locale. True costs over the life cycle of the product must be considered so that improvements can be measured properly. Ideally, any effort should lead to a companywide policy of environmental stewardship.

REFERENCES

1. B. Allenby, "Industrial Ecology: The Materials Scientist in an Environmentally Constrained World," *MRS Bulletin* (March 1992), 46–51.
2. T. Graedel, B. Allenby, and P. Linhart, "Implementing Industrial Ecology," *IEEE Technology and Society Magazine* (Spring 1993), 18–26.
3. J. Cross, "Pollution Prevention and Sustainable Development," *Renewable Resources Journal* (Spring 1992), 13–17.
4. S. Weissman and J. Sekutowski, "Environmentally Conscious Manufacturing: A Technology for the Nineties," *AT&T Technical Journal* (November/December 1991), 23–30.
5. J. Fiksel, ed., *Design for Environment: Creating Eco-Efficient Products and Processes*, McGraw-Hill, New York, (1996), 3–10.
6. T. Graedel and B. Allenby, *Design for Environment*, Prentice-Hall, Upper Saddle River, NJ, 1996.
7. "EPA Manual for Waste Minimization Opportunity Assessment," 2–88/025 (April 1988).
8. D. Navin-Chandra, "Overview of the Green Engineering Project," unpublished paper, updated February 9, 1993.
9. T. Holloway, "Life Cycle Assessment: A New Yardstick?" *Process Engineering* (1992), 19–21.
10. T. Bishop, "Waste Away," *Engineering* (January 1992), 21–22.
11. M. Hundal, "DFE: Current Status and Challenges for the Future," *Design for Manufacturability 1994*, J. Mason, ed., American Society of Mechanical Engineers, 1994.
12. D. Ufford, "Design for Environment—Getting Started," *Nepcon West '94 Proceedings*, Reed Exhibition Companies, 1994.
13. D. Peck, "Rework and Repair of TAB and FPT Devices," *Circuits Assembly* (May, 1993), 42–54.
14. J. Owen, "Environmentally Conscious Manufacturing," *Manufacturing Engineering* (October 1993), 44–55.
15. M. Robins, "The History of Pb-Free," *Electronic Packaging and Production Supplement*, **40**(6), 7–10 (June 2000).
16. Commission of the European Communities, "Proposal for a Directive of the European Parliament and of the Council on Waste Electrical and Electronic Equipment. Document No. 500PC0347(01), June 13, 2000.
17. L. Hester, "Administration Promotes Rule on Lead Emissions Information," U.S. Environmental Protection Agency Environmental News Release, R–056, Washington, DC, April 17, 2001.
18. Environment Agency, "The Challenge to Establish the Recycling-based Society," The Design Agency of Japan, Tokyo, Japan, 2000.
19. European Union (EU), "Directive 2002/95/EC of the European Parliament and of the Council of 27 January 2003 on the Restriction of the Use of Certain Hazardous Substances in Electrical and Electronic Equipment," *Official Journal of the European Union*, **37**, 19–23 (2003).
20. European Union (EU), "Directive 2002/96/EC of the European Parliament and of the Council of 27 January 2003 on the Waste Electrical and Electronic Equipment," *Official Journal of the European Union*, **37**, 24–38 (2003).

21. U.S. Environmental Protection Agency, "Guidance to Hazardous Waste Generators on the Elements of a Waste Minimization Program," *Federal Register*, **58**(102), 31113–31120 (May 28, 1993).
22. J. Erdmann, "Getting the Lead Out of Electronics," [Online], Available: http://www.sai-global.com/NEWSROOM/TGS/2005–06/LEAD/LEAD.HTM.
23. B. Vigon, D. Tolle, B. Cornaby, H. Latham, C. Harrison, T. Boguski, R. Hunt, and J. Sellers, *Life-Cycle Assessment: Inventory Guidelines and Principles*, U.S. Environmental Protection Agency, February 1993.
24. S. Moore, and P. Brunner, "Material Flows: A View from the South," *Journal of Industrial Ecology*, **7**(2), 7–9 (2003).
25. H. de Haes, R. Heijungs, S. Suh, and G. Huppes, "Three Strategies to Overcome the Limitations of Life-cycle Assessment," *Journal of Industrial Ecology*, **8**(3), 19–31 (2004).
26. "Introduction to the ISO 14000 Standards," *Environmental Resource Center*, September 1998, [Online], Available: http://www.ercweb.com/ISO/ISO.HTM.
27. S. Aasness, "Chromium Plating," *Design for the Environment/Engineering Update*, October 1993.
28. R. Breitengross, "Pollution Control and Hazardous Materials Minimization in a Printed Wiring Board Shop: A Case Study," *IPC Technical Review* (April 1993), 21–28.
29. J. Fellman, "Plating and Stripping Tin and Tin-Lead Deposits with Methanesulfonic Acid Baths that Are 'Environmentally-Friendly'," IPC Technical Paper presented at the IPC Fall 1993 Meeting.
30. L. Lambert, *Soldering for Electronic Assemblies*, Marcel Dekker, Inc., New York, 1988, pp. 237–263.
31. "Lemon Juice Gives Green Solder Flux," *The Engineer* (March 5, 1992), p. 34.
32. L. Beradinis, "The Packaging Hurdle," *Machine Design* (April 25, 1991), 55–65.
33. J. Martin, "The Impact of Carbamate on Heavy-Metal Reduction," *Printed Circuit Fabrication* (May 1991), 38–40.
34. J. Price, "Balancing Cleaning Options," *Electronic Packaging and Production* (February 1994), 119–120.
35. L. Guth, and J. Morris, "No-Clean Soldering Processes," *AT&T Technical Journal* (March/April 1992), 37–44.
36. T. Suzuki, K. Hanamura, and N. Kanemaki, "Environmentally Friendly HDI Materials," *Future Circuits International*, **7**, 63–69 (2000).
37. B. Richards and K. Nimmo *Lead-free Soldering: Update 2000, Department of Trade and Industry, Middlesex, UK,* URN 00/750, 2000.
38. JEDEC, *Moisture/Reflow Sensitivity Classification for Non-hermetic Solid State Surface Mount Devices*. IPC/JEDEC J-STD-020B, Electronic Industries Association, Arlington, VA, 2002.
39. R. Parker, "The Next No-lead Hurdle: The Components Supply Chain," *CircuiTree*, **13**(8) (2000).
40. M. Pecht, and A. Govind, "In-situ Measurements of Surface Mount IC Package Deformations during Reflow Soldering," *IEEE Transactions on Components, Packaging and Manufacturing Technology—Part C*, **20**(3), 207–212 (1997).
41. J. Raby and R. Johnson, "Is a Lead-Free Future Wishful Thinking?" *Electronic Packaging and Production*, **39**(8), 18–20 (August 1999).

42. W. Trumble, "A Case Study: A Printed Circuit Assembly with a No-Lead Solder Assembly Process," *Green Electronics/Green Bottom Line*, Lee H. Goldberg and Wendy Middleton, eds., Newnes, Boston, 2000, Appendix A.
43. Japan Electronics and Information Technology Industries Association, "Reports on Lead-Free Surface Finish of Semiconductor Terminals: Technical Standards Committee on Semiconductor Device Package," EE-13/EIAJ, June 2000, pp. 15–26.
44. G. Swan, A. Woosley, N. Vo, T. Koschmieder, and T. Chong, "Development of Lead-Free Peripheral Leaded and PGBA Components to Meet MSL3 at 260°C Peak Reflow Profile," *Proceedings of IPC SMEMA Council APEX 2001, LF2-6*, January 2001, pp. 1–7.
45. K. Barrett, "Learning to Drop Lead." September 18, 2000, [Online], Available: http://209.67.253.180/enews/Issue/RegisteredIssues/2000/09182000/z74f-1.asp.
46. W. Leonhard, A. Gemmler, W. Heck, M. Jordan, and W. Gust, "Lead-free Solders for Flip-chip Technology," *Micromaterials Conference Proceedings*, Berlin, Germany, April 17–19, 2000.
47. D. Frear, J. Jang, J. Lin, and C. Zhang, "Pb-Free Solders for Flip-chip Interconnects," *JOM*, **53**(6), 28–38 (June 2001).
48. H. Manko, "Lead Poison, Solder and Safety in the Workplace," *Electronic Packaging & Production* (February 1994), 93–96.
49. M. Pecht, R. Agarwal, P. McCluskey, T. Dishongh, S. Javadpour, and R. Mahajan, *Electronic Packaging: Materials and Their Properties*, CRC Press, Washington, DC, 1999.
50. K. Barrett, "Learning to Drop Lead. 2000;" 2003, [Online], Available: http://www.reed-electronics.com/electronicnews/index.asp?layout=article&articleid=CA50368.
51. D. Romm and D. Abbott, "Pb-free Solder Joint Evaluation," *Surface Mount Technology*, **12**(3), 84–88 (1998).
52. N. Lee, "Lead-free Soldering for Chip Scale Packages," *Chip Scale Review*, **4**(2), 42–43 (2000).
53. M. Pecht, *Product Reliability, Maintainability, and Supportability Handbook*, CRC Press, Boca Raton, Florida, 1995.
54. D. Shangguan, "Environmental Leadership in Electronics Manufacturing: Lead-Free and Beyond," *Proceedings of the 12th IEEE International Symposium on Electronics and the Environment*, Scottsdale, AZ, 2004, pp. 33–39.
55. R. Ciocci, "Handling the Migration to Lead Free," *IEEE Transactions on Components and Packaging Technologies*, **24**(3), 536–538 (2001).
56. D. Das, N. Pendse, M. Pecht, L. Condra, and C. Wilkinson, "Understanding Electronic Part Data Sheets for Parts Selection and Management," *IEEE Circuits and Devices*, **16**(5), 26–34 (2000).
57. B. Levine, "Facing a Lead-free Future," *Semiconductor Manufacturing*, **4**(9), 30–54 (2003).
58. M. Pecht, A. Dasgupta, J. Evans, and J. Evans, *Quality Conformance and Qualification of Microelectronic Packages and Interconnects*, John Wiley, New York, 1994.
59. K. Kima, S. Huha, and K. Suganumab, "Effects of Intermetallic Compounds on Properties of Sn–Ag–Cu Lead-free Soldered Joints," *Journal of Alloys and Compounds*, **352**(1–2), 226–236 (2003).
60. B. Trumble, "Printed Circuit Assembly with No Lead Solder Assembly Process," *Proceedings of the 5th IEEE International Symposium on Electronics and the Environment*, San Francisco, CA, 1997, pp. 25–27.

61. T. Fryer, "Japan Progresses Down the Lead-free Path," *Electronics Manufacture and Test*, **4**, 25–30 (2001).
62. M. Pecht, L. Nguyen, and E. Hakim, *Plastic-Encapsulated Microelectronics*, John Wiley, New York, 1995.
63. S. Crum, "Components, Materials Re-Examined for Lead-Free Compatibility," *Electronic Packaging and Production Supplement*, **40**(6), 13–16 (June 2000).
64. G. Diepstraten, "Analyzing Lead-free Wavesoldering Defects," *Surface Mount Technology*, **14**(10), 48–52 (October 2000).
65. *Green Products by Design: Choices for a Cleaner Environment*, Office of Technology Assessment Publication, September 1992, pp. 3–20.
66. Brian T. Oakley, "Total Quality Product Design—How to Integrate Environmental Criteria into the Product Realization Process," *Environmental TQM*, 2nd ed., J. Willig, ed., McGraw-Hill, New York, 1994, Ch. 25.
67. J. Bartelo, "The Effect of Temperature Range During Thermal Cycling on the Thermomechanical Fatigue Behavior of Selected Pb-Free Solders," *APEX 2001 Proceedings*, San Diego, CA, January 14–18, 2001.
68. S. Dunford, P. Viswanadham, and P. Rautila, "On the Road to Lead Free," *Circuits Assembly*, **12**(4), 34–40 (April 2001).
69. J. Hwang and Z. Guo, "The Effects of Lead Contamination on Lead-Free Sn/Ag/Cu/In Solder," *Chip Scale Review*, **5**(5) (July 2001).
70. R. Comizzoli, J. Landwehr, and J. Sinclair, "Robust Materials and Processes: Key to Reliability," *AT&T Technical Journal* (November/December 1990), pp. 113–128.
71. IPC. *IPC Roadmap: A Guide for Assembly of Lead-free Electronics*, Draft IV, June 2000.
72. R. Gagliano and M. Fine, "Growth of η Phase Scallops and Whiskers in Liquid Tin-Solder Copper Reaction Couples," *JOM*, **53**(6), 33–38 (June 2001).
73. R. Schetty, "Minimization of Tin Whisker Formation for Lead-free Electronics Finishing," *Journal of the Institute of Circuit Technology*, **27**(2), 17–20 (February 2001).
74. B. Hampshire and L. Hymes, "Shaving Tin Whiskers," *Circuits Assembly*, **11**(9), 50–54 (September 2000).
75. M. Robins, "Conductive Adhesives Show Alternative Solutions without the Lead," *Electronic Packaging and Production Supplement*, **40**(6) (June 2000).
76. P. McCluskey, D. Das, J. Jordan, L. Condra, T. Torri, J. Fink, and R. Grzybowski, "Packaging of Electronics for High Temperature Environments," *International Journal of Microcircuits and Electronics Packaging*, **20**(3), 409–423 (Third Quarter 1998).

CHAPTER 9
DISASSEMBLY FOR END-OF-LIFE ELECTROMECHANICAL PRODUCTS

Liu Zhifeng, Gao Yang, Chen Qing, and Hong C. Zhang
Hefei University of Technology

1	INTRODUCTION	211
2	PRODUCT DISASSEMBLY: RESEARCH ACTIVITIES AND OVERVIEW	214
	2.1 Proposed Methods for Evaluating Disassembly	214
	2.2 Disassembly Process Layout	218
	2.3 The Programming of Disassembly	224
	2.4 Active Disassembly	225
	2.5 Automatic Disassembly	227
	2.6 Disassembly Tool	228
	2.7 Disassembly System Control	228
3	RESEARCH AND METHODOLOGIES IN DISASSEMBLY STUDY	229
	3.1 Disassembling Techniques[3]	229
	3.2 Disassembling and Recovering Models[48]	231
	3.3 Composition of Disassembling System[50]	236
	3.4 The Disassembly Evaluation[3]	237
	3.5 Method of Estimation	247
	3.6 Development of Three-Dimensional Environment for Disassembly[54]	252
	3.7 Study of Disassembly Tools	255

1 INTRODUCTION

Since the 1960s, the world economy has been developing faster than ever before. This rapid expansion brings great pressure to bear on and is having serious effects on human development and lifestyle. Traditional electronic products are sought after and produced in enormous quantities. The rapid advancement of technology effectively shortens the useful life of these products causing enormous consumption of natural resources. And because the design of traditional electronic products rarely considers product disposal, the old and useless equipment creates enormous danger to the environment. Take, for example, the computer. According to the U.S. Environmental Protection Agency (EPA), more than 300 million computers were projected to be discarded in the United States in 2004. It was expected that the waste, which contains harmful elements such as lead, mercury, and cadmium, of 8.5 million of these would end up in landfill.

In order to solve these fundamental problems, we must develop a "green" approach to the making of electronic products. Such an approach must systematically consider, from the beginning, the impact on the environment of both the manufacturing process and the product itself, and it must do so without sacrificing product function, quality, and cost efficiency. It should strive to minimize the negative impact of the product on the environment throughout the entire life cycle and to have the highest utilization ratio of recovery. Consequently, the final dismantlement of electronic products occupies a very important position in green manufacturing system.[1]

Recycling and reuse are the last two key segments of the lifespan of products, and they are the most difficult segments to address. The degree to which these two segments are fully realized influences resources and the environment directly. The primary factors complicating recycling or reuse are as follows:[2]

- The product design is not suitable for dismantling and recycling.
  - *Methods of joining*. Generally, the connection methods for component parts are chosen in order to simplify assembly and safe connection; therefore, there are connections that cannot be loosened, and so they are difficult or impossible to dismantle.
  - *Material variety*. Economical efficiency and best performance dictate the material selection, which generally results in adopting a large number of different materials, many of which are nonrecoverable and carry a high cost for dismantlement.
  - *Process design*. The processes selected to provide optimal function and assembly efficiency can create a large number of unnecessary dismantlement steps.
- Normal daily use of the product may make recycling or reuse of parts impossible owing to disintegration, corrosion, contamination, or other damaging forms of physical or mechanical change.
- Dismantled products lack the intact product information necessary for recycling or to ensure proper disposal.

There are three reasons to undertake dismantlement:

1. *Reutilization of the products*. Reutilization encompasses both direct and indirect methods. Direct reutilization is suitable when the parts are expensive to produce, take a long time to be improved, or have a long performance life, such as the bottles used for beverages, which can be reused directly for holding other liquids. Indirect reutilization is suitable for parts that cannot be reused directly as they are but that, after some treatment, can be used in other ways. For example, some parts of cars can be reconfigured for use in tractors, and discarded mainboards of computers can be reconfigured for use in game machines.
2. *The reclamation of components and parts*. This is especially suitable to electronic products because they contain numerous materials and obsolescence is very rapid. Special methods are often necessary to retrieve component parts.

3. *Recovery of material.* When the material used to produce a product or component is expensive, but the price of the single part is low because of large-scale production and short product lifespan, it is often more efficient to recycle the material.

Dismantling can happen at the product level or at the part level. Based on the desired result, dismantling can be performed in any of three ways: destructive dismantlement, partial destructive dismantlement, and nondestructive dismantlement. Destructive dismantlement focuses on separating the parts, and the level of damage is not important. Partial destructive dismantlement allows for only some cheaper parts of the product to be damaged (perhaps, for instance, by cutting connections using a bluster cut, a high-pressure water cut, or a laser cut). Other parts must be separated intact without damage. Nondestructive dismantlement is the ultimate form of disassembly. It requires that all the parts of the product be separated intact (for example, loosening screws, disconnecting plugs, and removing soldering).

Many modern mechanical machines are electromechanical products or multidisciplinary technology-intensive products. Some products have both parts that will cause pollution to the environment and parts that will not cause pollution to the environment, which creates complications to disassembly. Even when the product is capable of being recycled or reused it may not be economically efficient to do so. Parts may be difficult or problematic to reuse, and disassembly may require disproportionate amounts of time and energy. Regardless of the issues surrounding the wasting of materials, improperly disposed of materials create serious environmental pollution. Figure 1 shows the life cycle of the materials.

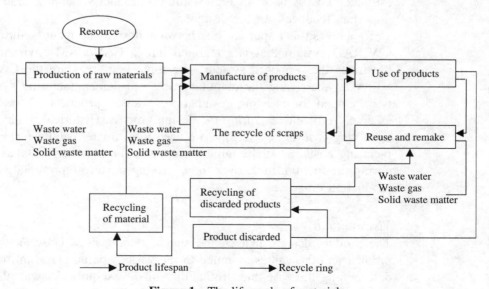

Figure 1 The life cycle of materials

We can see from it that a lot of waste material (such as waste water, waste gas, and solid waste) is created when products are manufactured, used, and discarded. Disassembly is essential to saving resources, protecting the environment, and realizing sustainable development.[3]

2 PRODUCT DISASSEMBLY: RESEARCH ACTIVITIES AND OVERVIEW

2.1 Proposed Methods for Evaluating Disassembly

Das

Das has employed a tool bag that calculates a product's disassembly effort index (DEI) score and evaluates the disassembly difficulty, accessibility, and reclaimability of products. This score provides an estimate of disassembly cost and return on investment.[4]

Mok

Mok has proposed an intelligent Virtual Assembly and Disassembly (VIRAD) system, which can be integrated into a CAD/CAM environment, for designers to evaluate products for assembly and disassembly during the design process.[5] The VIRAD system uses a hierarchical model of a Generic Assembly and Disassembly (GENAD) workcell to generate merging trees for simulating assembly and disassembly processes. In a GENAD workcell, every object is either a part for making a product or a handler to facilitate the workcell operations. Each workcell operation is associated with a cost for product assembly and disassembly. The estimated cost is used for representing a product's manufacturability, providing important feedback for the designer.

To address this important link between design and manufacturing, we propose a VIRAD system that is integrated into a CAD/CAM environment. Figure 2 illustrates the proposed system. The three key areas are the CAD system, the VIRAD system, and the manufacturing and demanufacturing systems. VIRAD is developed for assisting designers to evaluate products for assembly and disassembly efficiencies during the design process. Historical production data such as tooling cost and cycle time are used to estimate manufacturing and demanufacturing costs. Special equipment updates the historical database continuously during production, thus, over time, making it more representative of the actual production processes.

Viswanathan

The product design, that is, the spatial relationships between the parts of the product and the values assigned to the various parts, plays an important role in determining the ease and profitability of disassembly. Viswanathan defined the concepts of Desired Value Precedence (DVP), Best Value Precedence (BVP),

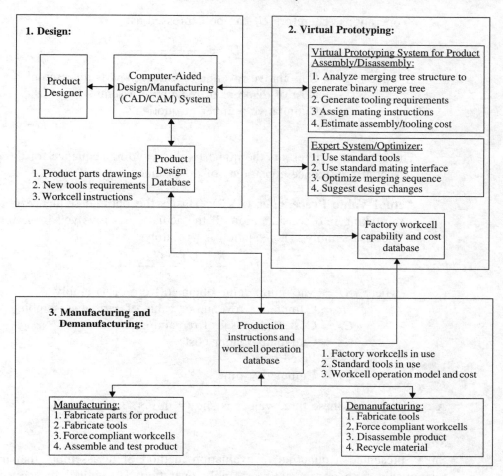

Figure 2 Products virtual assembly/disassembly block graph.

and Actual Value Precedence (AVP) and analyzed the similarities among them.[6] These concepts allows us to determine whether we should adjust a product's design based on various combinations of the spacing and value. Viswanathan's comparative concepts are defined as follows:

Desired Value Precedence (DVP): This is the ideal sequence where the parts are disassembled in descending order of their values, unhindered by the spatial constraints.

Best Value Precedence (BVP): This is the sequence where the parts are removed considering the spatial precedence constraints but neglecting the disassembly costs. So, it would be the sequence along which an idealized return U_i is maximized assuming unit time is taken for all disassembly operations. Therefore U_i can be expressed as:

$$U_i = \frac{R - C_d - t'}{t'}$$

where, R = Cumulative revenue or value of parts disassembled
C_d = Cost of disposal of remaining assembly
t' = Cumulative number of quotients

Therefore this represents the highest value-yielding sequence for the given spatial precedence constraints of the product.

Actual Value Precedence (AVP): This is the optimal sequence obtained by taking into consideration all the actual costs involved in performing the disassembly. The equation is as follows:

$$U = \frac{R - C_d - C_p - C_f}{t}$$

where, U = Maximum value obtained from disassembly
R = Cumulative revenue or value of parts disassembled
C_d = Cost of disposal of remaining assembly
C_p = Cumulative labor cost
C_f = Fixed costs
t = Cumulative time

A comparison of these three values is shown in Fig. 3.

Suga

Suga[7] presented a quantitative evaluation method of disassembly that has no relationship with an actual disassembly operation. He quoted the concept of

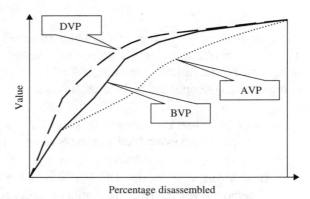

Figure 3 Value precedence comparisons: Desired Value Precedence (DVP), Best Value Precedence (BVP), Actual Value Precedence (AVP).

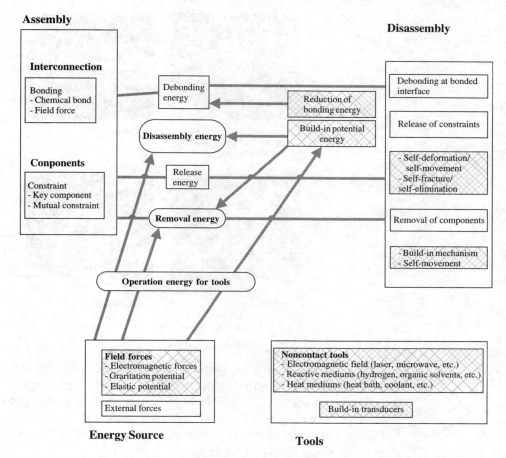

Figure 4 Active assembly and disassembly process.

active disassembly, which was first described at Delft University (Figure 4). He theorizes that all parts of future products be designed to be removable. The energy needed to disassembly the product is built into the product itself. All that is needed is to deliver a small amount of power or energy to the device to release the assembly. No tools are necessary for the disassembly process. Two parameters quantitatively describe the disassembly performance: disassembly energy and disassembly entropy. Disassembly energy is the sum of the energy needed to release all the connections. Disassembly entropy is determined by the number of types of connection and the different disassembly directions required.

Additional Evaluation Methods
Further perfecting of virtual prototyping techniques creates a new environment and methods for evaluating disassembly. Krause presents a virtual environment

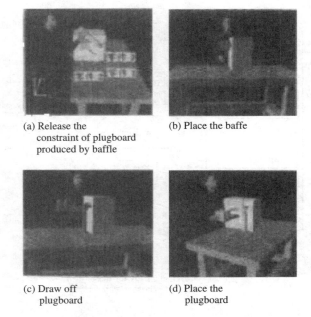

(a) Release the constraint of plugboard produced by baffle
(b) Place the baffe
(c) Draw off plugboard
(d) Place the plugboard

Figure 5 Virtual human disassembly.

simulating highly qualitative interactions such as touching, pressing, pulling, gripping, and handling of virtual tools, work pieces, or other components, including the precise movement of flexible objects. These simulations help to formulate statements about product assembly and disassembly.[8]

Figure 5 shows a virtual environment presented by Liu Jihong for demonstrating human disassembly of an object.[9]

Building-Block Design
Building-block design of products will reduce the amount of disassembly work and improve the disassemblability of products. Researchers at Tokyo University pointed out the characteristics of building–block design: mature techniques, upgradable function, long lifetime, easily guaranteed quality, easily cleaned and repaired.[10] Fumihiko et al.[10] and Zhou and Zhu[11] also study modular design disassembly.

2.2 Disassembly Process Layout

Disassembly Process Layout Based on Nondirected Graph
Zhang and Kuo[12] offer that the key problem with product recycling is the disassembly process and the recycling strategy itself. A complete disassembly module should address three aspects: disassembly analysis, database and database management, and optimization of reclaiming self-cost (Fig. 6). The first aspect is

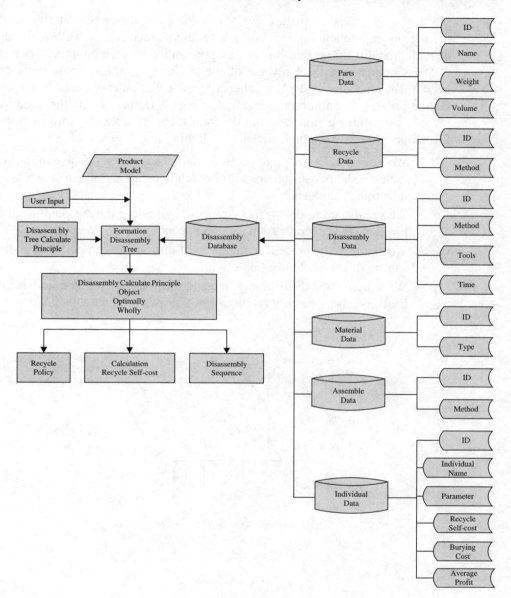

Figure 6 Disassembly model.

disassembly analysis, which analyzes the structure and disassembly sequence of the product. Structure mainly involves parts' physical dimensions and the assembly relationship among parts. The second aspect is database and database management. After analyzing the structure of product, we must analyze the nature of the materials. The material database supplies the data for the materials'

220 Disassembly for End-of-Life Electromechanical Products

physical characteristics (nature, weight, volume, virulence of material, etc.) and for the proper reclamation method (reuse/remanufacture), as well as the degree of reclaimability). The third aspect is optimization of reclaiming self-cost.

Disassembly is not the reverse of the assembly process. Some parts do not need to be disassembled. The schematic of a computer chassis based on the factors previously outlined is presented in Fig. 7. In the picture, the nodal points include parts information, and the diagram includes fastening information. The formation of a disassembly sequence is divided into four steps:

1. We use graphs to analyze the relationship of parts assembly, which includes the part-fastener assembly relationship and the disassembly optimization relationship.
2. We predigest the disassembly process by finding the cut point to divide a product into several assembly components.
3. We evaluate the precedence relation of disassembly to ensure that the part can be disassembled in this stage.
4. We structure the disassembly tree and formulate the disassembly sequence based on the assembly relationship of waste products.

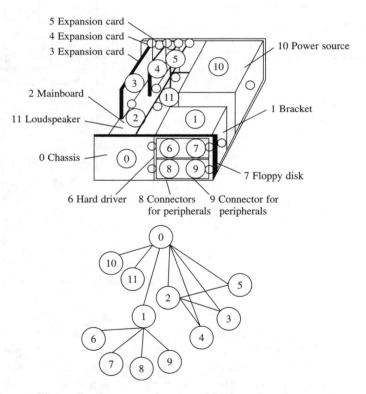

Figure 7 Computer chassis and its fastening diagram.

Following these steps allows us to arrive at the best disassembly sequence aimed at the least reclaiming self-cost.

Disassembly Process Layout Based on Petri Net

Suzuki has presented the concept of the Disassembly Petri Net (DPN).[13] In DPN, the warehouse represents products, parts, or subassemblies; transition represents the disassembly method (Fig. 8). The primary researchers of DPN are Moore and Zussman.

The process layout offered by Moore[14] is that we extract the Disassembly Precedence Matrix (DPM) based on geometrical relationships from the CAD modeling of products and turn it into DPN automatically. We use a reachability tree to analyze the DPN and arrive at a feasible Disassembly Process Plan (DPP). We then determine the best DPP according to a self-cost equation. The factors that affect the disassembly self-cost are time delays created when tools and disassembly directions are changed and when noxious parts are disassembled. Moore has proven that the DPN structured by itself is active, limiting, and reversible.

Zussman[15,16] pointed out that the problem with disassembly layout is determining the best sequence for the disassembly operation. Zussman used a simple but effective method to solve uncertainty issues in the disassembly operation by assigning different a priority to each stage of the disassembly operation of components. The priority determined the value produced in the disassembly operation:

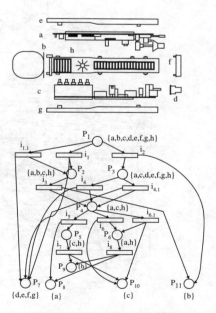

Figure 8 Radio and its DPN.

the more valuable, the higher the level of priority. When one disassembly operation fails, the system will choose the transition of next priority. This strategy also brings mission success rate into the disassembly operation and integrates it into the decision-making process of DPN.

Disassembly Process Layout Based on AND/OR Graph

Zussman has designed a recovery graph that is a variation of the AND/OR graph data structure.[17,18] The recovery graph is an AND/OR graph in which every point and every AND relation line is added to determine recovery value. For points, the recovery value should recover directly or disuse the benefit/self-cost instead of disassemble this point; for the AND relation line, recovery value is the self-cost divided components assembly into subassembly components.

Kang[19] uses extended process graphs to research the formation of disassembly sequences for parallel disassembly aimed at creating the highest total profits. The basis for disassembling AND/OR graph models is integration, and the starting point is disassembling the AND/OR graph completely. Guidelines are used to screen for errors and to correct the sequence. The final layout model describes the possible disassembly processes for the product (Fig. 9).[20]

Lambert also used AND/OR graph models as adjunct support when he developed the disassembly linear layout.

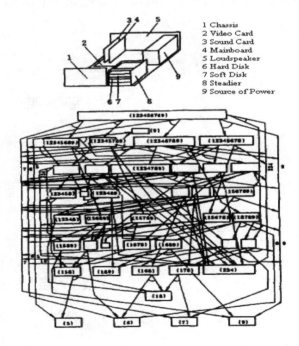

Figure 9 Computer chassis and its disassembly AND/OR graph.

2 Product Disassembly: Research Activities and Overview

Disassembly Process Based on Directional Graphs

Using directional graphs to describe the sequence of disassembly is rare. However, Cai Xuyuan[21] offers a disassembly logic network method, which includes nodes for disassembly, activity of disassembly, unit, logic relation, and disassembly path.

Disassembly Process Based on Multiple Graphs

Kanai[22] proposed four different graphs to uniformly model the product and the steps of its recycling process (Fig. 10a). He also suggested rules for planning disassembly, shredding, transforming the graph's structure, and changing the graph's topology and the attributes of nodes. The four graphs are as follows:

1. The product configuration graph is a set of directed trees. It represents parent-child relationships between subassemblies and parts, or between a fragment and its constituents. Shown in Fig. 10a, a root node indicates a subassembly or a part or a fragment, and its terminal nodes indicate parts or fragment constituents. A subassembly changes into a fragment when shredding has been completed. A fragment consists of several fragment constituents each of which is composed of a single material.

2. The product connection graph is a set of networks and represents connective relationship between parts or between fragment constituents, as shown in Fig. 10b.

3. The process graph is a Petri net (bipartite directed graph). Shown in Fig. 10c, it represents the sequence of disassembly, shredding, and material sorting performed on a retired product and also shows the inputs and outputs of each activity.

4. The retrieval condition graph represents the pairings of a part or a subassembly with one of the categories reuse, recycle, or disposal. The term

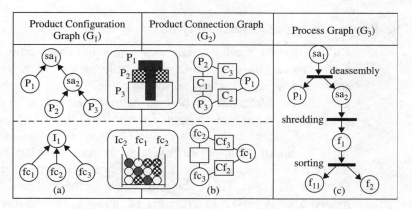

Figure 10 Graph-based models of product and processes.

recycle means horizontal and cascade material recycling in our study. It is a simple extension of the product configuration graph.

Touzanne[23] proposed three kinds of disassembly processes for differing levels of precision:

1. *Product level.* At the product level the process describes the part through product structure tree.
2. *Handling level.* At the material handling level, the process is adopted from a function graph showing the product and the logical operations; it does not to consider equipment and time.
3. *System level.* At the system level, a Petri net is used to express the process of disassembly; it includes equipment, space, and process time information.

Disassembly Process Based on Other Graphs

Gungor used a priority connection matrix based on the geometry between parts to build the best sequence of disassembly.[24]

Johnson used income loss to determine the end of the disassembly process thereby minimizing the disassembly cost and maximizing the profit of recycling materials.[25]

Xue Shanliang constructed the best economy optimization method of configuration based on the product bill of materials (BOM) tree of an AND/OR graph.[26] Hesselbach[27] proposed a four-part method that included classifying the character of the product and of the environment of the disassembly and then building a model of the product and disassembly process; dividing products into disassembly families based on models of the product; creating a strategy for the disassembly process; and finally, simulating the process for special disassembly needs to gain the best programming of the disassembly process. These methods are integrated into the software of LaySiD.

2.3 The Programming of Disassembly

Xu[28] proposed the automatic generation of disassembly motions. First use a partial medial axis to create all possible paths within the space of the parent subassembly, using graph- searching to determine the best path. Next, using the geometric relationship of the best path for the movable subassembly, program the feasible motion. He proposed the method that fixes on a partial medial axis when the parent subassembly is 2-D polygon. A set of key nodes are fixed when geometric shapes are given, with the set of key nodes being based on the restrictive relationships of the geometry. Partial medial axes can be used to create all paths within the parent subassembly. The relation between partial medial axes is shown in Figs. 11, 12, and 13. The figures complexity increases with the increasing of information of partial medial axis.

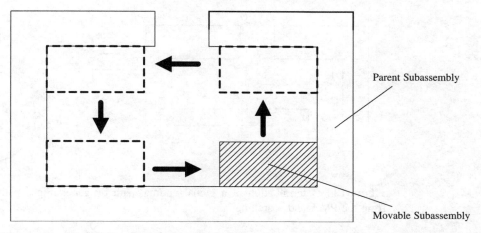

Figure 11 If there is a free space within the parent subassembly, the disassembly motion of the movable subassembly is iterative back and forth within the free space.

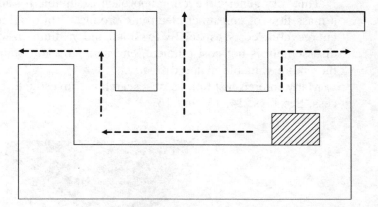

Figure 12 If there are multiple paths available for disassembly motion, an optimal path must be determined based on the degree of difficulty for each path.

Determining the best path through a graph-searching technique optimizes the path of disassembly. This method has been used to automatically create assembly designs based on geometric modeling and evaluation.

2.4 Active Disassembly

Recently, more and more people pay attention to Active Disassembly using Smart Materials (ADSM). ADSM uses smart materials to make joining components (such as bolts, screw bushings, or clips). These joins can actively release when the environment is changed. For example, joins can release in different temperatures. When a product comes to the end of its useful life, we can make the

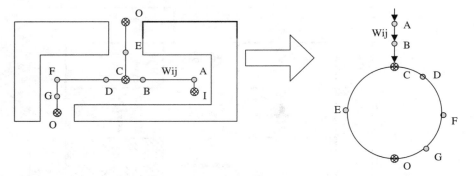

Figure 13 Determination of a global optimal path by generating a path graph abstracted from the PMA and searching.

parts and materials automatically disassemble.[29,30] ADSM can provide support to DFD.[31,32,33]

This very generic recycling approach is suitable for automated disassembly of a mass flow of consumer electronic products from various categories. It aims to cut recycling costs, especially for small and medium-sized products where manual disassembly is not cost efficient, and also aims to obtain less contaminated and thus more valuable material fractions.

Many foreign institutions of research are investigating ADSM with some success. See Figs. 14, 15, and 16.

Figure 14 Using ADSM to disassemble Sony digital walkman.[24]

Figure 15 Using SMP to make bracket of LCD.[24]

Figure 16 Result of using ADSM to disassemble Motorola mobile telephone.

Between 2000 and 2001, Nokia Research Center, Helsinki University of Technology, and the University of Art and Design Helsinki investigated recycling mobile phones through ADSM.[34] This research aimed at exploring a mechanism triggered by heating, to simplify disassembly and the recycling process of mobile phones. This project includes four phases: concept design, product design, detailed design/modeling, and testing/finishing. In concept design, it mainly considers the ability of disassembly and the recycling of product. In product design, it mainly considers the following factors to optimize the design of the product and to make and test the product model: product function (reliability), triggering temperature, disassembly time, product size (weight and volume), product structure, and cost of life cycle (material cost, product cost, assembly and disassembly cost). Through testing, the disassembly time is 2 seconds, which is 1/50 of manual disassembly time. It improves remarkably the efficiency of disassembly and achieves both economy and high efficiency of mobile phone recycling.

2.5 Automatic Disassembly

The development of automation and the use of robots is advanced, but has only been applied to assembly processes. Automated disassembly is still undeveloped. The aim of automated disassembly is to reduce the cost of disassembly and improving work conditions within the factory.

Lee[35] proposed a model of automated disassembly, using the information of product design and feedback sensors, that fixes on the next part needed to be disassembled. It minimizes the mount of process to get to the target part. This model can test the difference between old product and new product design, including evaluating the lack of a part, replacing a part, adding a new part, addressing a damaged part.

Knoth[36] proposed the concept of disassembly families to identify manual disassembly functions from automated and semi-automated work areas and systems. The disassembly family gathers together all products requiring similar processes of disassembly.

David adopted a method of document analysis for querying workers and observing operations of disassembly, thereby obtaining and providing information for disassembly automation.[37]

2.6 Disassembly Tool

Uchiyama[38] employed an image processing system and robotic automatic cutting system that used a supersonic wave sensor to disassemble air conditioners and washing machines. Geskin adopted high-pressure waterjets to separate the faceplate glass from the funnel of CRTs.[39]

Zuo[40] designed a low-cost flexible disassembly method. The screwnail indentation process overcomes geometrical uncertainties of dismantling products. The separating process damages only a portion of the part's surface.

Aria proposed a method that can separate copper and aluminum from air conditioner heat exchangers. It can improve disassembly of air conditioner heat exchangers through the use of a rolling mill and deals with product crushing, separating, and sorting.[41]

2.7 Disassembly System Control

Information Flow Control

It is important for disassembly systems to manage the uncertainties surrounding a product's nature by obtaining, storing, and sharing information regarding its kind, brand, and type.

Ranky[42] developed object component oriented requirements analysis of disassembly line manager and operator user needs. This software enables shared analysis and design of critical aspects for demanufacturing processes via the Internet or company intranet. Sanou explored an artificial intelligence system that can identify classes of product entering the disassembly line by measuring size and weight.[43]

Stobbe[44] and Ishii[45] adopted industrial image processing software and automated systems to identify and collect information for the disassembly process.

Logistic Control

Because disassembly time for various products is uncertain, Yuji[46] explored a flexible transfer system for materials and products. This system provided an adjustable transportation management model to handle work flow regardless of the nonuniform work time.

The Maintenance of Disassembly Equipment

Kouikoglou[47] researched a disassembly system with finite buffers and unreliable equipment, with the objective of maximizing system throughput while controlling costs through balancing the rate of machining and repairing of all parts.

In summary, many scholars have done extensive research into the field of disassembly, including evaluation of disassembly, programming of disassembly processes, programming of disassembly sequences, disassembly techniques, disassembly tools, and disassembly systems, and have done so with a great deal of accomplishment. The development of disassembly techniques will guide and perfect product design, solve the problem of disassembly difficultly, and consequently help to resolve threats to the environment and the problem of limited resources created by end-of-life products.

But there are still many shortfalls in the research, so the following aspects need be further researched and developed.[1]

1. By approaching disassembly from perspective of the life cycle of the product and integrating the DFD rules into the product design process, it will be able to provide designers with easily implemented design standards and easily operated design tools.
2. By creating either general or product-special disassembly sequence evaluation systems, disassembly process programming could include both economic factors and environment factors to optimize the disassemble process, thereby obtaining maximum economic profitability and environmental benefit.
3. Considering the great diversity of variety, brand, model, and such of discarded products in the disassemble line, basic sorting according to product disassembly characteristics before disassembly is required. To get better sorting efficiency, classification techniques at the automation level, such as adopting vision-based identification systems, should be improved.
4. Since there are many uncertain factors in disassembly, disassembly processes should be flexible. Such flexibility needs to address obtaining in real time each product's information, transferring this information to the necessary handlers, and adjusting the disassembly techniques.
5. To improve disassembly efficiency, new disassembly techniques and special disassemble tools should be developed. The automation of disassembly processes, such as that developed in Japan to automatically position and cut apart refrigerator compressors, also should be improved.
6. The industrialization of mechanical and electronic products disassembly, such as HP's disassembly line for their modular ink cartridges and Japan's AEHA (Association Electric Home Appliances) home appliances recycling factory, should be encouraged.

3 RESEARCH AND METHODOLOGIES IN DISASSEMBLY STUDY

3.1 Disassembling Techniques[3]

Disassembling techniques establish appropriate strategies, determines the optimal disassembling route, and selects the best extraction tool for either automated or manual disassembly.

Establish Disassembling Strategy

There are two strategies for disassembly: one is that the product is disassembled into separate parts, meaning it disassembles from the top to down; the other is selectively disassembling the product according to the current condition of the product. Generally speaking, technique, economical efficiency, and environmental synthesis determine the disassembly approach. It is impossible to salvage recycle value for each part of a specific product. And the cost of disassembling some parts is far greater than the value of the recovered material or part. Thus, we often adopt selective disassembly. Partial disassembly is the most common disassembly option, which requires rational disassembly of objects and determination of appropriate end points according to the current condition of the product.

Determine Disassembly Objects. According to the material value of parts, the ability of disassembled parts to be reused, and the environment effects of failure to recover, disassembled objects are categorized as one or more of the following:

1. *Optimum value parts.* These parts are usually made from expensive materials or manufactured with high cost.
2. *Recyclable parts.* These parts have long operational life span and can be reused in the same or different products without degradation of performance.
3. *Parts with serious repercussion for environment.* These parts are manufactured with hazardous materials or include ingredients harmful to the environment. These must be disassembled and recovered in spite of cost.
4. *General parts material.* These materials are sorted and recycled according to material composition and melted down and refined as raw materials.

Determine End Point. According to general rules for disassembled and recycled products, high-value parts should be disassembled first. The further a product is disassembled, the lower the recovered value from recycling can be. However, the cost of disposal of the discarded product also continues to decrease, thereby offsetting the lower return and simultaneously enhancing recycling efficiency because of reduced surplus waste material.

End point is the point at which disassembly stops. It is a constraint in the disassembly process. Based on disassembly object categories, end point can be after the most valuable parts are recovered, after the constituent having the most deleterious effect on the environment is neutralized, or after reusable parts have been recovered. In all cases, the controlling factor of the disassembly process is economic. When the recycling process becomes uneconomical, stop disassembling (see Fig. 17). In Fig. 17, $f1$ shows the disassembly cost, including cost of labor and equipment costs; $f2$ is the cost of processing waste material, which includes transportation expense and burying expense; $f3$ is the value recovered from the recycled material. The D-value is the difference between recycling expense and

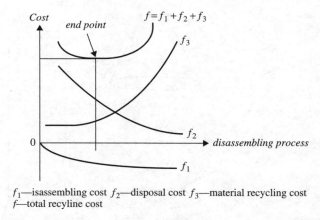

f_1—isassembling cost f_2—disposal cost f_3—material recycling cost
f—total recyline cost

Figure 17 Determining end point where recycling becomes economically unviable.

recycling value. F totals $f1$, $f2$, and $f3$. On curve f, the lowest point shows the lowest recycling cost, namely the best recycling benefit; this point is end point. As Fig. 17 shows, the lowest point is the most economical level of recycling. Under the premise of recycling the high-value parts first, disassembling beyond this point returns a negative value. In this situation, disassembly should stop.

3.2 Disassembling and Recovering Models[48]

The construction of disassembling and recovering models and the formation of the disassembling sequence are key aspects to the study of disassembly and recovery. The choice of the models to be disassembled and recovered has a great effect on the economic efficiency and the feasibility of the disassembly process. At the present time, the methods of study for disassembling and recovering models, both in China and around the world, are mainly as follows: AND/OR diagrams, nondirected graphs, Petri nets, logic graphs, and disassembly trees. The following discussion looks at different presentations of models, using the product description illustrated in Fig. 18 as an example.

AND/OR Graph Model

Figure 19 is an AND/OR graph model. The model shows every possible dissembling sequence, when there are comparatively fewer nodes. However, the number of cut sets when using AND/OR relations between parts and components is fairly large making them difficult to screen. This problem solution will become overwhelming and untenable when there are a large number of parts and components.

Undirected Graph Model

Figure 20 depicts an undirected graph model of the model flashlight. This method illustrates directly the connections between the parts and components of a product, which can be obtained directly from the computer-assisted design (CAD)

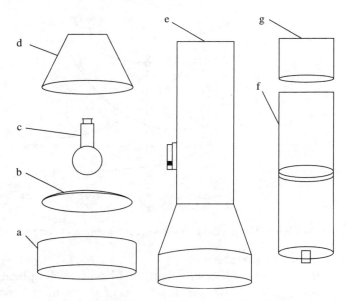

Figure 18 Structure diagram of flashlight.

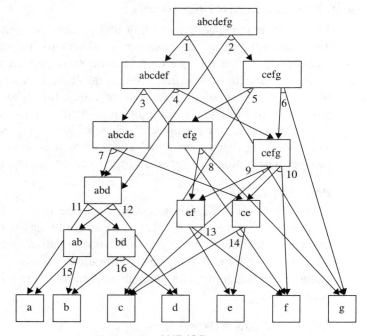

Figure 19 AND/OR graph model.

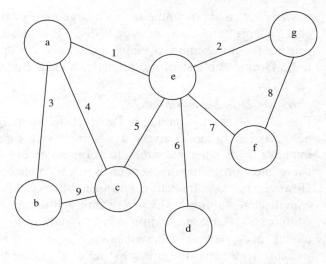

Figure 20 Undirected graph model.

system drawings. The structure is very simple, but it can show the procedural order for the disassembly process.

Directed Graph Model
Figure 21 is the directed graph model of Fig. 18. This model shows the constraint relationships among the parts and components, as well as the disassembling

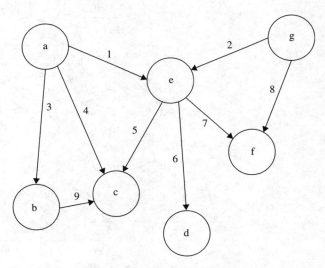

Figure 21 Directed graph model.

procedures. Its shortcoming is that a large number of parts and components generates a large number of nodes and edges. This results in a model with great depth causing the disassembly planning model and the optimizing planning algorithm to be fairly complicated. Additionally, it cannot address certain circumstances.

Petri Net Disassembly Model

Figure 22 shows the quintuple Petri net disassembly model. Based on the Petri net disassembly model, every disassembly task can be diagrammed. It also can integrate with other resources to form a comprehensive model, for example, using linear programming techniques to obtain the optimum path for disassembly. However, because the Petri net disassembly model has numerous connections and complicated algorithms, its practicality will diminish as the number of parts and components increase. Additionally, an endless loop can occur when the base model encounters a circular situation.

Table 1 identifies the strengths and weakness of the various models. Common shortcomings are as follows:

- When abstracting models from a product, each part and component is seen as the same. The models do not fully consider the assembly relationships between parts and components; nor do they take into consideration

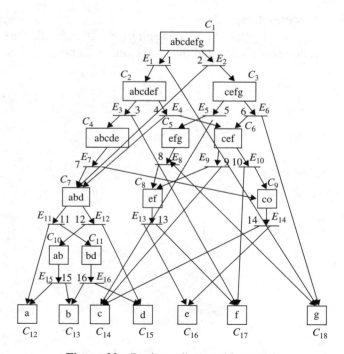

Figure 22 Petri net disassembly model.

Table 1 Merits and Shortcomings of the Models

Model	Merits	Shortcomings
Connection graph (Undirected graph)	• Can obtain the information on direct linkages directly from CAD drawings • Can use graph theory searching algorithm to obtain the optimum disassembly path	The model building of disassembly tasks is difficult
Directed graph	Can illustrate the disassembly sequence	The searching range of nodes and edges is relatively large
AND/OR graph	• Contains all the possible disassembly sequences • Can display the concurrency in disassembly strategy • Obtains the optimum path by heuristic reasoning	• Weak in integrability • Cut-set pattern interference • Cannot represent self-disassembly
Petri net—Method 1	• Can present every disassembly task • Easy to integrate with other resources to form a comprehensive model • Allows derivations of characteristics • Can display the concurrency of disassembly tasks	Needs connecting information for parts and components to build models from graph
	Obtains the optimum path by linear planning	
Petri net—Method 2	• Allows for EOL values • Easy to realize dynamic planning	• Nonlinearity • Difficult derivation

the experience level of operators. Consequently, many infeasible strategies appear, and even pattern failure is likely to occur.

- Models do not show clearly the differences of constraint relations between parts and components. These variations add to the difficulty of the disassembly process, making it hard to automatically determine the optimum sequence for disassembly from the models.

Therefore, a more efficient disassembly model and disassembly sequence algorithm is urgently needed. Disassembly is the process of removing the constraints between parts and components in order to separate them. It requires building models to show clearly the direct and indirect constraint relations and to filter out the infeasible disassembly sequences to find the optimum disassembly pattern. Below is a model of a classified directed confinement graph and the its applications.[49]

3.3 Composition of Disassembling System[50]

A disassembly system encompasses disassembly analysis, disassembly sequencing, disassembly evaluation, and the creation of system software to guide original product design. In the creation of a disassembly system, we can perform a virtual disassembly using the design model of the product. Then we can produce the disassembly sequence and the disassembly route and assess the disassembly of the product, pointing out design deficiencies and defects and providing revision suggestions for the design according to the information obtained. Using virtual disassembly, it is not necessary to manually perform a real entity prototype disassembly. Designers can analyze the design plan with the help of a computer-aided designing system, improving the design until a satisfactory result is achieved. Based on the complexity of the design, simplify the procedure as follows while researching and analyzing:

- Disassemble the primary product completely, but do not perform alternative disassembly, namely disassembling subparts piece by piece.
- Do not consider destructive dismantlement in the disassembly; that is, do not dismantle connection structures between parts that require destructive actions, for example, welds, rivets, etc.
- Dismantle only those parts capable of being disassembled through linear motion, sequentially and without backtracking.
- Disassemble parts serially, that is one part at a time; do not take in to account parallel or simultaneous disassembly of multiple parts.

Composition of Design System for Disassembly
Design systems for disassemble include the following four modules:

System Setup. There are two basic methods for inputting design information and setting up product modeling: (1) reading the data interchange file from the CAD system and (2) inputting interactively by users. Each of these methods has deficiencies. The first method generally establishes product modules only with regard to particular fields, and because of the incompleteness of product information from the CAD system, it does not provide complete design information necessary to the design of disassembly. The second method requires users to answer a large number of questions. The process is tedious and prone to mistakes. It is best to combine the two methods, capturing data input such as the geometric model of the product and assembly relations from the CAD system first and then interactively inputting other design information, for instance, physical attributes, etc. After intact design information is acquired, establish the product model design for disassembly.

Analyze Disassembly. Disassembly analysis includes producing an optimized disassembly array and then determining the best sequence for disassembling the parts and components. Generally there are several options for when and how to remove a part. It is necessary to select the easiest and fastest method. Using graph theory, search for feasible solutions. Narrow the search field using appropriate criteria to find the optimum solution.

Assessing Disassembly. Assessing disassembly generally includes quantifying disassembly costs and calculating disassembly time. Disassembly costs refers to expenses related to the disassembly. Because current operations are mainly executed manually, this mainly points manpower costs. Disassembly time refers to the time necessary to complete the disassembling operation, including dismantling time, separation time, and time spent addressing adjoining structures. Using statistical analysis for common connective structures, create an objective and comprehensive assessment calculation formula. Higher disassembly costs and longer disassembly times indicate greater complexity within the structure and portend poor disassembly performance of the product.

Provide Design Improvement Suggestions. The primary function of this module is to offer suggestions for the improvement of product design based on the knowledge gained through the analysis. It is necessary to analyze a large number of the connection structures in order to form more complete knowledge base. Figure 23 illustrates a functional module of this disassembly design system, including mutual relationships.

3.4 The Disassembly Evaluation[3]

The disassembly evaluation is a dynamic process that reiterates valuation, modification, revaluation, and remodification to a design project until the design

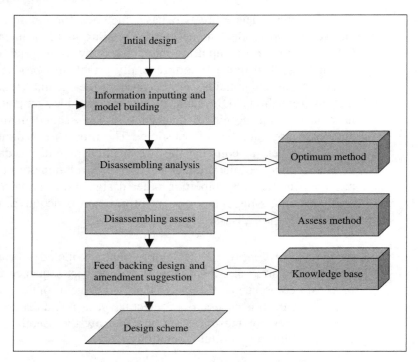

Figure 23 Composition of disassembling system.

objective is met. The primary valuation problems are how to valuate products' disassembly, which index to valuate, and what criterion to measure. So, the objectives of a valuation system are to put forward a complete disassembly valuation index system and to establish a set of valuation criteria.

The valuation index system must consider a wide variety of factors, including scientific, objective elements, such as the primary components of system, costs, resource usage, and measurable environmental impact. It must also address subjective aspects, such as actual possibility, intangible costs, immeasurable environmental impacts, and so on. Additionally, it should be capable of comparing the value of the specific projects. It also should allow us to perform proper cross-checks to confirm accuracy.

Generally, the disassembly valuation begins with two aspects: the first one is the product's physical disassembly; the other one is the relative time, expense, and so on. These indexes contain a fixed-quantity index and a fixed-capability index.

Relative Index for Objective Aspects
The relative index for disassembly mainly contains disassembly cost, disassembly time, disassembly resource expense, and the environmental impact of the disassembly.

Disassembly Expense. Products' parts have different join structures and so does the disassembly. The disassembly expense manifests different value. The disassembly expense is the cost that is directly attributed to the disassembly, including human expense, investment expense, and so on. Investment expense includes the disassembling tools expense, supporting and holding equipment expense, equipment and work space expense, and part conveyance mechanism expense, disassembly operating expense, the removed materials' identification and sorting expense, and transportation and storage expenses. The human expense consists mainly of workers' wages. The disassembly expense is the primary index for measuring economic viability of the structure's disassembly. If a part's disassembly expense is high, its return and reuse value is low. When a part's disassembly cost is higher than its discard cost, it has lost any return or reuse value. We can see, the less disassembly expense is, the higher the part's return and reuse is.

We can use following formula to express the disassembly expense:

$$C_{\text{disa}} = K_1 \sum_i C_1 t_i / 60 + K_2 \sum_i C_2 S_i$$

In this equation:

C_{disa} = Total disassembly expense
K_1 = Labor cost coefficient; takes into consideration the type of labor expense by disassembly method (manual or automatic) and the worker's experience level, and speed
K_2 = Tool expense coefficient; takes into consideration various disassembly tool expenses based on disassembly method
i = Disassembly time
C_1 = Current labor cost of disassembly operation f (RMB db)
t_i = Time of disassembly operation j (mh)
C_2 = Current tool cost consumption of disassembly operation j
S_1 = Tool utilization rate of disassembly operation f

Disassembly Time. The disassembly time is the time it takes to take apart a specific joint. Some parts of the product may consist of many joints, so this part's disassembly time is the total time to complete the separation of all the joints. It contains basic disassembly time and auxiliary time. The basic disassembly time includes the actual time it takes to loosen the joint and the time it takes to separate the related joints and parts that have been prepared for disassembly. Auxiliary time is the time needed for supplemental work to complete the disassembly, for example, preparing the disassembly tool or the time it takes for a worker's arm to approach the disassembly position, and so on. The longer disassembly time required, the worse is the product's disassemblability. Long disassembly times indicate complicated structures.

The disassembly time can be calculated using the following formula:

$$T_{\text{disa}} = \sum_{i=1}^{n} t_{di} + \sum_{i-1}^{m} N_{fi} t_{fi} + t_a$$

In the formula:

T_{disa} = Total disassembly time (min)
t_{di} = Disassembly time of single part i (min)
N_{fi} = Number relative fixed parts
N = Number of total parts
m = Number of linked parts
t_{ri} = Removal time of fixed part (min)
t_a = Auxiliary time (min)

Each different joint has its own computation for disassembly time. For example, a single screw joint can be computed by following formula:

$$T = T_1 + T_2$$
$$T_1 = L/(\text{np})$$
$$T_2 = K_t T_1$$

In the formula:

T_1 = Screw's disassembly time (min)
T_2 = Other time, including separating joined parts and auxiliary time (min)
T = Total disassembly time (min)
L = Length of screw (mm)
n = Number of screw
p = Pitch of screw (mm)
K_t = Correction coefficient, the screws portion of the auxiliary time included in the basic disassembly time

Obviously, choice of disassembly tool, workers' expertise, and complexity of the structure all impact the disassembly process, and the correction coefficient can be adjusted for practical application.

The disassembly time can be determined by practical research and analysis of disassembly data. Table 2 lists typical disassemble times determined by the accumulative method.

Energy Consumption in the Disassembly Process. Obviously, the disassembly of products consumes energy. Useful energy is found in two forms: manpower and external power sources (i.e., electrical energy, heat energy, etc.). Energy consumption in the disassembly of parts is an index that is indicative of the part's disassembly feasibility. The less the energy consumption, the better is the part's disassemblability. There are many different joining possibilities in mechanical

Table 2 Mechanical and Electrical Joint Structures Disassembly, Installation and Replacement Times

Standard Serial Number of Time	Item fixed part	Standard Time		
		Disassembly	Install	Replace
1	Standard screw	0.16	0.26	0.42
2	Hex screw	0.17	0.43	0.60
3	Moorage screw	0.15	0.20	0.35
4	Fast fixed part (1/4 circle)	0.08	0.05	0.13
5	Fast fixed part (less 1 circle)	0.06	0.06	0.12
6	Dish screw	0.06	0.08	0.14
7	Machine screw	0.21	0.46	0.67
8	Nut bolt	0.34	0.44	0.78
9	U gasket Button lock	—	0.27	
10	Ring lock	0.03	0.03	0.06
11	Pinchcock lock	0.04	0.03	0.07
12	Dish lock	0.05	0.05	0.10

and electrical products, for example, fit, welding, and mechanical connections such as screw connections and snap connections. Each method has a different disassembly energy usage calculation. The disassembly energy consumption of mechanical connections includes the screw's releasing energy, the spring distortion energy of snap connections, the friction energy of linking elements, and so on. Separating methods for chemical connections can also be considered, where the consumption energy is melting energy, rupture energy, or dissolving energy. In the following discussion, we will analyze screw and snap connections, giving the computing methods for disassembly energy.

SCREW CONNECTION. The screw connection is accomplished through a screw down force Fd to make the screw tighten by the corresponding screw-down moment. Figure 24 shows a screw connection structure. The screw-down moment is in direct proportion to the screw-down force and the screw diameter d and can be shown by following formula:

$$M = 0.2 \text{Fd} \times 10^{-3}$$

In the formula:

M = Screw-down moment (N·m)

0.2 = Moment coefficient; value is relative to friction coefficient and screw's minor diameter

From mechanical knowledge, the screw connection's axial force equals 10 percent of the screw-down moment. Because this axial force is on the screw's loosening direction, the screw's loosening moment needs 80 percent of screw-down

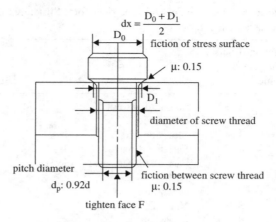

Figure 24 Screw connection.

moment. So, the screw's loosening energy E_1 can be computed by following formula:

$$E_1 = 0.8M\theta$$

In the formula:

θ = Revolving angle in axial stress loosening direction (rad.)

So, the screw's disassembly energy is loosening screw energy E_1.

SNAP CONNECTION. Figure 25 is a diagram of a snap connection. Disassembly energy of the snap connection, E_2, can be defined as the strain energy that changes the height of the pin fitting in the sleeve. Using the mechanical properties of the material, we can treat the snap connection as a cantilever. The strain energy of

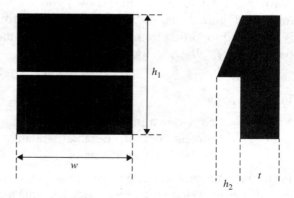

Figure 25 Snap connection structure.

this cantilever can be calculated in the following formula:

$$E_2 = 1/8(\text{Ewt}^3 h_2^2 / h_1^3)$$

In the formula:

E = Young's modulus of material
h_1 = Height of the snap connection
h_2 = Height of the pin
t = Thickness of the snap connection
w = Width of the snap connection

Up to this point, we have calculated the disassembly energy of a single connecting unit. If a component waiting for disassembly has more than one connecting unit, the disassembly energy for that component will be the sum of all connecting units. The disassembly energy of other connecting modes are shown in Table 3 but will not be discussed here.

Influences to the Environment of the Disassembly Process. The impact on the environment from disassembly processes occurs mainly in the form of the variety and quantity of contamination discharged into it and in the form of noise. Hazardous materials in the disassembly process, such as materials containing harmful or poisonous ingredients, should be disassembled with care using protective procedures. Attention must be paid to safety measures, and the removed components must be properly classified and stored to prevent environmental pollution or a cross-contamination of other parts. Other kinds of materials in the everyday environment, such as gasolines and lubrications used with automobiles, should also be properly collected and handled to prevent spills and overflows from polluting the work-yard and environment or polluting water resources through runoff. Tables 4 and 5 present standard pollution levels for noise and exhaust emissions.

Table 3 Connection Type[51]

Connection Type	Connection Index	Unidirection or Equity
Injection molding	260	Equity
Welding	220	Equity
Bolting connection	210	Equity
Screw fastening	210	Unidirection
Buckle	200	Unidirection
Light force fit	90	Unidirection
Clearance fit	50	Unidirection
Loose fit	20	Unidirection
Lid	10	Unidirection
Spacing	5	Unidirection

Table 4 Noise Environment Standards[52]

Noise Extent	Mark
Operating noise <65 dB*	0
65 dB ≤ Operating noise < 75 dB	0.3
75 dB ≤ Operating noise < 85 dB	0.5
85 dB ≤ Operating noise < 95 dB	0.7
Operating noise ≥ 95 dB	1

*The noise environment standard value of industrial areas is 65 dB.

Table 5 Exhaust Emission Standards[52]

Exhaust Emission Amount*	Mark
Exhaust emission amount < 350 μg	0
350 μg ≤ Exhaustemission amount < 700 μg	0.3
700 μg ≤ Exhaust emission amount < 1000 μg	0.5
1000 μg ≤ Exhaust emission amount < 1500 μg	0.7
Exhaust emission amount > 1500 μg	1

*Primarily the emission amount of CO_2.

Target Correlations for Subjective Aspects
As discussed in the preceding section, the relative index correlates with objective things such as disassembly process time and energy. However, actually designing a unique, all-encompassing analysis is always key to evaluating the potential success of a disassembly. Accounting for intangibles is the most significant component in estimating disassembly value. The advantages and disadvantages of disassembly capability for products are not often evaluated by qualitative analysis. Here, we try to qualitatively evaluate disassembly capabilities. It must be noted that by definition the nature of this indexing is not quantitative.

Reach Capability. Reach capability includes three aspects: vision, or the ability to see clearly all that needs to be seen; entity, or the physical ability to reach everything that needs to be touched or handled; and spatial, or having enough space in which to actually perform the disassembly. Regardless of whether disassembly is done by hand or is automated, sufficient space is needed to perform the disassembly actions. Take a shared-space valuation for example. Manual disassembly requires space to allow the operator to use disassembly tools conveniently and smoothly. Automated disassembly requires space to allow the disassembly device to approach the disassembly part conveniently and to execute the separation procedure in the prescribed disassembly direction. The necessary space is relative to the linking structure, the dimension of the disassembly part, the method of disassembly, the disassembly tools, the directional flow of the separation, and so on.

For disassembly by hand, we also should consider what we know about both human capabilities and the engineering of machines. For example, the size of a screw may be too big to disassembly by hand or a workpiece might be too lightweight or fragile to withstand the stress of automated disassembly. When a bolt dimension is too big and using a wrench requires too much strength or expends too much energy, we should use an automated machine or another type of tool.

In initial product design, bolts of differing dimensions should allow for screw disassembly, by leaving the minimum space required to conveniently apply a wrench to the bolt. When designing products using standard values, all of the parameters should allow for at least the minimum standard value. In order to pick up linking parts conveniently, they should have enough separating space in the axial direction.

Degree of Standardization. The degree of standardization is another measure used with product disassembly. A standardization coefficient is the primary tool used to measure the degree of a product's standardization. Standardization coefficients can be based on the number of standard parts, the number of general parts and adapted parts, the total number of parts, or total number of different kinds of parts. Some parts in a product can be very difficult to classify, especially in very small electric subassemblies. These parts should be given special attention during product design, to ensure adherence to the appropriate standards. For example, small rivets should utilize standard lengths, diameters, and materials, as should gaskets and screws. These parts should be designed to work with standard tool types, thereby avoiding numerous specialized tools. Standardization makes purchasing replacement parts easy and exchanging the parts convenient. Generally, the bigger the standardization coefficient, the better it is for cutting cost of design, manufacture, and disassembly and the more easy it is to apply advanced methods and technology.

The product standardization coefficient includes both a standardized pieces coefficient and a standardized type coefficient.

The product standardized pieces coefficient = (number of standard parts + the number of general parts + the number of adapted parts number)/total number of parts × 100%.

The product standardized type coefficient = (the number of types of standard parts + the number of types of general parts + the number of types of adapted parts)/total number of parts × 100%.

Disassembly Direction. The direction of disassembly is very important aspect in the operation of automated disassembly tools. Generally, a knock-down subassembly must be dismantled by maneuvering parts in a specific direction or series of directions, called the disassembly direction (DD) scope. For analysis purposes, subassembly parts can be identified and labeled using letters and numbers such. A generic application of this strategy would be that the direction of disassembly for part C_i can be denoted as DD_i. For example, in Fig. 26a C_1's DD can be called DD_1, indicating only one disassembly direction. When researching disassembly direction scopes, we can map them to a Gaussian sphere to provide specific descriptions (namely, the unit radius sphere). Disassembly direction scopes can occur in following configurations:

1. Mapped to a point (0 dimension) on the Gaussian sphere—single direction (e.g., DD_1 of Figs. 26a and 26b)
2. Mapped to an arc (1 dimension) on the Gaussian sphere—series of directions (e.g., DD_1 which abuts the shaded circular area of Fig. 26c)
3. Mapped to a surface on the Gaussian sphere—series of directions (e.g., DD_1 in Fig. 26d)

In some instances, the disassembly direction scope may have multiple dimensions. In this case, we will need a way to make the various dimensions comparable. We can divide the Gaussian sphere into quadrangles of equal area in order to unify points, arcs, and surfaces.

Different DD_i corresponding to G can be calculated using C_i. In Fig. 26a, C_1's disassembly direction is in one direction only. So, the value G is 1 because overall disassembly direction requires only one quadrangle. In Fig. 26b, the disassembly value G is 2 because C_1 can be removed in either of two directions (the overall DD_1 requires two quadrangles). In Fig. 26c, DD_1 is made up of

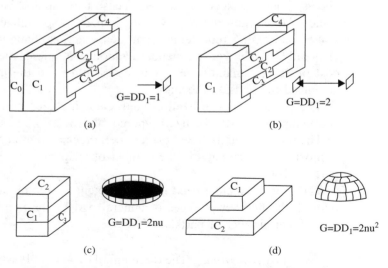

Figure 26 Numerical expression of disassembly direction scope.

many directions whose end points describe a circle on the Gaussian sphere. For overall disassembly direction scope, a quadrangle whose width is 1/u is made up of a line whose total length is 2. The corresponding G_1 is 2u, that is, the overall disassembly direction scope requires 2u quadrangles. In Fig. 26d, G_1 is a hemispheroid that converts to single quadrangle area of $2\pi u^2$. Here, u (where $u > 1$) indicates the potential number of disassemble directions available to the operator; usually $u = 10$.

During the assembly design process, the sequence of disassembly should be considered. After determining its disassembly sequence, we can compute each subassembly's composite disassembly direction G_i for evaluation.

At the original design stage, this methodology expresses whether disassembly will be easy or difficult for the specified subassembly and can be used to evaluate disassembly capabilities of various design projects. The bigger the value of G_i, the easier disassembly will be.

To obtain a complete disassembly sequence, the total of all subassembly G's can be expressed by Q as follows:

$$Q = \sum_{i=1}^{m} G_i$$

Generally, the higher the value of Q, which means the disassembly direction scope is bigger, the easier disassembly is. Designing special disassembly equipment based on these findings is economical because it can be made to install and disassemble in a series of directions. A single complex machine can be used many times to accomplish different sub-disassemblies.

Complexity of Design. Using **AND/OR** charts to evaluate the complexity of a product's design is somewhat subjective. However, they can be used to abstract the product into structural modules (physics) for analysis in order to assign them a quantity.

3.5 Method of Estimation

The assessment of disassemblability requires numerous estimations. It attempts to provide suggestions for improvements to designers and can aid in developing simulations to test design results.

Disassemblability Degree[52]

Disassemblability degree, DD, is the degree of disassemblability of products. DD comprises the sum of all the estimations applied to the product. The formula for DD is:

$$DD = \sum_{i=1}^{m} w_i a_i + \sum_{i=m+1}^{n} w_j(1 - a_j)$$

In the formula:

m = Number of estimations identified for the product; the larger the number, the better the disassemblability of the product
n = Number of estimations identified for parts and subassemblies
i = Subscript assigned to identify individual parts; $1, 2, \ldots, n$
a_i = Value of the ith estimation
w_i = Weighting factor assigned to the ith estimation

The range of DD is [0,1]: 0 denotes that it cannot be disassembled; 1 denotes perfect disassemblability. A discrete value is assigned to each estimate based on its relation to the disassembly process, with product-level disassemblies being assigned higher values than subassembly-level disassemblies. Subassembly estimations address aspects such as size of the fastening piece, reachability, and exhaust gas and noises pollution, and the values of these estimates should be adjusted by a factor of $(1 - a_i)$. We can weight the importance of the estimates by applying an analytic hierarchy process, a fuzzy analytic hierarchy process, etc. The formula given for DD provides total disassemblability of a product, including the disassemblability of its subassemblies.

Disassembly Evaluation Chart[52]. The methodology of AND/OR charts cannot derive the estimations we need. To do so, we need to use a disassembly evaluation chart. Table 6 outlines the basic structure of a disassembly evaluation chart. The following is a brief explanation of each of the categories listed in Table 6.

1. The Part Number is the number assigned to each part of the product. Identical parts that are disassembled simultaneously or parts with the same disassembly characteristics, for example, three identical screws used to

Table 6 Basic Structure of Disassembly Evaluation Chart

1		Part Number
2		Minimum Theoretical Number of Parts
3		Repetitive Operation Times
4		Type of Disassembly Tasks
5		Direction of Disassembly
6		Disassembly Tools
7	Difficulty Grade:	Reachability
8	• Simple	Localization Requirement
9	• A little difficult	The Value of Disassembly Force
10	• Difficult	Additional Disassembly Time
11	• Very difficult	Special Disassembly Problem
12		The Summation of Difficulty Grades
13		The Summation of Difficulty Grades and Repetitive Times
14		Annotation

fasten a single part, can be assigned the same number. When a single part appears as multiple components of product, we can differentiate the parts with subscripts.

2. Minimum Theoretical Number of Parts is the theoretical minimum the number of parts that comprise the product as determined by Design for Assembly (DFA) analysis. Here we estimate each step of the disassembly process to ascertain which parts need to be separated. If, for example, a part functions differently from other parts, is made of a different material, or needs to be removed in order to get to another part, assign it a value of 1; if not, assign it a value of 0. Every part marked as 1 needs to be estimated. For components, the value filled into the chart is decided by the following disassembly operation. If the component needs further disassembly, mark it with 1; if it does not need further work, mark it 0.

3. Repetitive Operation Times records the number of times an action needs to be repeated in order to complete each disassembly task. Here we mainly consider identical parts needing to be disassembled synchronously, For example, three identical screws holding a single part in place require the screw removal action to be repeated three times.

4. Type of Disassembly Tasks describes the mechanical operation necessary to complete the disassembly, for example, push/pull, move, incision, rap, etc. A single disassembly process may require several modes of operation to remove one part. For instance, the operation "loosen the screw" contains the operation "remove the screw." In this case, we only denote the first operation performed, so in the preceding case just denote "loosen the screw." For estimation purposed, concrete tasks of the disassembly operation can be derived from trial disassembly of the product.

5. Direction of Disassembly describes the axes of direction that humans or automated disassembly tools use to access parts for disassembly. The reference frame usually used aligns the Z-axis with the face of the workbench where the product is positioned. Once established and throughout the disassembly process, this reference frame remains constant. It does not change as the disassembly process progresses, Even though one disassembly action usually involves several movement directions. We must record the order of all actions in the chart.

6. Disassembly Tools lists the type of tools needed in the disassembly process to complete the all the disassembly tasks, for example, Phillips head screwdriver, flathead screwdriver, wire-cutter, etc. If the disassembly operation is finished by handwork alone, it need not to be recorded.

7. Reachability is uses a scale to describe the degree of difficulty that an operator or disassembly tool has in accessing (reaching) the part for disassembly. It mainly addresses whether there is proper space to perform the

disassembly action and the difficulty in handling the parts and components during the disassembly process.

8. **Localization Requirement** of disassembly describes the exact localization, or turning degrees, of the operator's arms or of the disassembly tools necessary to complete the disassembly tasks. For example, compared with a simple operation such as grasp or pull, positioning a Phillips head screwdriver into the head of a screw and loosening it demands higher localization precision.

9. **The Value of Disassembly Force** is a measurement of the force needed to complete the disassembly action. For example, the value of disassembly force to remove a force-fit part such as a snap connection is much greater than that needed to remove a clearance-fit part such as a bolt. Separating fitted parts or destructive disassembly of parts requires greater force.

10. **The Additional Disassembly Time** addresses the extra time it takes to dissemble a part owing to some factor that is integral to the nature of the part, causing the disassembly process to be more difficult or take more time. For instance, it more time to disassemble a long screw than a short screw. Note that this additional disassembly time is not accounted for in other places, so it is added here. If it has been taken into consideration in another part of the chart, it should not be relisted here.

11. **Special Disassembly Problems** are any issues that have not been considered elsewhere or have not been included in the chart. Examples of a special problem include stuck wires or part or when the exact location of the join is not known.

12. **The Summation of Difficulty Grades** is the sum of the difficulty grades assigned to the items 7 through 11 above.

13. **The Summation of Difficulty Grades and Repetitive Times** is the product of the values entered in items 3 and 12, and combines the repetitions of a certain disassembly task with its total difficulty grade.

14. **The Annotation line** is used to further explain any special tasks listed above, for example, special tools that are needed or the complexity of circumstance requiring special disassembly.

Using the outline and definitions discussed above, we can estimate the difficulty of disassemblability by completing the disassembly evaluation chart using data derived from the desired product.

Fuzzy Analytic Hierarchy Process (FAHP)[53]. Gradation analysis is an analytic hierarchy process that synthesizes qualitative analysis and quantitative analysis. The synthesized evaluation of various principles is achieved through setting up an object hierarchy and constructing a judgment matrix, which provides the context for selecting the best principle. In fuzzy gradation analysis, a fuzzy judgment matrix replaces the traditional judgment matrix, thereby avoiding the judgment inconsistency issues found in the analytic hierarchy process. The specific construction process of fuzzy judgment matrix is $A^{(K)}$.

Elements that have relationship with each other will be considered to have greater importance than those without relationship. If element C has relationship with element $a_1, a_2, \ldots, a_n$, then the judgment matrix can be shown as follows:

C	a_1	a_2	...	a_n
a_1	a_{11}	a_{12}	...	a_{1n}
a_2	a_{21}	a_{22}	...	a_{2n}
...	...	...	...	...
a_n	a_{n1}	a_{n2}	...	a_{nn}

The notation a_{ij} indicates that the elements a_i and a_j have a fuzzy relationship and that this relationship carries a value that "... is more important than..." when element a_i is compared with a_j.

When comparing any two principles there will be degrees of importance. Standardizing these degrees in the form of a graded scale 0.1 to 0.9, such as that shown in Table 7 is useful.

Table 7 Gradation of Relative Importance on a Scale of 0.1 to 0.9

Gradation	Definition	Explanation
0.5	Equal importance	In comparing the two elements, they are of equal importance.
0.6	Somewhat important	In comparing the two elements, one is somewhat more important than the other.
0.7	Obviously important	In comparing the two elements, one is obviously more important than the other.
0.8	Very important	In comparing the two elements, one is much more important than the other.
0.9	Extremely important	In comparing the two elements, one is extremely more important than the other.
0.1, 0.2, 0.3, 0.4	Cross compare	If the judgment r_{ij} results from comparing element a_i to the element a_j, then the result from comparing the element a_j to the element a_i is $r_{ji} = 1 - r_{ij}$.

Confirming the Weight of the Evaluating Indicator.

$$w_i = \frac{\sum_{j=0}^{n} a_{ij} - 0.5n + a}{2a_n}$$

$$(0 < a \leq 0.5)$$

In the FAHP, weight

weight vector $W = [W_1, W_2, \ldots, W_n]^T$

$$w_i = \frac{2a + \sum_{j=1}^{n}(a_{kj} - a_{jk})}{2a_n}$$

When judgment matrix A doesn't satisfy the criteria of FAHP, it can be adjusted to do so.)

3.6 Development of Three-Dimensional Environment for Disassembly[54]

SolidWorks

SolidWorks is a Windows-based software for three-dimensional design having the following characteristics:

1. It is a characteristic-based design software capable of making detail and partial design drawings of complicated parts more clearly by using its characteristic manager.
2. SolidWorks adopts a related-whole technique, so the updating among parts and three-dimensional models and the engineering drawings is executed simultaneously.
3. SolidWorks is a dimension-driven system; the designer can change the dimension or the shape of one part by specifying the dimensions and the geometric relationships among all related parts are automatically adjusted.
4. SolidWorks is a set of CAD/CAM/CAE; it gathers design, analysis, processing, and data management into an integrated whole.
5. SolidWorks can carry out dynamic stress tests automatically, and it has a robust ability to sculpt complicated curved surfaces.
6. SolidWorks has a strong dynamic simulation function, so it can carry out the simulation of assembly and disassembly processes.
7. SolidWorks has a variety interfaces that allow it to integrate with different CAD system.
8. SolidWorks contains an application programming interface (API), which allows the user to customize the characteristic design system.

For supplemental development needs, SolidWorks allows any OLE- or COM-supported programming language to be used as developer's tools. Developers can

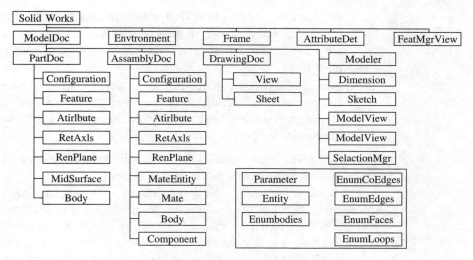

Figure 27 SolidWorks API objects.

create supplements in two forms. One form automatically develops and creates EXE files; the other, based on COM, creates *.dll files and then registers them as plug-ins for SolidWorks.

SolidWorks API objects are divided into several layers (see Fig. 27). Every object has its own attributions, methods, and events.

Application of SolidWorks in DFD System

SolidWorks, as the design platform used in the development of a DFD system, offers several useful features as described here:

- File sharing allows remote viewing and access to files by many designers and from many locations. Web folders can be stored on the main server in the design department. Permissions can be set to allow various levels of access, and files can be downloaded to a local hard disk. Any local computer with access and running SolidWorks can be used to open and work on files.
- Messaging software allows expedited information exchange among departments. Using the design guide module, designers can access the real-time estimates, design suggestions, re-engineering requests. They can modify the design proposal by accessing Internet or intranet database information, and update the design by refreshing the browser window.
- SolidWorks is capable of extracting information from existing parts, including characteristic information such as dimensions, materials, and assembly information. It saves the information into a local database. The information then can be uploaded synchronously to a full-system database and used be all departments through a shared-distribution database.

- By distilling the real-time design information from the designer and searching the related design-help information guide, the browser window can display the design department's general design guidelines or project-specific guidelines.
- Since SolidWorks is a dimension-driven system, it can use the API to carry reconfigurations throughout the design of related dimensions and shapes, for example, when disassembly evaluations result in suggestions for an improved design.

Using the distilling of parts information in a DFD system as an example, we analysis how SolidWorks works in the system.

The basic contributing unit of a product is one part, so the key input of product information consists of distilling the characteristics of each part. SolidWorks is a CAD system based on characteristics. Since the system interfaces with other CAD systems, it is easy to integrate isomeric systems. Since the system provides API functions for developers, users can develop their own applications to distill the desired information from parts and the assembly.

SolidWorks provides developers with powerful secondary interfaces using OLE objects. Some commonly used OLE objects are SldWorks, ModelDoc, PartDoc, AssemblyDoc, DrawingDoc, Sketch, Dimension, etc. OLE-supported developer tools such as Visual C++, Visual Basic (VB), Delphi, etc. can be used to develop controls for distilling additional part information though OLE objects.

Next we discuss using VB 6.0 as the developer's tool to develop a disassembly-oriented system. Through the API interface function of SolidWorks, we can call the objects to create a control that will distill the part's geometric and material information and the connection information of each part in the assembly. Also a temporary database is created to save the distilled information.

Take the bill of materials (BOM) in a schedule drawing as an example. The BOM contains the materials and quantity information for the parts. its content is based on everything visible on the schedule drawing. The main steps of the distilling process include:

```
retval=DrawingDoc.GetCurrentSheet0; \\Get the Drawing activated at present
retval=View.GetBomTable0;      \\Get the BOM Table in the View
retval=BomTable.Attach30; \\Activate BOM
retval=BomTable.GetTotalRowCount0;  \\Get Total Line Count of BOM
retval=BomTable.GetTotalColumnCount0;  \\Get Total Column Count of BOM
retval=BomTable.GetHeaderText(col); \\Get the Text of Header
void BomTable.Detach0; \\Release BOM
```

Figure 28 diagrams the process used to distill the information on create parts, assembly, and characteristic objects by SolidWorks.

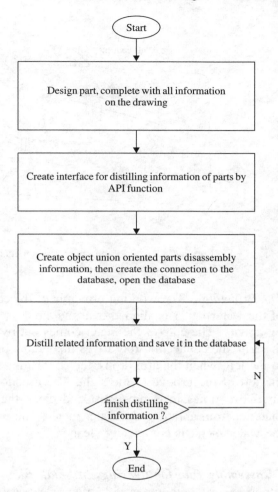

Figure 28 Work route for distilling information of parts.

3.7 Study of Disassembly Tools

Research of Equipment Used for Disassembling the Body and Core of Water Meters [55]

The Structure. The equipment is mainly composed of a base plate, a structure for disassembling, and a structure for cleaning. The structure for disassembling includes the rotating shaft bearing, the arm, the arm bracket, the lead screw, the sleeve, and the anvil. The structure for cleaning includes the sliding plate, the sliding block, the plywood, the long shaft, the short shaft, the brush, and the connecting rod (see Fig. 29).

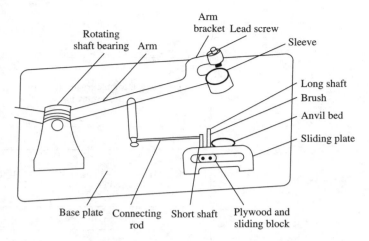

Figure 29 Shield and core of water meter disassembling device.

The Principle. The working principle of the equipment is as follows: The body of the water meter is aligned over the anvil, is lifted by the arm, and is dropped by gravity. The core of the meter comes out by tapping. In the process of lifting the arm, the connecting rod drives the sliding block from the right of the groove to the left; when the arm drops, the connecting rod drives the sliding block from the left of the groove to the right. These actions cause the brush to sweep the anvil two times to remove rust and clean the meter, preventing the corroded material from abrading the water meter. The motion continues until the core of the water meter is completely cleaned.

Disassembly Tool for Bearings in Small Blind Bores[56]

When repairing equipment, it's often necessary to disassemble the bearings in it. Since the fit of bearings is an interference fit, it is difficult to disassemble them; this is especially true for those in small blind bores. Fig. 30 shows a bearing in a blind bore. Such pieces are almost impossible to disassemble using current

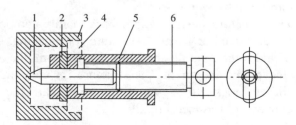

Figure 30 Bearing disassembly: (1) pin roll, (2) sliding block, (3) workpiece, (4) bearing, (5) thread sleeve, (6) lead screw.

standard tools. In order to solve this problem, a simple piece of equipment for disassembling bearings is researched. Its principle is as follows.

The sliding block is positioned in the groove of the thread sleeve so that it can slide freely. Initially, the two sliding blocks rest inside the thread sleeve. The sliding blocks together with the thread sleeve move through the bearing, and then a cotter pin is inserted into the thread sleeve. The front end of the cotter pin presses against the workpiece. The sliding blocks can go to only a certain height, which is the thickness of the bearing inner race. The thread sleeve is screwed down, tightening the cotter pin against the workpiece and moving the sliding blocks closer to the bearing inner race. When the thread sleeve is screwed in enough, the sliding blocks will drop into place allowing the thread sleeve to be pulled out with the sliding blocks removing the bearing from the workpiece.

Tool for Disassembling Bearing's Outer Shell[57]

When repairing the grinding tools in a grinding machine, we sometimes find that the bearing shell is damaged and that the ball bearings fall out. When the grinding tool is disassembled, the outer shell of the bearing remains in the shaft. Because the outer shell of the bearing is placed in an offset in the shaft, it's very difficult to disassemble it with standard methods (see Fig. 31). To solve this problem, we use a tool that can disassemble the outer shell of the bearing easily (see Fig. 32).

The Structure of the Tool. Measure the size of the bearing's outer shell as d; make the size of the outer arc of the bulking slug as $d_{-0.15}^{-0.10}$, and mill a 3 mm opening at the halfway point of the bulking slug. Different bulking slugs can be machined to suit to bearings with different specs.

The Operation Method.

1. Put the key into the bore of the bulking slug. Fit cylinder head screws into the six squares to seat the key;

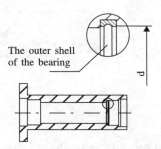

Figure 31 Outer shell of bearing.

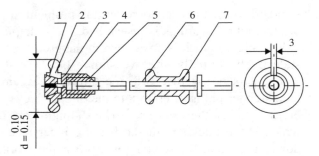

Figure 32 Structure of tool used to disassemble bearing out shell.

2. Position the bulking slug in the ball bearing groove of the bearing's outer shell. Adjust its position and tighten the inside six square screws to embed the outer circle of the bulking slug in the ball bearing groove.
3. Twist the key and tension bar, using a block to tap the step in the tension bar repeatedly until the bearing's outer shell can be removed.

This popular tool can also be used to disassemble bearings in other situations.

Design for Tool to Disassemble 1E Series Electric Motor[58]

The Principle. According to the design of the 1E electric motor top cover, we can use the single knock-down part—the screw thread openings that align the top cover and outer cover—to disassemble the top cover and outer cover at the same time. As Fig. 33 shows, the tool has two main parts: the sleeve and the mandrel. The mandrel is connected to the sleeve by screw threads, and the sleeve is positioned outside the top cover of the motor. As the mandrel is screwed in, the tip presses against the core of the motor. Continuously twisting the mandrel

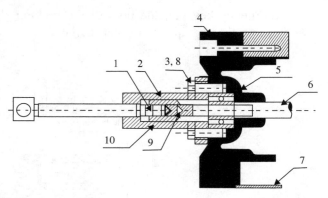

Figure 33 Design for tool to disassemble IE electric motor: (1) mandrel, (2) sleeve, (3) bolt, (4) cover, (5) bearing, (6) rotor, (7) stator frame, (8) gasket, (9) center, (10) roof filament.

allows the sleeve to draw up the top cover. Ultimately, the top cover and the bearing will be disassembled at the same time.

Design for Sleeve. The sleeve (shown in Fig. 34) is aligned directly on the top cover. Bolts can be used to affix it to the cover by milling holes on the underside of the sleeve. In order to make the sleeve versatile and multipurpose, we should consider several factors, such as the maximum diameter of the electric motor core, the minimum head size of the screw in the outer cover, the space around the outer cover for turning the bolts and nuts, and the distance that the mandrel must travel. After calculating, we design three sorts of sleeves, as shown in Table 8.

Design for Mandrel. The design for the mandrel is controlled by the dimensions of the sleeve, the depth of the core, and the distance that the top cover moves. Three different mandrels are designed for use with the three sizes of sleeves.

One final refinement is added to the design of the mandrel. There is a B center hole at one end of the 1E electric motor shaft and a C center hole at the other end. In order not to damage the end faces of the shaft in the process of disassembling it, a copper finial is added to the tool. It is affixed on the end of the mandrel

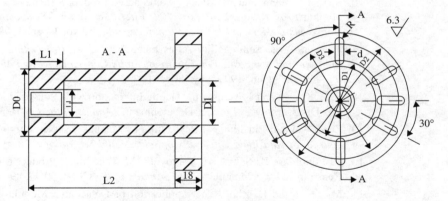

Figure 34 Design for the sleeve.

Table 8 The Specs of Different Sleeves

Electromotor Types	Dimension									
	D0	D1	D2	D3	D4	L1	L2	d	R	M
IE80–IE112	φ52	φ33	φ102	φ87	φ61	25	125	7	3.5	M20*1.5
IE132–IE200	φ83	φ64	φ200	φ160	φ96	50	230	10	5	M27*2
IE225–IE315	φ110	φ90	φ300	φ250	φ140	50	300	12	6	M30*2

such that the mandrel and the copper finial slide in opposite directions during the disassembly process. This prevents the end faces and the center holes of the shaft from being damaged. The convenience and effectiveness of this tool gives it perfect applicability.

REFERENCES

1. Gao Jiangang, Wu Ying, Xiangdong, Liu Xueping, "The Overview of Electromechanical Products Disassembly Research," *China Mechanical Engineering*, 40(7), 7 (2004).
2. Liu Guangfu and Liu ZhiFeng, "The Research of Disassembly Design System of Industry Product," *China Mechanical Engineering*, 40(7), (1998).
3. Liu Guangfu, Liu Zhifeng, and Li Gang, *Green Design and Manufacturing*, Mechanical Press, Peking, 2000.
4. S. K. Das, P. Yedlarajiah, and R. Narendra, "An Approach for Estimating the End-of-Life Product Disassembly Effort and Cost," *International Journal of Production Research*, 38(3):657–673 (2000).
5. S. M. Mok, C. H. Wu, and D. T. Lee, "A Hierarchical Work Cell Model for Intelligent Assembly and Disassembly," in *Proceedings, 1999 IEEE International Symposium on Computational Intelligence on Robotics and Automation*, Monterey, California, 1999, pp. 125–130.
6. S. Viswanathan, and V. Allada, "Value-based Product Structure Evaluation for Disassembly," in *Proceedings, EcoDesign '99: First International Symposium on Environmentally Conscious Design And Inverse Manufacturing*, Tokyo, Japan, 1999, pp. 778–783.
7. T. Suga, "Disassemblability Assessment for IM," in *Proceedings, EcoDesign '99: First International Symposium on Environmentally Conscious Design and Inverse Manufacturing*, Tokyo, Japan, 1999, pp. 580–581
8. F. L. Krause and J. Neumann, "Haptic Interaction with Non-rigid Materials for Assembly and Disassembly in Product Development," *Annals of the CIRP*, 50(1), 81–84 (2001).
9. Zhou Wei and Liu Jihong, "Realization of Manual Disassembly in Virtual Environment," *Journal of Huazhong University of Science and Technology*, 28(2), 45–47 (2000).
10. K. Fumihiko, K. Satoru, H. Tomoyuki, M. Takefumi, "Product Modularization for Parts Reuse in Inverse Manufacturing," *Annals of the CIRP*, 50(1), 89–92 (2001).
11. M. Zhou and H. Zhu, "The Application of Detachable Modular Design in Packaging-Related Machinery," Light Industry Machinery, Co., Ltd., 2001(4), 2–4.
12. H. C. Zhang and T. C. Kuo, "A Graph-based Approach to Disassembly Model for End-of-Life Product Recycling," in *1996 IEEE CPMT International Electronics Manufacturing Technology Symposium*, Austin, Texas, 1996, pp. 247–254.
13. T. Suzuki, T. Zanma, A. Inaba, and S. Okuma, "Learning Control of Disassembly Petri Net—An Approach with Discrete Event System Theory," in *Proceedings of the 1996 IEEE International Conference on Robotics and Automation*, Minneapolis, Minnesota, 1996, pp. 184–191.
14. K. E. Moore, A. Gungor, and S. M. Gupta, "Petri Net Approach to Disassembly Process Planning for Products with Complex AND/OR Precedence Relationships," *European Journal of Operational Research*, 135(2), 428–449 (2001).

15. E. Zussman and M. Zhou, "A Methodology for Modeling and Adaptive Planning of Disassembly Processes," in *IEEE Transactions on Robotics and Automation*, 15(1), 190–194 (1999).
16. E. Zussman, M.-C. Zhou, and R. Caudill, "Disassembly Petri Net Approach to Modeling and Planning Disassembly Processes of Electronic Products," in *Proceedings of the 1998 IEEE International Symposium on Electronics and the Environment*, 0ak Brook, Illinois, 1998, pp. 331–336.
17. Y. Pnueli and E. Zussman, "Evaluating the End-of-Life Value of a Product and Improving It by Redesign," *International Journal of Production Research*, 35(4), 921–942 (1997).
18. E. Zussman, A. Kriwet, and G. Selinger, "Disassembly-Oriented Assessment Methodology to Support Design for Recycling," *Annals of the CIRP*, 43(1), 9–14 (1994).
19. J-G. Kang, D-H. Lee, P. Xirouchakis, and J. G. Persson, "Parallel Disassembly Sequencing with Sequence Dependent Operation Times," *Annals of the CIRP*, 50(1), 343–346 (2001).
20. J. Gao, G. Duan, and J. Wang, "The Method of Disassembly AND/OR Graph in the Recovery Evaluation of Products Disassembly," *Chinese Journal of Mechanical Engineering*, 38(supplement), 26–31 (2002).
21. X. Cai, Y. Hu, and J. Hu, "The Confirmation of Maximum Return in the Process of Products Recovery, *Mechanical Science and Technology*, 20(1), 80–83 (2001).
22. S. Kanai, R. Sasaki, and T. Kishinami, "Representation of Product and Processes for Planning Disassembly, Shredding, and Material Sorting Based on Graphs," in *Proceedings of the l999 IEEE International Symposium on Assembly and Task Planning*, Porto, Portugal, 1999, pp. 123–128.
23. F. Touzanneand and C. Perrard, "Representation of Disassembly Processes in Order to Allow Time Evaluation of Their Performances," in *Proceedings of the 1999 IEEE International Symposium on Assembly an d Task Planning*, Porto, Portugal, 1999, pp. 135–140.
24. A. Gungor and S. M. Gupta, "Disassembly Sequence Planning for Products with Defective Parts in Product Recovery," *Computers and Industrial Engineering*, 35(1–2), 161–164 (1998).
25. M. R. Johnson and M. H. Wang, "Planning Product Disassembly for Material Recovery Opportunities," *International Journal of Production Research*, 33(11), 3119–3142 (1995).
26. Xue Shanliang, Ye Wenhua, and Wang Ningsheng, "The Optimization Research of Products Disassembly Configuration." *Mechanical Science and Technology*, 21(3), 400–402 (2002).
27. J. Hesselbach and K. Westernhagen, "Disassembly Simulation for an Effective Recycling of Electrical Scrap," in *Proceedings, EcoDesign '99: First International Symposium on Environmentally Conscious Design and Inverse Manufacturing*, Tokyo, Japan, 1999, pp. 582–585.
28. Y. Xu, R. Mattikalli, and P. Khosla, "Generation of Partial Medial Axis for Disassembly Motion Planning," in *Proceedings, 1991 IEEE International Conference on Systems, Man, and Cybernetics*, Charlottesville, Virginia, 199l, pp. 997–1003.
29. www.edw.com.cn/news/show.aspx?ClassID=60&articleID=9951.
30. www.sciencetimes.com.cn/col36/col78/article.htm1?id=49399.
31. Yao Dekang and Cheng Guoxiang, *Intelligence Material*, Chemical Industry Press, Beijing, 2002.

32. Yang Dazhi, *Intelligence Material and Intelligence System*, Tianjin University Press, Tianjin, 2000.
33. Du Shanyi, Leng Jinsong, and Wang Dianfu, *Intelligence Material System and Structure*, Science Press, Beijing, 2001.
34. Roope Takala and Pia Tanskanen (Nokia Research Center Helsinki, Finland), "Concept of a Mobile Terminal with Active Disassembly Mechanism," in conference publication of the International Electronics Recycling Congress, Davos, Switzerland, 2002.
35. K. M. Lee and M. M. Bailey-van Kuren, "Modeling and Supervisory Control of a Disassembly Automation Workcell Based on Blocking Topology," *IEEE Transactions on Robotics and Automation*, 16(1), 67–77 (2000).
36. R. Knoth, M. Brandstotter, B. Kopacek, and P. Kopacek, "Automated Disassembly of Electronic Equipment," in *Proceedings of the 2002 IEEE International Symposium on Electronics and the Environment*, San Francisco, California, 2002, pp. 290–294.
37. B. T. David, N. Boutros, K. Saikali, et al., "Automation of Disassembly Processes and Its Information Systems," in *Proceedings, EcoDesign '99: First International Symposium on Environmentally Conscious Design and Inverse Manufacturing*, Tokyo, Japan, 1999, pp. 564–569.
38. Y. Uchiyama, R. Fujisawa, Y. Oda, et al., "Air Conditioner and Washing Machine Primary Disassembly Process," in *Proceedings, EcoDesign '99: First International Symposium on Environmently Conscious Design and Inverse Manufacturing*, Tokyo, Japan, 1999, pp. 258–262.
39. E. S. Geskin, B. Goldenberg, and R. Caudill, "Development of Advanced CRT Disassembly Technology," in *Proceedings of the 2002 IEEE International Symposium on Electronics and the Environment*, San Francisco, California, 2002, pp. 249–253.
40. B. R. Zuo, A. Stenzel, and G. Scliger, "Flexible Handling in Disassembly with Screw Nail Indentation," in *Proceedings of the 2000 IEEE International Conference on Robotics and Automation*, San Francisco, California, 2000, pp. 3681–3686.
41. Y. Arai, Y. Oda, A. Sunarni, et al., "The Separating Method of Copper and Aluminum from Heat Exchanger in Air Conditioner," in *Proceedings, EcoDesign '99: First International Symposium on Environmentally Conscious Design and Inverse Manufacturing*, Tokyo, Japan, 1999, pp. 824–829.
42. P. G. Ranky, R. J. Caudill, K. Limaye, et al., "A Disassembly User Requirements Analysis Method, Tools and Validated Examples," in *Proceedings of the 2002 IEEE International Symposium on Electronics and the Environment*, San Francisco, California, 2002, pp. 128–132.
43. M. Sanou, A. Fujita, and H. Hoshina, "Environmental Assessment on Operating Condition of the Demonstration Plant for Recycling Post-Use Electric Home Application," in *Proceedings, EcoDesign '99: First International Symposium on Environmentally Conscious Design and Inverse Manufacturing*, Tokyo, Japan, 1999, pp. 285–287.
44. I. Stobbe, H. Griese, H. Potter, et al., "Quality Assured Disassembly of Electronic Components for Reuse," in *Proceedings of the 2002 IEEE International Symposium on Electronics and the Environment*, San Francisco, California, 2002, pp. 299–305.
45. T. Ishii, K. Kai, K. Shinoda, et al., "Integrated Recycle Plant for Electric Home Appliances—Automated Recognition and Sorting of Product Types, in *Proceedings, EcoDesign '99: First International Symposium on Environmentally Conscious Design and Inverse Manufacturing*, Tokyo, Japan, 1999, pp. 263–267.

46. Y. Inada, H. Yamamoto, S. Shibabta, et a1., "Advanced Transportation in the Dismantling/Recycling Factory-Development of the Flexible Transfer System," in *Proceedings, EcoDesign '99: First International Symposium on Environmentally Conscious Design and Inverse Manufacturing*, Tokyo, Japan, 1999, pp. 576–579.
47. V. S. Kouikoglou, "Optimal Rate Allocation in Unreliable Assembly/Disassembly Production Networks with Blocking," *IEEE International Conference on Robotics and Automation*, 2000, 16(4), 429–343.
48. Li Jianfeng, Chen Jian, Li Fangyi, Wang Yuling, "The Model of Electromechanical Products Disassembly and the Creation of Disassembly Lists," *The University of Shandong College Journal*, 34(5), (2004).
49. Li Jianfeng, Li Fangyi, Wang Jingsong, and Duan Guanghong, "Research of the Green Design Method in Present Condition and Outlook—the DFX Method Research," *The Aviation Manufacturing Technique*, 2004(11).
50. Jiang Dandong and Liu Guangfu, "The Structure of Disassembly Design Dystem," *Hefei University of Technology Journal* (natural science version), 22(4), (1999).
51. Zhang Jun, Hou Wenle, Ni Junfang, and Cai Jianguo, "A Research of Products Disassembly Modeling and Arithmetic," *Computer Application and Software*, 1999(6).
52. Pan Xiaoyong, XiangDong, and Duan Guanghong, "Electromechanical Products Disassembly Evaluation," *Applied Science and Technology*, August 2005.
53. Liu Zhifeng, "The Green Product Comprehensive Evaluation and an Analysis Method Researches," doctoral dissertation.
54. Xu Hongxiang and Chen Zhiwei, "The Application of SolidWorks in Disassembly-Oriented System," *Technique Normal School of Jiangsu Journal*, 10(2), (2004).
55. Yu Hong, "The Develop of Body, Core of the Water Meter Disassembly Equip," *China Calculate*, 2005(7).
56. Jiang Junbo and Wang Yujie, "A Bearing Disassembly Equip in Blind Hole," *Manufacture Technology and Machine Tools*, 2005(7).
57. Yang Xuefeng, Yang Weidong, and Zhao Bin, "A Convenience and Practical Bearings Outer Shell Disassembly Tool," *Nicety Manufacture and the Automation*, 2001(2).
58. Guo Qinghua, "The Design of the Disassembly Tools for 1E Series Electrical Engineering," *Flameproof Electrical Engineering*, 2004(3).

CHAPTER 10
INDUSTRIAL ENERGY EFFICIENCY

B. Gopalakrishnan, D. P. Gupta, Y. Mardikar, and S. Chaudhari
Industrial Assessment Center, Industrial and Management Systems Engineering, West Virginia University, Morgantown, West Virginia

1	**INTRODUCTION**	**265**	4.1 Furnaces, Ovens, Boilers, and Steam Systems	280
	1.1 Energy Usage in Manufacturing Industry	266	4.2 Motors	283
	1.2 Benchmarking Industrial Energy Consumption	266	4.3 Lighting	285
	1.3 Potential for Industrial Energy Conservation	266	4.4 Chillers	286
	1.4 Initiatives of the U.S. Department of Energy for Energy Efficiency	267	4.5 Heating, Ventilation, and Air Conditioning Equipment	286
			4.6 Air Compressors	287
	1.5 Economics of Energy Efficiency and Importance to Company's Bottom-Line	268	4.7 Combined Heat and Power (CHP)	288
	1.6 Environmental Benefits of Energy Efficiency	269	**5 CASE STUDIES OF DEVELOPMENT OF ENERGY-EFFICIENCY MEASURES**	**288**
2	**LITERATURE REVIEW**	**269**	5.1 Glass Manufacturing	289
3	**DATA ANALYSIS OF ENERGY EFFICIENCY MEASURES**	**271**	5.2 Plastics Manufacturing	289
			5.3 Metals Manufacturing	290
	3.1 Utility Bill Analysis	271	5.4 Chemicals Manufacturing	290
	3.2 Analysis of IAC Database	277	**6 CONCLUSION**	**291**
4	**ENERGY-EFFICIENCY MEASURES IN MAJOR ENERGY CONSUMING EQUIPMENT**	**280**	**7 ACKNOWLEDGMENT**	**291**

1 INTRODUCTION

Environmentally conscious manufacturing (ECM) has made significant gains on account of advances in energy efficiency. Energy efficiency relates to reducing wasted energy, hence reducing energy consumption. Utilization of fossil fuels adversely affects the greenhouse gases released into the atmosphere and results in undesirable quantities of emissions. Increasing energy efficiency will reduce the unwanted environmental effects produced by manufacturing.

1.1 Energy Usage in Manufacturing Industry

U.S. manufacturing plants, mines, farms, and constructions firms currently consume about 25 quads (quadrillion British thermal units, or Btu) of energy each year, which is about 33 percent of the nation's total consumption of energy.[1] A few of the cost-effective methods for improving energy efficiency are general housekeeping and maintenance programs, energy management and accounting programs, improved methods, and procedures for existing production methods and product changes. There are several technologies associated with heat recovery, high efficiency motors, variable speed drives, and cogeneration that have applications in many industries. Potential areas can be identified for energy conservation measures, namely lighting, boilers, motors and pumps, destratification of air, insulation of steam lines and process equipment, air and steam leaks, and compressors.

An important factor for industrial energy utilization and efficiency is the corporation's internal culture and external relationships. Capital investment in modern equipment usually enhances energy efficiency even when efficiency is not the primary purpose of the investment. An operations manager's key priorities are retaining efficiency of the production line. In this context, reducing energy costs tends to be a low priority.

1.2 Benchmarking Industrial Energy Consumption

The energy costs at a facility can be controlled by effective energy analysis, diagnostics, and management. Energy management is an important aspect for various manufacturing and service operations across U.S. industries to reduce operating costs.[2] The management in manufacturing companies is often interested in focusing on efficiency targets and comparing their performance with respect to established industrial practices. In this process, *benchmarking* is often used to identify performance levels with respect to the competition or the industry on an average. In terms of energy consumption in a manufacturing facility, benchmarking can be used effectively to judge whether there is an opportunity to implement significant energy efficiency measures. As an example, the energy use for a particular mill can be compared with that for similar mills or with that for a model mill representing the current best practice.[3]

1.3 Potential for Industrial Energy Conservation

A significant potential indeed exists for energy reduction in the manufacturing community. Energy efficiency and subsequent reduction can be achieved by focusing on the manufacturing processes, utilities that support these processes, and total productive maintenance procedures. Achieving energy efficiency depends on the technical aspects of operations, as well as the necessity for a cultural change in the viewpoint of management, in day to day operations.[4]

Industry's options for reducing energy costs were summarized in a study.[5] This study primarily identified energy efficiency measures that yield energy and environmental benefits for large volume, commodity and process industries. Prioritization of these opportunities based on various criteria incorporated in the study suggested 5.2 quadrillion Btu—21 percent of primary energy consumed by the manufacturing sector. These savings equate to almost $19 billion for manufacturers, based on 2004 energy prices and consumption volumes.[5]

1.4 Initiatives of the U.S. Department of Energy for Energy Efficiency

The major initiatives taken by the U.S. DOE toward industrial energy efficiency include the establishment of industrial assessment centers, BestPractices, plant-wide energy assessments, industrial systems initiatives, and Industries of Future (IOF) program. In order to help industries identify and then capture EEM (energy efficiency measures) and EEO (energy efficiency opportunities) as of 2006, the USDOE EERE/ITP has supported the establishment of 26 industrial assessment centers (IAC) throughout the country. The IAC are located in universities and avail themselves of engineering students under the guidance and supervision of professors to conduct facility resource assessments. The assessments then help industries identify measures they can take to harness EEM and EEO. The IAC have been funded by the U.S. Department of Energy (USDOE), Office of EERE, Industrial Technologies Program (ITP) since 1976.

The focus of the IAC has been toward reducing industrial energy consumption, mainly for small and medium-sized facilities. The no-cost industrial assessments performed by the IAC are available to manufacturing facilities in the SIC 20 to 39, provided that they have no in-house professional staff to perform the assessment, have gross annual sales below $100 million, have fewer than 500 employees at the plant site, have annual utility bills more than $100,000 and less than $2.5 million, and be located within 150 miles of the IAC. The energy assessment process includes the analysis of the utility costs and on-site data gathering using effective instrumentation. The discussion with the plant personnel adds to the practicality of the recommendations and the use of the proper support data. One additional benefit from the program is that the data generated by the assessments provides a unique opportunity to quantify the state of energy, waste and productivity management in small and medium sized manufacturing facilities and the potential of the assessment process to improve efficiency. Since 1981, the data have been compiled from the assessment performed under this program.[6]

Another initiative started by U.S. DOE is BestPractices, a program area within the Industrial Technologies Program (ITP) that supports ITP's mission to improve the energy intensity of the U.S. industrial sector through a coordinated program of research and development, validation, and dissemination of energy-efficient technologies and practices. BestPractices develops, implements, and disseminates best practices in energy management. The manufacturers' participation in BestPractices program can earn the organization technical, financial, and engineering personnel support for achieving energy efficiency. The BestPractices program has provided invaluable software for industrial use known as decision tools for industry. These decision tools include tools such as Process Heating Assessment Tool (PHAST), Steam System Assessment Tool (SSAT), Steam System Scoping Tool (SSST), AirMaster+, MotorMaster+, PEP, and 3Eplus.

Plant-wide energy assessments help large-sized industries to save energy. Plants are selected on a competitive basis and agree to share costs of energy assessment spaced throughout a year. A maximum award of $100,000 can be granted to a plant in this type of assessment, and equivalent or greater cost is shared by the selected company.[7] Since this assessment is spaced throughout a year, it is a comprehensive energy consumption study and leads to significant EEM.

Industrial energy systems account for 80 percent of energy consumed by industry. Industrial energy systems is an initiative by which research and development efforts are focused on improvements in basic systems such as steam, process heating, compressed air, motors, pumps, fans, and combustion. This area provides potential of 10 to 20 percent reduction in energy usage, which converts approximately to 1.6 to 3.2 quads.[8]

The U.S. DOE has developed the Industries of Future (IOF) program to develop energy-conservation opportunities in energy-intensive industries such as aluminum, steel, chemicals, metals casting, glass, mining, forest products, and petroleum and refining. Each project under this category meets rigorous criteria of industry participation, cost sharing, and relevance to industry-defined priorities. To date, the savings from this initiative have totaled to 1.6 quadrillion Btu, or $6.5 billion.[9]

1.5 Economics of Energy Efficiency and Importance to Company's Bottom-Line

Most manufacturing facilities seek a two-year payback on investment in terms of any projects. It is important that the projects that promote energy efficiency be practical and that they afford economic desirability in terms of implementation by the manufacturing facilities. U.S. manufacturing must extract the highest percentage of productive energy from each Btu of energy produced, especially since it has a high per-capita consumption of energy in the industrial sector. This is necessary for competing in the global marketplace. A significant portion of the operating costs of any manufacturing or service industry is in the form of energy costs. Hence, energy conservation and management is an important activity within any manufacturing or service organization, as it can reduce operating costs. The ability of a company to compete for business in an international scale is very important, especially in today's global business environment. Energy management is an important function within any manufacturing or service organization as it pertains to operating costs. Any reduction in operating costs is bound to increase the competitive edge of the industry.

1.6 Environmental Benefits of Energy Efficiency

Burning fossil fuels generates CO_2, NOx, SOx, and other gases that create an environmental problem. For each kWh of electricity that is reduced, the CO_2 emissions are reduced by as much as 2.19 pounds. Similarly, for every MCF (1,000 cubic feet) of natural gas that is not burned, the CO_2 emissions are reduced by approximately 113 pounds. This indicates that significant emissions reductions are possible due to energy efficiency measures implemented in manufacturing facilities.

2 LITERATURE REVIEW

Since the mid-1990s, manufacturing plants in the United States have been undergoing a major shift with measures to reduce energy consumption. The U.S. Department of Energy's Office of Energy Efficiency and Renewable Energy (EERE) has encouraged the efforts through its various programs.

A key feature of any energy conservation program is the energy assessment.[10] Management commitment is the most important thing in an overall sense.[11] The methodology for conducting an energy assessment that attempts to obtain a comprehensive view of the energy utilization in all forms is important.[12] It addresses the issue that many of the energy assessments conducted by companies have the potential to overlook substantial energy conservation opportunities. It describes how plant performance is estimated based on limited data. According to the Energy Auditor's Handbook,[13] the energy audit is a brief, on-site survey and analysis of a plant and its energy usage patterns, identification of opportunities to save energy through the implementation of operational and maintenance changes, and an assessment of its need for implementation of energy conservation measures. Energy auditing is the first phase in improving the energy efficiency of a facility.[14] Preparation of the energy management report, implementing, and monitoring are the other three phases. The implementation of energy efficient technologies among manufacturing firms can lead to short-term as well as long-term benefits.[15] There is also an environmental benefit from projects, the reduction of fossil-fuel consumption. For example, the Pacific-Northwest region pursues energy conservation as a source of energy.[16] When energy problems caused by the rapid increase in demand in the face of dwindling fuel supply first became apparent, the immediate response was to seek new supplies and alternative fuels.[17]

A rational approach to targeting energy costs and outlining the key strategies in energy assessment has been specifically determined for iron foundries.[18] Large metal-casting foundries are excellent targets for energy saving opportunities and should be targeted for continuous technological improvements. The nature of energy conservation opportunities that exist in glass manufacturing facilities are varied.[19] The energy management program pursued by a specialty chemical manufacturer has shown significant benefits.[20] The strategy used in a chemical plant to reduce energy consumption is unique in terms of process and utility parameters.[21] Effective instrumentation is needed to facilitate a chiller plant energy efficiency monitoring as a prerequisite to any efficiency improvement initiatives for centrally air-conditioned buildings.[22]

As mentioned earlier, the U.S. DOE Industries of the Future (IOF) program includes industries such as agriculture, aluminum, chemical, forest products, glass, metal casting, petroleum refining, and steel, and these industries consume around 33 percent of all energy used in the United States.[23,24] These industries have developed strategies to forecast growth and have identified the tools that are needed to increase their competitiveness. The plant-wide assessment at a paper mill identified five opportunities resulting in overall savings of over $1.5 million.[25] Assessments at the world's leading aluminum manufacturing plant identified $60 million in annual savings opportunities companywide.[26] Energy-saving opportunities included improved heat recovery, better furnace operations, and development of process-energy use targets.

Since DOE began the plantwide energy assessments program, several plants have submitted successful proposals.[27] The total energy savings that has been realized through this program has been significant and has saved millions of dollars.[27] Only when the energy problems faced in the plant are reviewed in a total perspective do the solutions and the various conservation opportunities become visible. This is called the *systems engineering approach*. In a systems approach, the systems are evaluated macroscopically first and then micro studies are initiated only within the framework of the macro study to ensure the effectiveness of the assessment.

The effectiveness with which energy resources are converted into work is often termed as *energy efficiency*.[5] One example of a metric used often in evaluating heat producing systems is thermal efficiency. Factors that have an effect on thermal efficiency include completeness of combustion, heat balance, and operating system characteristics.[28] The manufacturing facilities have varied overall energy efficiency that depends on operating-system characteristics, production aspects, and technological considerations. Efficiency can be influenced by the design aspects of the manufacturing equipment, as well as by the extent of use. In the manufacturing sector, these energy losses amount to several quadrillion

Btus (quadrillion British Thermal Units, or quads) and billions of dollars in lost revenues every year.[29]

3 DATA ANALYSIS OF ENERGY EFFICIENCY MEASURES

This section describes the energy cost components for the energy bills and the analysis of the IAC database. Every manufacturing facility needs to understand how it is being billed for electricity and natural gas. Many facilities do not understand their rate structure, especially when it comes to charges for electrical billing demand. In terms of natural gas, many facilities may be placed in a position to pay high rates based on the type of contract they negotiate.

The Industrial Assessment Center database is extensive and contains elaborate information on all energy assessments conducted by the various IAC since 1981. The analysis of the database can reveal interesting energy profile and savings potential for various types of manufacturing facilities. The database is also a treasure house of knowledge in terms of energy conservation opportunities and the economic impact potential of these opportunities in terms of industrial implementation.

3.1 Utility Bill Analysis

The utility bill for a facility may have different components based on the need of the processes. Typically, the facilities have electricity and natural gas as major energy sources, and some of them may have other utilities such as oil, coal, wood, and water. Since electricity and natural gas are common components in most utilities, this section will be focused on discussion of these utilities. In general, all the facilities will have a contract with the utility company that will dictate the price per unit of electricity and natural gas. Many a time, the facility may have two or more rate schedules to choose from. Based on the data collected from different facilities, analysts may be able to study monthly or yearly usage pattern to find significant savings in terms of utility charges. In some cases, the facility may have done the analysis and chosen the best possible rate but over time, because of change in the usage pattern, the current rate schedule may not be optimal. In other cases, the utility company may decide the rate based on the maximum need of the facility without considering the usage pattern, which may result in more utility cost to the facility. The following section will discuss some key components in electricity and natural gas energy bills.

Components in Electricity Energy Bills
The required energy cost mainly has three components[30]: (1) fixed costs (consumer/customer charges, administrative costs), $/month; (2) electricity costs—that is, the real cost of electricity that is consumed by the process (variable costs); and (3) demand costs—that is, the cost of maintaining a level of energy to run

the operation (investment costs), $/kW. Fixed costs (consumer/customer charges, administrative costs) have no direct relation to the amount of energy used during the billing period. They cover the expenses in readings, accounting, and billing by the power supplier company, which is fixed in any billing period. In general, this cost component is insignificant in comparison with the energy and demand charges. The major components in the electricity bills are electricity (or usage) and demand cost.

Energy Charges

Energy charge is based on the direct consumption of the electricity in terms of kWh (kilowatt-hours) during the electricity consumption period. The kWh value is multiplied by the energy charges per unit ($/kWh) for the total bill in the billing cycle. These charges may vary based on the service provider, voltage, and energy consumption during each billing cycle.[31–35] A summary of different rate structures may be given as a flat rate for each kWh consumed by the facility, a variable rate based on the time of day during which the electricity is consumed, a variable rate based on the time of the year during which the electricity is consumed, a flat or variable rate with low power factor penalty, or a flat or variable rate based on the total amount of power (kVA or kW) consumed.

Demand Charges

This charge is to compensate the utility company for the capital investment required to serve peak loads, even if that peak load is used only for partial operating period. The demand is measured in kW (kilowatts) or kVA (kilovolt-amperes). These units are related to the energy (kWh) consumed in a given time interval of the billing period. The demand periods vary with the type of energy demand. The high fluctuating demand has a short demand period, which can be as short as five minutes, but is generally 15, 30, or 60 minutes long.[35–37] The utility companies use the period with the highest average demand for determining the demand charges in any billing cycle.

It should be noted that not all the utility companies charge their customers based on energy and demand. Also, there is no specific ratio or number of utility companies that charge based on energy only and do not include demand in their bills. The calculation of the demand can be explained with a simple example. Assume that the demand pattern for any particular process is as given in Table 1. The average demand charged to the facility for this 15-minute interval can be calculated as follows:

$$\text{Demand charged (kW)} = ((10 \times 2) + (12 \times 4) + (2 \times 7) + (10 \times 2))/15$$

$$= \text{kW}$$

Similarly, the demand will be calculated for each of the 15-minutes intervals. Finally, the facility will be charged for the maximum of all these calculated

Table 1 Example Demand for a 15-minute Interval

Demand (kW)	Time Units for This Demand
10	2
12	4
2	7
10	2
Total	**15**

values for each 15-minute interval during the billing period. It may be noted that in some cases, the demand rate is also a variable charge based either on the time of the day or time of year.[31–35] Finally, the energy charge may be based on the demand (kW or kVA).

Rate Schedules

We can compute an energy bill for a sample facility. The energy rate structure for the facility is as follows:

Minimum monthly charges $2.75/kVA of demand

Energy charges

    $0.10599 for first 30 kWh per kVA of demand

    Next 170 kWh per kVA of demand:

        $0.06965/kWh for first 2,000 kWh

        $0.05883/kWh for next 8,000 kWh

        $0.04867/kWh for next 90,000 kWh

        $0.04175/kWh for all over 100,000 kWh

    For kWh in excess of 200 times kVA:

        $0.03341/kWh for first 200,000 kWh

        $0.02855/kWh for all remaining kWh

Average fuel cost adjustment (FCA) $0.01902/kWh

Average tax on kWh   8.27% of the kWh cost

Note: The average FCA and tax rates may be calculated from the energy bills or can be obtained directly from the utility company.

Example of the Electricity Cost Calculation

As an example, for one of the billed months, the readings are assumed to be:

$$\text{Total energy} = 201,800 \text{ kWh}$$
$$\text{kVA recorded} = 416 \text{ kVA}$$

Therefore, the cost of electricity can be calculated as follows:

Minimum monthly charges = $2.75/kVA × 416 kVA = $
Energy charges
  $0.10599 for first 30 kWh per kVA of demand
    = $0.10599/kWh × 30 kWh/kVA × 416 kVA   = $1,322.76
  Next 170 kWh per kVA of demand (or 170 kWh/kVA
                                        × 416 kVA   = 70,720 kWh)
    first 2,000 kWh : $0.06965/kWh × 2000 kWh   = $139.30
    (*Note: remaining kWh in this bracket is*
              70,720 − 2,000 = 68,720 kWh)
    next 8,000 kWh : $0.05883/kWh × 8,000 kWh   = $470.64
    (*Note: remaining kWh in this bracket is*
              68,720 − 8,000 = 60,720 kWh)
    next 90,000 kWh : $0.04867/kWh × 60,720 kWh   = $2,955.24
    (Note that only 60,720 kWh was available
       in this bracket)
    (*Note: remaining kWh in this bracket is*
              60,720 − 60,720 = 0 kWh)
    all over 100,000 kWh : $0.04175/kWh × 0 kWh   = $0.00
  For kWh in excess of 200 times kVA [201,800 kWh −
    (200 kWh/kVA × 416 kVA) = 118,600 kWh]
    (Note that only 118,600 kWh is available
       in this bracket):
    first 200,000 kWh = $0.03341/kWh × 118,600 kWh   = $3,962.43
    (*Note: remaining kWh in this bracket is* 118,600 −
       118,600 = 0 *kWh*)
    all remaining kWh: = $0.02855/kWh × 0 kWh   = $0.00
  Sub-total cost = $1,322.76 + $139.30 + $470.64
                 + $2,955.24 + $0 + $3,962.43 + $0
                 = $8,850.37
Average fuel cost adjustment
    (FCA) = $0.01902/kWh × 201,800 kWh   = $3,838.24
Average tax on kWh = 8.27% of $8,850.36   = $731.92

Therefore, the total cost (TC) for the month is calculated as

$$TC = \text{Subtotal cost} + \text{FCA} + \text{Tax}$$
$$= \$8{,}850.37 + \$3{,}838.24 + \$731.91$$
$$= \$13{,}420.53$$

It may be noted that the FCA rate may change from month to month. The same calculation method will be used to calculate the cost savings from any recommendation. The energy savings will be calculated in terms of kWh and kVA (derived from the kW savings as ratio of kW to the average power factor of 0.9788).

Example of Energy Cost Savings Calculation

The energy cost savings calculations are based on the average monthly value of energy and demand usage. As an example, for one of the manufacturing facilities, based on the energy bills for 12 months in 2005, the average energy used was 222,533 kWh and the average demand usage was 475 kVA. Based on the values in the bills, the average FCA rate is assumed as \$0.01902/kWh and the average tax will be taken as 8.27 percent. As discussed earlier, the total average cost of the electricity per month is calculated as \$14,869.65. To calculate the savings from any recommendation, the savings in the energy and demand usage will be calculated in terms of kWh/yr and kW-month/yr. As an example, if the proposed savings are 10,000 kWh/yr and 50 kW-month/yr (or a reduction of 4.26 kW in the peak demand in each month), the new average usage can be calculated as follows:

$$\text{Proposed average kWh} = (\text{Current average} - \text{Proposed savings})/\text{Month}$$
$$= (222{,}533 \text{ kWh}) - [10{,}000 \text{ kWh/yr}/12 \text{ months/yr}]$$
$$= (222{,}533 - 833.33) \text{ kWh/month}$$
$$= 221{,}699.7 \text{ kWh/month}$$

$$\text{Proposed average kVA} = (\text{Current average} - \text{Proposed Savings})/\text{Month}$$
$$= (475 \text{ kVA}) - [50 \text{ kW/yr}/(12 \text{ months/yr}^*\text{PF})]$$
$$= (475 - 4.26) \text{ kVA/month}$$
$$= 470.74 \text{ kVA/month}$$

(*Note*: PF is the average power factor is used to convert the kW to kVA.)

Based on the proposed average kWh and kVA and the method of energy cost calculation discussed in cost calculation section, the new cost of electricity is calculated as $14,801.66. Therefore the savings can be calculated as

$$\text{Savings} = (\text{Current cost} - \text{Proposed cost}) \times 12 \text{ months/yr}$$
$$= (14,869.65 - 14,801.66) \times 12$$
$$= \$816/\text{yr}$$

Components in Natural Gas Bills

The utility bills for the natural gas are rather simple as compared to the electrical energy bills. In general, the units for the measurement of natural gas consumption are based on volume or energy. Some common units in these categories are:

Volume.

Cf: Cubic foot $\cong$ 1,030 Btu
Ccf: One hundred cubic feet
Mcf: One thousand cubic feet
Tcf: Trillion cubic feet $\cong$ one quad.

Energy.

DTH: One deca-therm = 10 TH $\cong$ 1MMBtu
TH: One therm $\cong$ 100,000 Btu
MMBtu: One million Btu
MBtu: One thousand Btu

The total cost of the natural gas for a billing period may have one or more components and depends on the supplier and amount of natural gas usage in the facility. Some common billing terms used in the natural gas bills are:

- *Gas usage rate:* Usage rate determined by the utilities ($/unit of gas)
- *Standby charges:* Charges to maintain a constant supply of gas to the facility ($)
- *Customer charge:* Covers the cost of billing, meter reading, and equipment ($)
- *Purchase gas adjustment (PGA):* To recover the cost (fuel cost) of purchasing natural gas ($/unit of gas)
- *Transportation charges:* To recover the cost of transportation of gas to the customer site ($/unit of gas)
- *Taxes:* Applicable taxes on the total bill for the facility ($/unit of gas)

An example of a natural gas bill is given in Figure 1. It may be noted that the standby charges for this facility is much more than the usage cost of the gas. This facility was using the gas company as a standby to their gas wells, from which they were getting the natural gas for their process use. It is evident that

ACCOUNT NUMBER	PREVIOUS BALANCE	PAYMENT RECEIVED	CURRENT BILLING	TOTAL DUE	TARIFF MCF	TRANS-PORT MCF	TOTAL MCF
	2,005.83	2,005.83	1,835.00	1,835.00	4.00	0.00	4.00

ACCOUNT NO.:		INVOICE DETAIL			BILLING MONTH: APR 05
CONTRACT	DESCRIPTION		BASIS		DOLLARS
	COMMODITY CHARGE		4.00	MCF	31.31
	CUSTOMER CHARGE		1.00	MON	18.38
	DELIVERY CHARGE		4.00	MCF	12.86
	GAS COST ADJUSTMENT		4.00	MCF	5.58
	STANDBY SERVICE		1.00	MON	1,661.82
	TAX SURCHARGE		1,729.95	DOL	1.21
	TRANSITION COST		4.00	MCF	0.00
	TAX SURCHARGE		4.00	MCF	0.03-
	PA SALES TAX — 100% TAXABLE		1,731.13	DOL	103.87
	TOTAL CURRENT BILL		4.00	MCF	1,835.00
INCLUDES	0.00 OF LATE FEES IN PREVIOUS BALANCE				0.00
	LATE PAYMENT CHARGES				0.00
	TOTAL DUE				1,835.00

Figure 1 Example of natural gas bill with high standby charges.

the usage pattern of the utilities should be considered properly to minimize the unnecessary charges.

3.2 Analysis of IAC Database

The IAC perform industrial assessments for small and medium-sized manufacturing companies to identify opportunities to save energy, reduce waste, and increase overall plant productivity. The program maintains a database of all the recommendations from assessments performed by the IAC centers and currently has a listing of 12,987 assessments and 97,205 recommendations.[38] This database is continuously updated after any of the centers upload the report from their assessment. The database has many features to find any recommendation that the user may be interested in. The assessments/recommendations can be filtered based on manufacturer codes (NAICS or SIC), IAC center, year of energy assessment, annual sales of the facilities, number of employees, annual energy cost, and state of facilities.

Recommendations
To facilitate the search for a particular recommendation, the database has assigned special codes, called as Assessment Recommendation Code (ARC), for each of the recommendations. The ARC contains six numbers with a format as X.YYYY.Z. The first number is designated for the recommendation type, the next four numbers detail the actual recommendation within the main category,

and the last number is the application of the recommendation (e.g., building, process, or other applications). All the recommendations are mainly categorized as energy management, waste minimization/pollution prevention, and direct productivity enhancements. Each of these categories have lists of subcategories based on the energy systems. The general categories of these recommendations are as follows.

Energy Management
The recommendations related to energy management deal with combustion systems, thermal systems, electrical power, motor systems, industrial design, maintenance and equipment control, lighting, space conditioning, administrative and other costs, and alternative energy sources such as wind and solar energy.

Waste Minimization/Pollution Prevention
This includes the recommendations related to reduction of waste and therefore pollution through operations, equipment, maintenance, raw materials use, water use, and disposal.

Direct Productivity Enhancements
The example recommendations in this category are manufacturing enhancement, purchasing, resource optimization, reduction of downtime, management practice, and other administrative savings. The top five recommendations in the list are shown in Table 2.

Implementation
The IAC program keeps a track of the recommendations implemented in the facilities, which are called six to nine months after the assessment date. This information is reported to the IAC database and is accessible to the public. The program has an excellent implementation record of more than 50 percent implementation rate for its recommendations. Table 3 lists the top five recommendations that were implemented the most by the manufacturing facilities. It may be noted that the same recommendations that were recommended the most have been implemented the most as well.

Table 2 Top Five Recommendations

Recommendation	Number of Times Recommended
Utilize higher-efficiency, lower-wattage lamps or ballasts	8,189
Eliminate leaks in inert gas and compressed air lines	4,828
Use most efficient type of electric motors	4,372
Install compressor intakes in coolest locations	3,860
Utilize energy-efficient belts and other improved mechanisms	3,227

Table 3 Top Five Implemented Recommendations

Recommendation	Number of Times Implemented
Utilize higher-efficiency, lower-wattage lamps or ballasts	4,700
Eliminate leaks in inert gas and compressed air lines	4,020
Use most efficient type of electric motors	2,877
Install compressor intakes in coolest locations	1,911
Utilize energy-efficient belts and other improved mechanisms	1,815

Energy Savings

Based on the analysis of the database, the total amount of energy savings from all the recommendations in terms of electricity is over 2 billion kWh and natural gas is over 38 million MMBtu. It is worth mentioning that the database has the records for many other fuel sources such as #2 oil, coal, and wood, but have not been considered here because electricity and natural gas are main energy sources in most of the industries visited through the IAC program.

Cost Savings

The IAC program has recommended savings over $1.5 billion. The average savings per assessment is estimated as $55,000. The database lists the energy savings for each recommendation in terms of the utilities (electricity, natural gas, oil, coal, and water, etc.) related to the process. The total annual cost savings from recommendations related to electrical energy exceed $120 million, and natural gas savings recommendations result in an annual savings of more than $180 million. Top ten recommendations and their cost savings are shown in Table 4.

Table 4 Top Ten Cost Savings Recommendations

Recommendation	Total Savings ($)
Add equipment/operators to reduce production bottleneck	88,846,940
Replace existing equipment with more suitable substitutes	44,651,442
Install automatic packing equipment	40,672,093
Use a fossil fuel engine to cogenerate electricity or motive power; and utilize heat	37,876,660
Utilize higher-efficiency lamps and/or ballasts	35,925,118
Use waste heat to produce steam to drive a steam turbine generator	24,659,543
Eliminate leaks in inert gas and compressed air lines/valves	24,504,452
Replace electrically operated equipment with fossil fuel equipment	23,947,776
Change procedures/equipment/operating conditions	23,336,815
Repair and eliminate steam leaks	22,959,756

Payback on Investment

Based on the total cost savings and the implementation cost of the recommendations, the average payback on the implementation cost for all the recommendations is estimated as 4 months. It may be noted that some of the recommendations do not require any implementation cost and thus result in an immediate payback, while others may have some implementation cost in terms of capital or labor cost investment. For most of the recommendations, the average payback is within two years.

Reduction in Emissions

The recommendations during the assessments not only save money for the facilities but result in reduction in emissions. Based on the analysis performed in the energy savings section, the reduction in CO_2 emissions through electricity and natural gas recommendations are estimated as approximately 9 billion pounds.

4 ENERGY-EFFICIENCY MEASURES IN MAJOR ENERGY CONSUMING EQUIPMENT

This section discusses a sample of the important energy-saving opportunities for major energy-consuming equipment commonly used in facilities (e.g., boilers, furnaces, ovens, electric motors, lighting, chillers, cooling towers, heating ventilation and air conditioning, air compressors, and combined heat and power systems). Description for assessing equipment, data collection requirements, related theory, necessary modifications in the equipment setup, implementation cost, and benefits of implementation are discussed further. U.S. DOE's BestPractices tools can be used to determine the energy savings in several areas of equipment usage.

4.1 Furnaces, Ovens, Boilers, and Steam Systems

Furnaces and ovens are equipment used to heat air, material, or other fluids. Common applications can be melting, annealing, preheating, heat treatment, and baking. Boilers produce hot water or steam under pressure in a closed vessel used in various processes.[39] Furnaces/ovens/boilers burn different types of fuel to generate heat, which is then used for the desired purpose. Following are some of the energy-saving opportunities that can be realized for furnaces/ovens/boilers.

Insulate the Bare Surfaces of Ovens, Furnaces, Boilers, and Steam Pipes

The bare surfaces of crucible furnaces, reverberatory furnaces, electric ovens, pit ovens, and boiler and steam pipes, if not insulated, will radiate significant energy due to lack of insulation. These surfaces can be insulated, thereby reducing the heat losses. Energy savings is a function of the difference in heat loss from the bare and insulated surface, area of the surface that needs insulation, annual operating hours of the equipment, and the efficiency of the heat supply. The DOE BestPractices tool 3E Plus can be used effectively for this energy-efficiency measure.[40]

Preheat Natural Gas Oven/Furnace/Boiler Combustion Air Using Hot Flue Gas

Significant amount of waste heat is available in the oven/furnace/boiler stack. This heat can be recovered by preheating the combustion air. A heat exchanger along with the necessary ductwork can be installed to preheat the intake air of the natural gas oven/furnace/boiler. The combustion air can be preheated and directed into the natural gas–fired oven/furnace/boiler air intake port using galvanized steel insulated ductwork. Warmer combustion air leads to increase in combustion efficiency. An increase in efficiency of approximately 1 percent is possible for every 40°F increase in combustion air temperature.[41] The existing combustion efficiency of the oven/furnace/boiler can be determined by using a stack gas analyzer. The energy savings to be realized by directing preheated air to the oven/furnace/boiler air intake is a function of the rating of the oven/furnace/boiler, the annual operating hours, and the existing and proposed combustion efficiencies of the oven/furnace/boiler.[42] The USDOE's BestPractices software tool PHAST can be used to determine the improvement in efficiency by preheating the combustion air.[40]

Recover Waste Heat from Gas Ovens/Furnaces/Boilers

Air-to-air type heat exchangers can be used to capture heat from the exhaust gases from the gas oven/furnace/boiler to heat the plant area during winter. By recovering the waste heat from the oven/furnace/boiler stack, the plant space heating system will be used for less time. The recovered heat would be used to heat the plant area during the heating season. The outlet temperature for the hot side of the gas oven/furnace/boiler can be determined using a stack gas analyzer or an infrared temperature gun in case the stack is not insulated. Outside fresh air or plant air can be used on the cold side of the heat exchanger. The potential annual space heating energy savings is a function of the exhaust gases' mass flow rate from the stack, specific heat of exhaust gases, temperature of exhaust gases from stack before heat exchanger, minimum practical exhaust temperature after heat exchanger,[41] operating hours during the heating season, and efficiency of the heating system.[42]

Adjust Air-Fuel Ratio on the Natural Gas Oven/Furnace/Boiler

Adjust the combustion system air-fuel ratio for the oven/furnace/boiler to reduce the amount of excess air passing through the oven/furnace/boiler and to improve the combustion efficiency of the oven/furnace/boiler. Based on the field combustion test conducted on the oven/furnace/boiler, it can be determined whether they need air-fuel ratio adjustment. Combustion test on the oven/furnace/boiler can be done using a stack gas analyzer. The optimum amount of O_2 in the flue gas of a natural gas-fired oven/furnace/boiler is 2.2 percent, corresponding to 10 percent excess air. In practice, a reduction in excess O_2 to about 3.5 percent (about 18 percent excess air) can be achieved. This estimate is based on discussions with several local oven/furnace/boiler adjustment contractors, who have indicated that a reduction in excess O_2 to less than about 3 percent is difficult for most oven/furnace/boiler. The corresponding CO_2 value for an O_2 value of 3.5 percent should be about 9.8 percent, as seen in Table 5, assuming no change in stack gas temperature before and after the adjustment. Combustion efficiency will increase due to the adjustments on the air-fuel ratio.[42]

The energy savings is a function of the rating of the oven/furnace/boiler, the annual operating hours of the oven/furnace/boiler, the firing factor of the oven/furnace/boiler, and the existing and proposed combustion efficiencies of the oven/furnace/boiler.[42] The USDOE's BestPractices tool PHAST can be used to determine the improvement in efficiency by adjusting the air-fuel ratio.[40]

Inspect, Repair, and Maintain Steam Traps

Steam traps are used to separate the steam from condensate, air, and other non condensable gases.[42] Institute a permanent steam trap management program. The program should include inspecting the steam traps, cleaning the traps, replacing the trap disc or lever if not working properly, and replacing the trap itself if it is defective. This program could be implemented by the plant maintenance staff. Steam traps are the key to an efficient steam system. The objectives of the steam traps are to remove condensate, air, and other noncondensable gases, and prevent steam loss. Efficient operation of any steam system requires well-designed trapping, which is periodically inspected and properly maintained. It is only in this way that condensate and air will be removed automatically as fast as they accumulate without wasting steam.[43] The traps can be tested using an ultrasonic testing device. Common outcomes of steam trap testing are continuous blowing, partial or complete blockage, or the trap functions properly. Energy savings is thus a function of the total steam loss from steam traps, heat

Table 5 Optimal Percentages of O_2, CO_2, and Excess Air in the Exhaust Gases

Fuel	O_2 (%)	CO_2 (%)	Excess Air (%)
Natural gas	2.2	10.5	10
Liquid petroleum fuel	4.0	12.5	20
Coal	4.5	14.5	25
Wood	5.0	15.5	30

content of saturated steam, specific heat of water at constant pressure, feed water temperature, boiler overall efficiency, and annual operating hours for which the steam source operates.[42] The USDOE's BestPractices tool SSAT can be used to determine the losses through steam traps.[40] The inputs to the tool are mainly the steam pressure, total number of steam traps in use, number of steam trap failures, boiler fuel source, and the number of steam trap failures after repairs. Based on these inputs to the tool, savings in fuel and makeup water cost can be determined.

Other Steam-related Energy-efficiency Measures

The other steam-related energy-efficiency measures can be listed as follows: (1) install reverse osmosis unit to reduce boiler blow-down rate; (2) inspect and repair steam leaks; (3) implement a condensate recovery system; (4) install waste heat boiler to produce steam; (5) install de-aerator unit to reduce steam system corrosion; and (6) install turbulators on fire tube boilers.[40]

4.2 Motors

Electric motors are used to convert electrical energy into mechanical work.[42] There is a drop in motor efficiency with its use. The following energy-saving opportunities discuss ways to reduce transmission losses and improve motor efficiency.

Implement a Motor Management System

Implement a motor management system (MMS) such as MotorMaster?™ to help document motor inventory and to identify/analyze motor-driven systems for various energy-conservation opportunities. Implementing the motor management system requires the facility to first obtain MotorMaster™ software, which can be downloaded at no charge from the U.S. Department of Energy (USDOE's) Web site.[40]

After obtaining the software, the user can develop an inventory of motors organized by facility, department, and process. Information such as nameplate data, load profile, field measurements, and energy cost can be stored for each motor so that motor performance can be tracked. A motor management system is designed to assist the facility to reduce energy costs through maximizing production efficiency, minimizing energy consumption, correcting for power factor, understanding utility billing statements, and establishing a preventative and predictive maintenance program. MotorMaster™ software (Figure 2) is capable of analyzing the energy and cost impacts of various hypothetical situations that occur either before and/or after a motor failure. These situations include downsizing to a smaller motor, rewinding a failed motor, replacing a failed motor with same size motor, and replacing a failed motor with a larger spare motor.

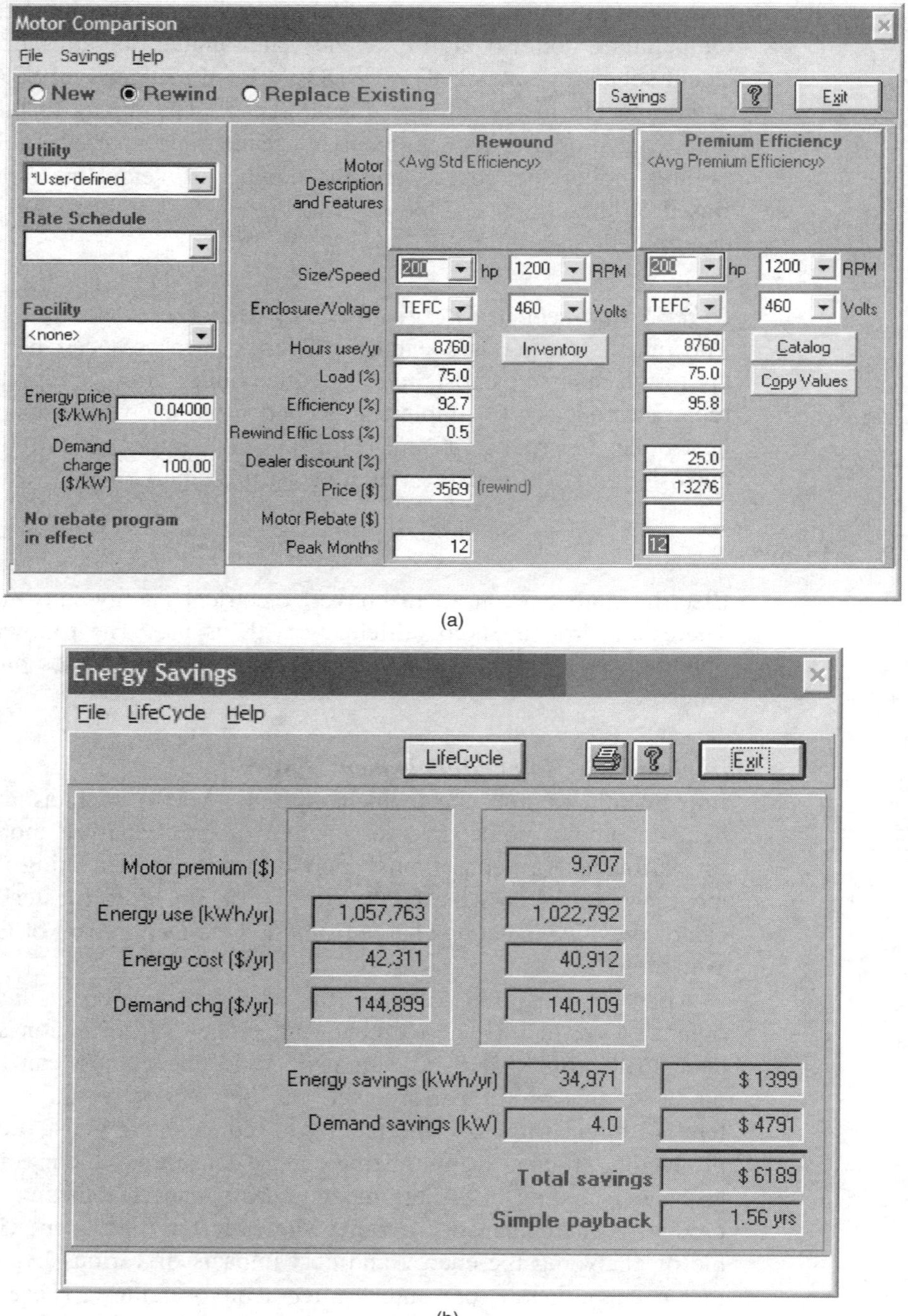

Figure 2 U.S. DOE's BestPractices Tool MotorMaster™ for determining energy savings through use of energy-efficient motor.

Other Motor-related Energy-efficiency Measures

The other important energy-efficiency measures as related to motors are: (1) perform vibration analysis on equipment; (2) replace drive belts on motors with energy-efficient cog belts; (3) maintain properly sized motor system; and (4) install variable-speed drive on the motors.

4.3 Lighting

Lighting systems use considerable amount of electrical energy to produce desired illumination levels.[39] Lighting technology has made significant developments with time and provides state of the art energy solutions.[42]

Replace the 400 W Metal Halide Lamps with 360 W Metal Halide Lamps

The lighting energy usage can be reduced by replacing the 400 W metal halide lamps with 360 W metal halide lamps. Because of more efficient 360 W bulbs, the illumination level practically remains the same with the reduction in power used. This is an option for reducing the amount of energy used for lighting. The replacement does not require the change of the fixture and the ballast. The lamps can be retrofitted with 360 W metal halide lamp, as the existing 400 W lamps burn out with time. Energy savings is a function of the difference in power used by the 400 W and 360 W lamps, the ballast factor of the lamp, the annual operating hours of the lamps, and number of similar lamps to be replaced in the facility.[42]

Replace the Existing T12 Lighting Fixtures with Magnetic Ballasts by T8/T5 Lighting Fixtures with Electronic Ballasts and Reflectors

Install T8/T5 lighting fixtures using electronic ballasts with reflectors in place of the T12 fixtures containing magnetic ballasts. This will reduce the lighting energy usage while maintaining the same lighting levels at the work surface. Lighting technology has evolved rapidly in recent years. In commercial buildings, significant reductions in energy use can be achieved by installing energy-efficient bulbs, fixtures, and controls. Retrofits to install new technologies such as electronic ballasts and specular reflectors are often cost-effective, providing a reasonable payback. T8/T5 bulbs provide almost the same light levels as T12 bulbs with the reduction in power consumption. Electronic ballasts consume less (approximately 5–10 percent) power as compared to the regular magnetic ballasts.[42] Hence, the energy savings is a function of the number of bulbs per fixture, the number of florescent fixtures in the facility, existing and proposed wattage of each fixture, the ballast factor, and the annual operating hours.[42]

Install Occupancy Sensors in Designated Areas

Occupancy sensors can be installed in some areas to reduce the electrical usage for lighting during unoccupied periods. If certain areas in the facility with less utility are heavily lit, then by wiring occupancy sensors into these areas, the lighting usage could be reduced during the unoccupied periods. Energy savings will result from reduced electrical usage for lighting. Some lights can be left on in the room for emergency purposes. Energy savings is a function of the total power consumption by the lights and the annual hours for which the lights can be controlled.[42]

4.4 Chillers

Chillers are used to produce chilled water for plant and process cooling. Mechanical chillers and absorption chillers are the two commonly used types.[39] Mechanical chillers use electric motors to drive the refrigerant compressors, whereas heat source is used in absorption chillers.

Increasing the chiller set point temperature would save electrical energy. Generally, the efficiency of chillers increases as the chilled water temperature increases. This is because in order to obtain lower-temperature chilled water, the refrigerant must be compressed at a higher rate, which in turn increases the compressor power requirements and decreases the efficiency of the chiller. There is approximately 1 percent increase in efficiency for each degree Fahrenheit increase in the chilled water set point temperature.[44] Before implementing this recommendation, it is advised to first check any effect of changing the set point temperature on the process. Thus, the energy savings can be realized by increasing the chiller set point temperature and is the function of the chiller capacity, increase in chiller efficiency due to increased set point temperature, chiller coefficient of performance, annual operating hours for the chiller, and operating load on the chiller.[42] The USDOE's BestPractices software tool CWSAT can be used to determine the energy savings from this recommendation.[40]

4.5 Heating, Ventilation, and Air Conditioning Equipment

Heating, ventilation, and air conditioning (HVAC) units are used to condition air (i.e., control temperature, humidity, and air flow as per the space requirements). Heating is provided to heat the air in the heating season, whereas air conditioning is used to cool the air in the cooling season. Ventilation involves air makeup units to maintain positive air pressure in the plant.

Older air conditioning units used for cooling were not made with installed economizers. Space cooling energy usage can be reduced by adding an economizer. Economizers are essentially a duct and damper system that allows fresh outside air to be used directly for space cooling whenever outdoor air has a lower total heat content, *enthalpy*, than return air. By using cool outside air whenever possible, the energy usage by the mechanical cooling units can be reduced significantly. To determine possible energy savings, the enthalpy of outside air should be known. The enthalpy is a function of the dry bulb temperature and the humidity. The higher the dry bulb temperatures and humidity levels, the larger the enthalpy value is. Using the ASHRAE Handbook of fundamentals, enthalpy values can be determined directly from wet bulb temperatures. Using TMY data and a bin analysis, 5°F wet bulb temperatures can be determined with their associated monthly hours. The enthalpy can also be determined directly from a psychometric chart. In order to use this, the user needs the dry bulb temperature and the humidity or both the wet bulb temperature and the dry bulb temperature. Energy savings is a function of the total Btu*hr per lbm of air for a month during the cooling period, total cubic feet per minute of the cooling capacity, density of air, coefficient of performance of the air conditioning unit, and the annual usage of the unit.[42]

4.6 Air Compressors

Air compressors are electric motor driven systems used to increase the air pressure by reducing its volume. Compressed air finds wide range of application in industries.

Use Outside Air for Air Compressor Intake

Duct outside air to the air compressor intake. This can be done by connecting one end of the pipe to the compressor intake. The other end is routed through the wall to the outside. The existing intake temperature at the compressor intake can be measured using an infrared temperature gun. On average, the outside air is cooler and therefore denser than the indoor air. The compressor's work for the usual operating conditions in manufacturing plants is proportional to the absolute temperature of the intake air. Being denser, cooler air has higher mass and thus compresses more mass of air in one compression cycle.[45] Therefore, by utilizing the outside air as supply to the air compressor intake, it can be compressed easily with reduced energy requirement. Energy savings is a function of the compressor motor capacity (horse power rating); fractional reduction in the compressor's work is due to lowering the compressor's intake air temperature; the horsepower reduction factor, based on operating pressure and power consumption at maximum pressure; operating load on the compressor; annual operating hours; and efficiency of the compressor motor.[42]

Repair Compressed Air Leaks

Repair compressed air leaks on a regular basis. Air leaks in the facility can be detected using an ultrasonic leak detector. The leak detector detects noises caused by air leak that are not audible to human ears in a noisy production setup. Output from the detector can be read from a visual analog (or digital) display or can be heard using earphones. Air is compressed from atmospheric pressure to the compressor discharge pressure at the expenditure of compressor work. Considerable volume of high-pressure air is lost through air leaks. At discharge pressure conditions, power is lost due to loss of compressed air volume through air leaks. Repairing air leaks result in reduced compressed air wastage and thus reduced artificial demand on the compressor. Air leaks are typically present near pipe joints, fittings, valves, and regulators. Energy savings is a function of the power loss due to leak, which is a function of the volumetric flow rate of free air exiting the leak and annual hours during which the leak occurs.[42] The USDOE's BestPractices tool AIRMaster+ can be used to determine the amount of energy lost through air leaks.[40]

4.7 Combined Heat and Power (CHP)

Combined heat and power systems generate heat and electricity as against traditional power plants only producing electricity. A CHP system typically involves a compressor, combustion chamber, gas turbine, and mechanical equipment or a generator coupled to the turbine shaft. The exhaust gases leave the turbine at an elevated temperature. Considerable heat is lost from the turbine exhaust. This heat energy is recovered and serves as a heat source for process heating, space heating, and boiler feedwater heating.[46]

A back pressure turbine can be installed at the boiler house to use the potential pressure difference between the turbine's inlet and exit steam as a prime mover or power source to any equipment using electricity. The high pressure steam produced by the boiler is often throttled at reducing stations to a lower pressure before it is used in the production process. There exists heat energy potential in this steam, which can be used by replacing the reducing station with a back pressure turbine. Therefore, the back pressure turbine will use the energy between the high-pressure steam at boiler operating conditions and the low-pressure steam being supplied to the plant. The back pressure turbine can therefore serve as a prime mover to any rotating equipment such as an electric motor. Also, a generator can be coupled to the turbine shaft to generate electricity, which can serve as a source of power for any equipment (electric motors) using electricity. The energy savings is a function of the mass flow rate of the steam, difference in enthalpies of the high pressure and low pressure steam, overall efficiency of the back pressure turbine, and the efficiency of the mechanical coupling between the back pressure turbine and the rotating equipment.[42]

5 CASE STUDIES OF DEVELOPMENT OF ENERGY-EFFICIENCY MEASURES

Various types of industries have implemented energy-efficiency measures to improve efficiency of their existing operations. This is especially true for manufacturing industries that have high energy intensity in their operations. These include manufacturing plants in glass manufacturing, plastics manufacturing, primary metals manufacturing, and chemical manufacturing industries. Some of the case studies that reveal typical energy-saving opportunities that existed in this industry segment and savings realized as a result of a particular measure are discussed here.

5.1 Glass Manufacturing

The manufacturing activities carried out in the glass-manufacturing facilities range from glass decorating using purchased glass to manufacturing glass from basic raw materials using the hand-blown or machine-based processes. Four factors that characterize the manufacturing facilities are energy consumption, the plant size, the number of employees, and the annual sales.[47] The industrial assessments conducted at the glass-manufacturing facilities revealed a variety of Energy Conservation Opportunities (ECOs). The manufacturing variety for glass was as follows; hand-blown glass, machine-made glass, pressed glass, flat glass, glass bending, cut glass, glass fabric production, window manufacturing, decorative glass manufacture, hermetic seal production, and marbles. Significant relationship exists between the annual energy usage and the plant size.[47] A list of ECOs was recommended for each plant assessed. The energy conserved was reported for each ECO. The ECOs were selected for each plant based on the percentage of energy savings they contributed with respect to the total energy savings recommended for that plant. In general, for each plant, ECOs that had energy savings of 20 percent or higher with respect to total recommended energy savings were selected for this list, as they were considered to be the most significant. The ECOs found in the glass-manufacturing facilities were adjusting air to fuel ratios, preheating combustion intake air, heat recovery and distribution, insulating hot surfaces, replacing motors with energy efficiency motors, modifying chiller operation, modifying heat output in process, improving compressor operation, and reducing infiltration and space-heating loads. A total of 280,000 MMBtu of energy savings were recommended at the 32 glass-manufacturing facilities as a result of these energy-efficiency measures.

5.2 Plastics Manufacturing

Energy assessment at a leading plastics manufacturer done by Industrial Assessment Center at West Virginia University has helped the company save nearly $100,000 per year in energy costs. The company is one of the largest plastic-packaging specialists in Europe and is expanding into the U.S. market. The plant manufactures injection-molded, rigid containers with open tops. The facility measures 187,000 square feet in size and operates continuously, 7 days per week. Energy costs before the implementation of energy-efficiency measures at the plant totaled approximately $760,000 per year, most of which was for electricity and the remainder for natural gas. Of the total 12 opportunities found feasible after study by assessment team, 6 were implemented at the manufacturing plant within first year. The implemented energy efficiency measures are briefly listed here. It was found that some bare surfaces of molding machines were radiating significant energy due to lack of insulation. However, by insulating these surfaces, the machines' heaters will operate less frequently, which will reduce energy consumption. USDOE-developed MotorMaster+ software was suggested to be used for motor management system. This software helped the company in evaluating economic decisions about buying new premium efficiency motors and rewinding or replacing a motor. Compressed air is used significantly in the plant. To help the company improve compressed air system energy efficiency, the assessment team suggested that the company reduce the compressor pressure to a safe minimum required by the system, repair compressed air leaks, and install engineered nozzles to reduce air consumption. As a result of the energy efficiency study and recommendations made by assessment team, the company is able to save 2.3 million kWh or 7,950 MMBtu annually. This translates into significant amount of energy costs savings of nearly $100,000.[48]

5.3 Metals Manufacturing

An energy-efficiency study carried out at a leading steel manufacturer in West Virginia has helped the company survive the market slowdown. The company recycles scrap steel from adjacent areas to produce steel sections, channels, and structural beams that are used in fabrication of frames, grousers, and welded frames for heavy equipments. The plant is approximately 40 acres in size and has been in operation since 1906. The plant operates approximately 8,160 hours annually. The energy consumption of the plant before energy assessment was 1,216,753 MMBtu/yr, or $10,029,682/yr. The assessment team found the following feasible energy-efficiency measures at the time of assessment: "ladling preheat energy reduction, hand-held cutting flame torches energy reduction, reduce infiltration in doors on furnace openings, preheat combustion air in ceramic recuperator, motor management, downsize motors, variable speed drives in several applications, cogged belt drives, improve power factor, insulate the furnaces and tundish dryer, synthetic lubricants, repair air leaks, outside air, lighting efficiency." The assessment recommendations made by assessment team helped the company save an estimated 139,843 MMBtu/yr. The total energy cost savings were approximately $1,034,547/yr.[49]

5.4 Chemicals Manufacturing

One of the leading chemical manufacturers in the world undertook a plant-wide energy assessment with West Virginia University. The objectives were to identify energy-saving projects in the company's utilities area, including boilers and associated steam systems, compressor systems, and electrical motor-driven pump systems. The project evaluation process was unique in that the company has obtained very favorable rates for electricity. Even so, the company found strong economic justification for several projects that would reduce either electricity consumption or fossil fuel consumption. The plant produces more than 1.2 billion pounds of chemical products each year. These materials are used in cars, appliances, furniture, home construction, steel manufacturing, food preparation, and many consumer products. The group operates around 350 companies worldwide and employs 117,000 workers. The energy-efficiency measures that were determined were replacing burner, condensate return, using VSDs on pumps, optimizing compressed air system, insulating steam system components, repairing steam leaks, adjusting the fuel/air ratio in one of the boilers, and repairing compressed gas leaks (air, NG, and nitrogen). The total cost savings were identified as 1,427,800 with and estimated implementation cost of $937,200 which results in a simple payback of less than 8 months.[50]

6 CONCLUSION

Industrial energy efficiency is necessary for several reasons. The energy efficiency will reduce operating costs and hence lead to manufacturing industry becoming globally competitive and contribute to increased profitability and growth in terms of jobs and market share. Energy efficiency will make the nation more secure and economically strong as it reduces the consumption of fossil fuels, hence playing a role in controlling the cost of energy to have a positive impact on every facet of the society. Energy efficiency will also enable the protection of the environment in terms of reduction in emissions. Pure and simply put, energy conservation makes sense.

Manufacturing facilities have the goal of putting the product "out the door" on a daily basis to survive and be profitable and they often do not have enough time and resources to ascertain whether they are using the least amount of energy possible per unit of product manufactured. The programs of the U.S. DOE's EERE and Industrial Technologies Program have gone a long way in helping industry and hence the nation to become energy efficient through a myriad of programs. Energy costs may or may not be a significant part of the product costs, hence not appearing in the top list of priorities of many companies. Rising fuel costs have changed some of that attitude. Aiding industry to save energy is a smart option that cannot simply be viewed as corporate welfare because the benefits are significant to the nation as a whole.

7 ACKNOWLEDGMENT

The authors wish to thank the U.S. DOE's Office of EERE and ITP program for supporting the Industrial Assessment Center (IAC) at West Virginia University since 1992. The development of this chapter by the director and the students of the IAC is a direct result of their experience gained at the IAC. We also thank the West Virginia Development Office's Office of Energy Efficiency for providing support to conduct energy assessments within the state of West Virginia for manufacturing facilities that do not qualify for an IAC assessment.

REFERENCES

1. B. Gopalakrishnan, R. Selvaraj, R. Turton, R. W. Plummer, and S. Sukumar, "A Systems Approach to Plant-Wide Energy Assessment," *Energy Engineering Journal*, **102**(5), 49–80 (2005).
2. B. Gopalakrishnan, A. Mate, Y. Mardikar, D. Gupta, R. Plummer, and B. Anderson, "Energy Efficiency Measures in the Wood Manufacturing Industry," Proceedings of the 2005 ACEEE (American Council for an Energy Efficient Economy) Summer Study on Energy Efficiency in Industry on CD ROM, ISBN 0-918249-54-6, West Point, New York, July 2005.
3. D. W. Francis, M. T. Towers, and T. C. Browne, "Energy Cost Reduction in the Pulp and Paper Industry—An Energy Benchmarking Perspective," Pulp and Paper Research Institute of Canada (Paprican), September 2004.
4. N. Sundarajan, "Analysis of the Trends in Energy Conservation Studies of the IAC Program," West Virginia University, Masters' thesis, 1999.
5. National Association of Manufacturers, "Efficiency and Innovation in U.S. Manufacturing Energy Use", http://www.nam.org/energyefficiency, Accessed in June 2005.
6. Department of Energy, BestPractices Program, http://www1.eere.energy.gov/industry/bestpractices/about_iac.html. Accessed in September 2005.
7. Department of Energy, Plant-Wide Assessments Program (PWA), http://www1.eere.energy.gov/industry/bestpractices/plant_wide_assessments.html. Accessed in February 2006.
8. Department of Energy, Energy Systems Program, http://www.eere.energy.gov/industry/energy_systems/. Accessed in December 2005.
9. Department of Energy, Industries of Future Program, http://www.eere.energy.gov/industry/technologies/industries.html. Accessed in April 2006.
10. U.S. Department of Energy, "Instructions for Energy Auditors," National Technical Information Service, Springfield, Virginia, Vol. I, September 1978.
11. B. L. Capehart, W. C. Turner, and W. J. Kennedy, *Guide to Energy Management*, 2nd ed., Fairmont Press, Atlanta, 1997.
12. C. Warfel, "An Energy Audit Method for Utilities and Industry," *Energy Engineering*, **90**(2) (1993).
13. A. Thumann, *Handbook of Energy Audits*, The Fairmont Press, Inc., 1979.
14. R. Hoshide, "Effective Energy Audits," *Energy Engineering*, **92**(6) (1995).
15. M. Brown, D. Smid, B. Matthews, M. McKeon, and S. Numminen, "Increasing the Implementation of Energy—Efficient Technologies among U. S. Manufacturers," *Energy Engineering*, **91**(6) (1994).
16. G. Spanner, D. Brown, G. Sullivan, and S. Riewer, "Impact Evaluations of Industrial Energy Conservation Projects in the Pacific Northwest," *Energy Engineering*, **90**(4) (1993).
17. A. Mate, "Energy Analysis and Diagnostics in Wood Manufacturing Industry," West Virginia University, Master's thesis, 2002.

18. W. Meffert, "Energy Assessments in Iron Foundries," *Energy Engineering*, **96**(4) (1999).
19. B. Gopalakrishnan, R. W. Plummer, and N. Alkadi, "Analysis of Energy Conservation Opportunities in Glass Manufacturing Facilities," *Energy Engineering Journal*, **98**(6) (2001).
20. F. Fendt, "A Highly Successful Holistic Approach to a Plant-Wide Energy Management System," *Steam Digest*, Office of Industrial Technologies, 2001.
21. J. Weber, "Celanese Chemicals Clear Lake Plant Energy Projects Assessment and Implementation," *Steam Digest*, Office of Industrial Technologies (2001).
22. F. W. H. Yik and J. Burnett, "An Experience of Energy Auditing on a Central Air-conditioning Plant in Hong Kong," *Energy Engineering*, **92**(2) (1995).
23. OIT Clearinghouse, "Vision: Results for Today. Leadership for Tomorrow," Office of Industrial Technologies, February 2001.
24. Energy Information Administration, "United States Background," http://www.eia.doe.gov/emeu/cabs/usa.html.
25. "Augusta Newsprint: Paper Mill Pursues Five Projects Following Plant-Wide Energy Efficiency Assessment," http://eereweb.ee.doe.gov/industry/bestpractices/pdfs/fp_cs_augusta_newsprint.pdf.
26. "Alcoa Teams with DOE to Reduce Energy Consumption," http://www1.eere.energy.gov/industry/bestpractices/pdfs/36152.pdf.
27. "Plant-Wide Assessments (PWAs)," http://www1.eere.energy.gov/industry/bestpractices/plant_wide_assessments.html.
28. "Role of Energy Efficiency: Understand the Value of Energy System Efficiency," http://www.eere.energy.gov/industry/energy_systems/pdfs/role.pdf. Accessed in December 2004.
29. Department of Energy, Industrial Technologies Program. Energy Use, Loss, and Opportunities Analysis, U.S. Manufacturing & Mining, http://www.eere.energy.gov/industry/pdfs/energy_opps_analysis.pdf. Accessed in December 2004.
30. C. F. Will, "Factors that Affect Your Plant Power Bill," Textile, Fiber and Film Industry Technical Conference, IEEE 1993 Annual, page 10/1–10/4. May 4–6, 1993.
31. "Industrial Rates, Utilities in Greater Cleveland," http://www.cose.org/pdf/FactSheets/Utilities.pdf. Accessed on November 22, 2005.
32. West Penn Power Company, "Schedule 20, General service," http://energy.opp.psu.edu/engy/ElecUtil/ElecRate/APS/PA20.pdf. Accessed on November 22, 2005.
33. Monongahela Power Company, "Schedule "B", General service," http://www.alleghenypower.com/Tariffs/WV/Wvmontariffs/WVMPRetailTariff.pdf. Accessed on November 22, 2005.
34. Allegheny Power, "Schedule "C", General and commercial service," http://www.alleghenypower.com/Tariffs/MD/Mdc.pdf. Accessed on November 22, 2005.
35. Monongahela Power Company, "Schedule "K," General power service rate," http://www.alleghenypower.com/Tariffs/WV/Wvmontariffs/WVMPRetailTariff.pdf. Accessed on November 22, 2005.
36. Buffington and Wolf, "Deregulation of electricity generation in Pennsylvania," http://www.age.psu.edu/extension/factsheets/h/H77.pdf. Accessed on November 22, 2005.

37. Monongahela Power Company, "Schedule LGS, Large general service," http://www.alleghenypower.com/Tariffs/WV/Wvmontariffs/WVMPlgs.pdf. Accessed on November 22, 2005.
38. IAC Database, http://iac.rutgers.edu/database/. Accessed on April 17, 2006.
39. Wikipedia, The Free Encyclopedia, http://en.wikipedia.org/wiki/Motors.
40. BestPractices Software Tools, http://www1.eere.energy.gov/industry/bestpractices/software.html. Accessed in April 2006.
41. Maple Dyer and Maxwell Dyer, "*Boiler Efficiency Improvement*," Boiler Efficiency Institute, Auburn, Alabama, 1981.
42. Industrial Assessment Center Reports, West Virginia University, 1992–2006.
43. "Energy Conservation Program Guide for Industry and Commerce," NBS Handbook 115, U.S. Department of Commerce/National Bureau of Standards, September, 1974
44. *Modern Industrial Assessments*, Rutgers University, 2004.
45. "BestPractices—Compressed Air Challenge," http://www.focusonenergy.com/data/common/dmsFilesStaging/B_GI_MKFS_CompressedAirBestPra.pdf. Accessed in April 2006.
46. C. B. Oland, "Guide to Combined Heat and Power Systems for Boiler Owners and Operators," Oakridge National Laboratory, 2004, http://www1.eere.energy.gov/industry/bestpractices/pdfs/guide_chp_boiler.pdf.
47. B. Gopalakrishnan, R. W. Plummer, and N. M. Alkadi, "Comparison of Glass Manufacturing Facilities based on Energy Consumption and Plant Characteristics," *Journal of Energy and Development*, **27**(1) (Autumn 2001).
48. Department of Energy, Industrial Technologies Program Factsheet, "Superfos Packaging: Plastics Manufacturer Saves $100,000 by Implementing Industrial Energy Assessment Recommendations," DOE/GO-102005-2169, September 2005.
49. B. Gopalakrishnan, and R. W. Plummer et al., Industries of Future, "Energy Assessment Report—Steel of West Virginia," West Virginia University, September 2003.
50. Department of Energy, Plant-Wide Energy Assessment Case Study, "Bayer Polymers: Plant Identifies Numerous Projects Following Plant-Wide Energy-Efficiency Assessment," DOE/GO-102003-1677, August 2003.

CHAPTER 11

INDUSTRIAL ENVIRONMENTAL COMPLIANCE REGULATIONS

Thomas J. Blewett
Community, Natural Resources, Economic Development University of Wisconsin Extension, Madison, Wisconsin

Jack Annis
College of Natural Resources University of Wisconsin–Stevens Point Stevens Point, Wisconsin

1 INTRODUCTION TO ENVIRONMENTAL REGULATIONS	295	
2 FEDERAL REGULATIONS AND THE UNITED STATES CODE	296	
3 OVERVIEW OF FEDERAL REGULATIONS	297	
3.1 National Environmental Policy Act of 1969 (NEPA); 42 U.S.C. 4321–4347	297	
3.2 The Big Three: The Major Environmental Laws Affecting Manufacturing	298	
3.3 Other Pollution Control Laws that Affect Industry	312	
3.4 Information and Communication Laws	315	
3.5 Other Federal Environmental Laws Affecting Manufacturers	316	
4 STATE REGULATORY REQUIREMENTS	317	
5 COMPLIANCE TOOLS FOR THE ENGINEER	319	
5.1 National Environmental Compliance Assistance Centers	319	
5.2 Environmental Management Systems (EMS)	320	

1 INTRODUCTION TO ENVIRONMENTAL REGULATIONS

In today's business world, consideration of environmental issues and laws have become an integral part of doing business in the United States and in international settings. One of the first steps for a business is to ensure that it is operating in compliance with environmental regulations. A company that wants to be competitive, and sometimes just to stay in business, must comply with environmental regulations and consider its options to operate in a more comprehensive environmentally conscious manner. Both of these dimensions can provide pathways to reduced financial liability and costs when considered together. The focus of this chapter is on the regulations.

The engineering staff of a company, just like every other business unit in a company, has a responsibility to be aware of, and comply with, environmental regulations as they may apply to the company overall and specifically as they relate to the products, services, operations and projects that engineering is responsible for (see Situation Example 1).

This chapter provides an overview of environmental regulations based on the federal perspective and contains many Web references. It is hoped that readers gain a better understanding of how environmental regulation impacts what companies and employees do in a manufacturing setting, and to appreciate the potential value of going beyond just compliance.

It would be difficult to think that anyone is not aware of some of the environmental problems that led to the need for environmental regulation in the United States. The Industrial Revolution brought many advantages that consumers enjoy today. However, with industrialization came polluted waters, unbreathable air, and contamination of the land that we are still dealing with through environmental remediation actions across the country. As the United States economy grew and convenience became a consumer demand, a throwaway society emerged, creating mountains of waste that led to the need for burial or other disposal for that waste. Reference to the term *deep pockets* in environmental parlance is a result of the billions of dollars spent for cleaning up the land, air, and water that were contaminated in part as a result of decisions made by companies many years ago regarding managing wastes and emissions from their manufacturing processes.

After ignoring or simply not understanding the impact of pollution for many years, public awareness began to force a government response and manufacturing accountability through the creation laws and regulations designed to prevent and even reverse the impact of the past practices. Now, every company in the United States is regulated under U.S. environmental law, and although the effects of poor environmental management are still with us, there has been, and continues to be, steady improvement.

The following sections review some of the things that manufacturing engineers need to know about environmental regulation: where the laws can be found, who enforces the laws, how to identify what applies to a manufacturer, and some suggested methods for approaching environmental regulatory compliance.

2 FEDERAL REGULATIONS AND THE UNITED STATES CODE

In the United States there are more than a dozen major environmental laws codified in the United States Code (U.S.C.). Once a law is passed by Congress it is published in the U.S.C, and this authorizes and requires the responsible agency to develop the specific regulations that affect a business or manufacturer. In most cases, the responsible federal agency for environmental regulation is the U.S. Environmental Protection Agency (EPA).

The federal environmental regulations established from these laws are published in the Federal Register as part of the Code of Federal Regulations (CFR). The CFR is divided into 50 titles, and most environmental regulations are contained in Title 40. Each volume of the CFR is revised once each calendar year, with Title 40 being revised every July 1.

It should be noted that during the process of developing rules from the laws there is always a lengthy process that provides the opportunity for those being regulated, and the public, to make comments and to challenge the new rule. Over the years this has included many court challenges, both to proposed and final rules.

The entire Code of Federal Regulations and other documents such as Federal Register notices are available in many libraries, as well as from the Government Printing Office (GPO). In this age of electronic resources, it is also possible to track rule making, make comments, check for currency, and, if needed, maintain all of 40 CFR and related references via the Internet or data files without creating paper files by using the EPA Electronic Code of Federal Regulations(e-CFR). The e-CFR contains the same files used by the GPO to print each year's paper edition of the CFR, but the e-CFR prototype is not yet considered the official, legal edition of the CFR.

Practicing engineers will probably only need to access Chapter 1 of CFR Title 40, or 40 CFR, which contains all of the environmental rules promulgated by EPA. Chapter 1 consists of subchapters A through U and parts 1–1068. The federal laws creating these regulations are presented in the following sections of this chapter.

3 OVERVIEW OF FEDERAL REGULATIONS

The following major environmental laws are most commonly referred to in the development and enforcement of United States environmental regulation found in 40 CFR. This section provides some history and key provisions of the laws, with an emphasis on the responsibilities of businesses, manufacturers, governments, and institutions affected by these laws. Individual states may vary in the implementation of federal environmental laws and may in some cases be more stringent than federal law. State agency contact information is provided later in the chapter to give a starting point in determining state specific regulatory requirements.

3.1 National Environmental Policy Act of 1969 (NEPA); 42 U.S.C. 4321–4347

The National Environmental Policy Act was a very important piece of legislation, as it was really the first major environmental law that put into place goals and policies of the federal government and recognized the comprehensive and essential linkages between human activities and adverse affects on the Earth's environment. It gave the government the authorization and responsibility to put into place practices that would protect and/or restore the environment.

A core provision of NEPA is the requirement of *environmental impact statements* (EIS) for projects or proposals by the federal government and its agencies, or by entities that require some federal approval, that have the potential to cause adverse impacts on the environment. The EIS must consider not only potential impacts, costs, and irreversible commitments, but also alternatives to the proposal. EIS reports have contributed to significant changes of some proposals and have even resulted in cancellation of projects.

NEPA is important from a historical sense because it set the stage for the major environmental protection legislation that was enacted in the 1970s and further refined in the 1980s. The following three major laws had a profound impact on industry and began the process of environmental protection and remediation.

3.2 The Big Three: The Major Environmental Laws Affecting Manufacturing

Although there are as many as a dozen environmental laws that might affect a company, the following three major pollution control laws and the resulting regulations are perhaps the most important. More attention is given to these laws because of their complexity and broad implications for regulated industries.

The Clean Air Act (CAA); 42 U.S.C. s/s 7401 et seq. (1970)

The Clean Air Act is the comprehensive federal law that regulates air emissions from area, stationary, and mobile sources throughout the United States. Sources of pollutants from individual industrial plants, or specific processes, power plants, and other facilities that have air emissions of pollutants are *stationary sources.* Pollutants that come from sources that are mobile, such as cars, trucks, trains, or planes, are called *mobile sources. Area sources* are considered regions, groups of small sources, states, or nonattainment areas of air quality under the National Ambient Air Quality Standards.

The CAA authorized the EPA to establish the *National Ambient Air Quality Standards* (NAAQS) to protect public health and the environment. The initial goal was to set and achieve NAAQS in every state by 1975. The setting of maximum pollutant standards was coupled with directing the states to develop *state implementation plans* (SIPs) applicable to appropriate industrial sources in the state. The act was amended in 1977 primarily to set new goals (dates) for achieving attainment of NAAQS, since many areas of the country had failed to meet the initial deadlines. The 1990 amendments to the Clean Air Act, in large part, were intended to meet unaddressed or insufficiently addressed problems such as acid rain, ground-level ozone, stratospheric ozone depletion, and air toxics such as *hazardous air pollutants* (HAPs). Some of the key elements implemented by the 1990 Clean Air Act Amendments included the following:

3 Overview of Federal Regulations **299**

- Establishing an air permit system throughout the United States, which regulated specific sources of pollutants with a single permit that addressed all types of pollutants based on the level of pollutant discharged to the air (see Title V Air Permit program described in 40 CFR Part 70 and Part 71)
- Providing EPA with the authority to enforce the permit conditions through fines and penalties without always having to go to court
- Providing for public participation in the process of how the CAA is carried out through making public comments and the opportunity to challenge any aspect of the law, rules, or permits developed or issued
- Recognition that interstate and international transport of air pollution affects air quality in the United States and needs to be addressed when considering a SIP
- Implementing market-based approaches that provide the opportunity to buy or sell air pollution credits that achieve net reduction and economic incentives for reducing pollution (e.g., cleaner fuels)
- Tying deadlines to SIPs and other activities that were thought to be more realistic than previously described in the CAA

Since 1990 there have not been new amendments to the CAA, but there has been a significant amount of regulatory activity through new air regulations, particularly in the individual states, that are trying to successfully meet their SIP obligations. Essentially, a state must meet its SIP requirements or EPA may be forced to step in and take over the air program.

For practicing engineers, this history leads to the point where it is important to be knowledgeable of the air permit system. Plant engineers must determine if their state is the *control authority* for air permits. Use the contacts provided in Table 1 to begin that process or review 40 CFR to see if the applicable state has this authority and responsibility.

In simplest terms, a facility will be issued an air permit based on the types and amounts of pollutants emitted to the environment, measured over periods of time and based on threshold quantities usually expressed in tons per year, or TPY. For example, a permit might be needed for a facility emitting 10 TPY of a specific *hazardous air pollutant* (HAP), *volatile organic compound* (VOC) or *particulate matter*.

The plant engineer may need to calculate the emissions, usually as *maximum theoretical emissions* (MTE), *potential theoretical emissions* (PTE), or the *actual emissions*. In many cases it will necessary to calculate all three emissions to obtain a permit. These emissions are calculated by first identifying the sources of emissions and what part or constituents of those emissions are regulated by air regulations—for example, a gallon of paint which contains VOC that will be applied by spray painting. The pounds of VOC in the gallon of paint are multiplied by various application factors such as transfer efficiency, how well the paint is applied, and the *potential, maximum*, and *actual* number of hours

Table 1 State Environmental Regulatory Agencies

Alabama Department of Environmental Management
http://www.adem.state.al.us/Regulations/regulations.htm
Alaska Department of Environmental Conservation
http://www.dec.state.ak.us/regulations/index.htm
Arizona Department of Environmental Quality
http://www.azdeq.gov/
California Environmental Protection Agency
http://www.calepa.ca.gov/LawsRegs/
Colorado Department of Public Health and Environment
http://www.cdphe.state.co.us/regulate.asp
Connecticut Department of Environmental Protection
http://dep.state.ct.us/index.htm
Delaware Department of Natural Resources and Environmental Control
http://www.dnrec.delaware.gov/info/Rules.htm
Florida Department of Environmental Protection
http://www.dep.state.fl.us/legal/Rules/rulelistpro.htm
Georgia Department of Natural Resources, Environmental Protection Division
http://www.gaepd.org/
Hawaii State Department of Health
http://www.hawaii.gov/health/environmental/
Idaho Department of Environmental Quality
http://www.deq.state.id.us/
Illinois Environmental Protection Agency
http://www.epa.state.il.us/
Indiana Department of Environmental Management
http://www.state.in.us/idem/index.html
Iowa Department of Natural Resources
http://www.iowadnr.com/epc/index/html
Kansas Department of Health and Environment
http://www.kdheks.gov/environment/index.html
Kentucky Department for Environmental Protection
http://www.dep.ky.gov/default.htm
Louisiana Department of Environmental Quality
http://www.deq.louisiana.gov/portal/
Maine Department of Environmental Protection
http://www.maine.gov/dep/index.shtml
Maryland Department of the Environment
http://www.mde.state.md.us/
Massachusetts Department of Environmental Protection
http://www.mass.gov/dep/dephome.htm
Michigan Department of Environmental Quality
http://www.michigan.gov/deq
Minnesota Pollution Control Agency
http://www.pca.state.mn.us/index.cfm
Mississippi Department of Environmental Quality
http://www.deq.state.ms.us/MDEQ.nsf/page/Main_Home?OpenDocument
Missouri Department of Natural Resources
http://www.dnr.mo.gov/
Montana Department of Environmental Quality
http://www.deq.mt.gov/index.asp

Table 1 (*continued*)

Nebraska Department of Environmental Quality
  http://www.deq.state.ne.us/
Nevada Division of Environmental Protection
  http://www.ndep.nv.gov/
New Hampshire Department of Environmental Services
  http://www.des.state.nh.us/
New Jersey Department of Environmental Protection
  http://www.state.nj.us/dep/rules/
New Mexico Environment Department
  http://www.nmenv.state.nm.us/
New York Department of Environmental Conservation
  http://www.dec.state.ny.us/
North Carolina Department of Environment and Natural Resources
  http://www.enr.state.nc.us/
North Dakota Department of Health, Environmental Health
  http://www.health.state.nd.us/EHS/
Ohio Environmental Protection Agency
  http://www.epa.state.oh.us/
Oklahoma Department of Environmental Quality
  http://www.deq.state.ok.us/
Oregon Department of Environmental Quality
  http://www.deq.state.or.us/
Pennsylvania Department of Environmental Protection
  http://www.depweb.state.pa.us/dep/site/default.asp
Rhode Island Department of Environmental Management
  http://www.dem.ri.gov/index.htm
South Carolina Department of Health and Environmental Control
  http://www.scdhec.net/environment/
South Dakota Department of Environment and Natural Resources
  http://www.state.sd.us/denr/denr.html
Tennessee Department of Environment and Conservation
  http://www.state.tn.us/environment/
Texas Commission on Environmental Quality
  http://www.tceq.state.tx.us/
Utah Department of Environmental Quality
  http://www.deq.utah.gov/
Vermont Department of Environmental Conservation
  http://www.anr.state.vt.us/dec/dec.htm
Virginia Department of Environmental Quality
  http://www.deq.state.va.us/
Washington Department of Ecology
  http://www.ecy.wa.gov/
West Virginia Department of Environmental Protection
  http://www.dep.state.wv.us/
Wisconsin Department of Natural Resources
  http://www.dnr.state.wi.us/
Wyoming Department of Environmental Quality
  http://deq.state.wy.us/

(See http://www.envcap.org/statetools/ for additional information.)

that paint will be applied to products each year. This must be done for each paint or source of emission in the plant. If a paint shop had one shift of 8 hours only, then one would use 8 hours times the number of days in operation each year to obtain the total amount of VOC expected. But if the shop *might* go to two shifts of 8 hours, there is the *potential* to emit at least twice the amount of VOC, which could trigger the need for a major source air permit. Depending on the situation of the company, the final air permit might include an agreed-upon *emissions cap* that restricts the amount that can be sprayed daily or each year based on these factors. These permits are sometimes referred to as *synthetic minor permits* because of the emissions cap. Generally, the regulatory permit reviewers and writers use a handbook called *AP-42* that contains many established factors when they review a shop's calculations for an air permit.

The following *criteria pollutants* are the common pollutants that are always regulated through the permit system and may also need to be calculated for the permit.

1. *Nitrogen dioxide:* A smog-forming chemical, commonly from burning of fuels. It is an ingredient of *acid rain.*
2. *Carbon monoxide:* Sources are commonly from burning of various types of fuels with extreme health effects in concentration.
3. *Sulfur dioxide:* An ingredient of acid rain. It commonly results from burning of oil, coal and industrial processes.
4. *Particulate matter:* Many sources generate dust, smoke, and soot and are regulated by particulate size, commonly referred to as PM-10.
5. *Lead:* This is being regulated out of products, but many sources remain.
6. *Ozone:* A major smog ingredient, ozone is created by the chemical reaction of nitrogen dioxide and volatile organic compounds (VOC).

Many VOC chemicals are also hazardous air pollutants (HAP). VOCs are not criteria pollutants but are heavily regulated because they contribute to the chemical reaction that creates smog. In addition to these criteria pollutants, there are extensive lists of HAPs and other toxic chemicals that are specifically regulated by exposure, time and thresholds (see Situation Example 2).

Essentially, the total amounts and types of all of the pollutants emitted from a facility determine the permit category or size. In most cases, this is considered the *operating permit*, and it may also be necessary to obtain a *construction permit* if a facility is installing any new equipment or processes, or making any process changes that affect the existing operating permit.

A facility will get an air permit based on the factors that have been discussed to this point, under the *Title V Permit Program*. EPA has two sets of regulations for the Title V Permit Program, as identified in 40 CFR 70 or 40 CFR 71. The 40 CFR 70 regulations are usually referred to as *Part 70* permits, and they are issued by state or even local authority as operating permits under their control authority. *Only EPA issues permits under 40 CFR 71, and only to major sources.* Currently, permits are normally renewed once every five years.

In writing an air permit, the EPA or the state agency may require a company to have *pollution controls* installed that remove or destroy an air pollutant from the air emission. Pollution control is a complex subject that includes any number of technologies with the goal of implementing *maximum achievable control technology*, or MACT for a particular pollutant. Section 112 of the 1990 amendments mandated the establishment of technology-based standards for 189 toxic substances regulated by MACT. The full complexity of this portion of the law is not addressed in this chapter.

Finally, either the state or the EPA will require submission of an annual report called the *air emissions inventory*. This is a method by which the regulatory agency can compare a facility's actual air emissions with its permit, and it also provides a means for calculating air emission fees. The fees are normally used to help pay for the state regulatory program costs.

Clean Water Act (CWA); 33 U.S.C. ss/1251 et seq. (1977)

The Federal Water Pollution Control Act was enacted in 1972 for the purpose of surface water-quality protection in the United States. The law became known as the Clean Water Act (CWA) following amendments that were enacted in 1977, with additional amendments in 1981, 1987, 1990, and 2002. Over the years the law has shifted from its initial emphasis on chemical degradation of surface water from point sources such as industrial plants and *publicly owned* (sewer) *treatment works* (POTWs) to a more comprehensive approach. Today the law is more systems oriented, with a watershed approach that addresses both point sources such as manufacturer discharges and nonpoint sources such as urban or farm run-off.

The major provisions of the law begin with the establishment of measurable water quality standards that are designed to meet the goals of the CWA. The standards include the establishment of a *designated use goal* for each water body, the water quality criteria that support that use, and *anti-degradation policies* that are designed to prevent a lowering of water quality in a water body. Water quality must be monitored on a regular basis to assure that degradation is not occurring. Water bodies that do not meet the water-quality standards continue to be monitored, but are subject to a number of additional provisions that are intended to restore the water quality to meet designated use levels.

Section 305(b) requires states, tribes, and territories to submit a biennial report on the condition of water bodies, including the designation of *healthy, threatened*, or *impaired* for each water body and the chemical, physical, and biological agents that have impacted each water body. Section 303(d) requires inclusion in the report of strategies to address the pollutants that would result in the affected water bodies meeting water-quality standards. One key strategy is the establishment of *total maximum daily loads* (TMDLs) that would limit pollutant loading of water bodies. These TMDLs are increasingly established on a watershed basis and may impact permit limits (see Situation Example 3) on point sources, such as through *National Pollution Discharge Elimination System* (NPDES) permits (Section 402), which apply to industrial sources and publicly owned treatment works (POTWs).

The NPDES program may be administered by the state or EPA, depending on whether a state has met EPA standards for program administration. Many companies have technology-based limits in their permits that, for existing dischargers, are known as the *best available technology economically feasible* (BAT). New sources are subject to *New Source Performance Standards* with limits that may be more stringent than BAT. Companies that discharge indirectly, through POTWs, are subject to *categorical pretreatment standards*. Any business or government must also comply with permit provisions of Section 401 when engaging in construction activities that may cause a water body to fail state water-quality standards.

Section 404 of the CWA addresses dredge-and-fill activities that result in discharges to U.S. waters, including tributaries of navigable waters and wetlands. This broad-reaching provision is administered by the Army Corps of Engineers with review authority by EPA for 404 permits. Section 404 covers a wide array of activities, including bridge building, dam and dike construction, alternations of water flows, and wetland filling. The 404 permit is also subject to review under other related federal environmental protection laws such as the National Environmental Policy Act, the Coastal Zone Management Act and the Wild and Scenic Rivers Act.

Nonpoint sources (Section 319 Nonpoint Source Program) have gained increasing importance and attention in recent years, as most point sources have been covered by permits and associated regulatory provisions. The CWA does not provide specific regulatory authority over nonpoint sources, but does provide incentives through funding to states, tribes, and territories to implement programs that support changes in practices that reduce nonpoint source pollution. A second funding provision of the CWA includes grants to states, tribes, and territories under Section 106 to implement programs in support of CWA goals. Since 1996, the CWA has provided a third source of additional funds to states to establish the Clean Water State Revolving Loan programs.

Resource Conservation and Recovery Act (RCRA); 42 U.S.C. s/s 321 et seq.(1976)

The Resource Conservation and Recovery Act (RCRA) was enacted in 1976 as an amendment to the 1965 Solid Waste Disposal Act. The initial focus of RCRA was on the establishment of a solid waste management system with statutes that provided a new national framework for disposal of municipal and industrial solid waste. The EPA developed the regulations (see 40 CFR, Parts 240–282) that established specific requirements to be in compliance with the mandates of RCRA. The Hazardous and Solid Waste Amendments (HSWA) are the 1984 amendments to RCRA that required phasing out land disposal of hazardous waste and the implementation of a comprehensive hazardous waste management system. The 1986 amendments to RCRA enabled EPA to address environmental problems that could result from underground tanks storing petroleum and other hazardous substances.

RCRA is today the most complex of the major federal environmental regulations. The EPA's *RCRA Orientation Manual* (http://www.epa.gov/epaoswer/general/orientat) provides an in-depth summary of the law. Three significant parts of RCRA have emerged from the initial act and subsequent amendments: *Subtitle C* addresses hazardous waste management, *Subtitle D* addresses state or regional solid waste plans, and *Subtitle I* addresses the regulation of underground storage tanks. In all, there are 10 subtitles in the RCRA act as it has been amended through 1996. The Federal Facilities Compliance Act of 1992 provided for EPA authority to enforce the RCRA regulations at federal facilities. The Land Disposal Program Flexibility Act of 1996 provided a mechanism for more regulatory flexibility in the use of land disposal for some types of waste. RCRA primarily focuses on active and future facilities and does not address abandoned or historical sites (see CERCLA).

Solid waste is defined in RCRA as garbage, refuse, sludges from waste treatment plants and water supply plants, sludges from pollution control facilities, nonhazardous wastes, and other discarded materials that include solid, semisolid, liquid, or contained gaseous materials resulting from industrial, commercial, mining, agricultural, and community activities. By definition, solid waste includes hazardous waste, which is regulated in Subtitle C. Hazardous wastes that are defined as exempt in Subtitle C are regulated under Subtitle D solid waste rules.

EPA guidelines for municipal solid waste management provide for a hierarchical approach, with the highest-priority management practice being *source reduction*, which emphasizes prevention of waste through all phases of product manufacturing and use. The successive priorities for municipal solid waste management are *recycling* to recover and reuse materials, *combustion* to reduce waste volume and recover energy, and *landfilling*. Many communities have adopted composting of organic matter, considered a form of recycling. Businesses and manufacturers that generate large volumes of compostable wastes have also adopted composing as a way to reduce waste.

The primary regulatory focus of Subtitle D is the establishment of criteria for solid waste disposal facilities that define environmental performance standards. These standards provide for the protection of endangered species, surface and ground water resources, and public health and safety. EPA established *technical criteria* to address seven aspects of municipal solid waste landfills, which may be publicly or privately owned:

1. *Location:* to avoid water resource contamination, unstable substrates, and other hazards such as attracting birds near airports

2. *Design:* including a landfill liner and leachate collection system
3. *Operation:* including control of disease vectors, daily cover, controlling explosive gases, and a prohibition on burning
4. *Groundwater monitoring:* monitoring wells to observe any potential failures in the liner system resulting in rainwater leakage or leachate
5. *Corrective action:* to be taken in the event of a failure in the liner and leachate collection system
6. *Closure and post-closure:* criteria for end-of-life management of the landfill facility
7. *Financial assurance:* plan to address the financial responsibilities of the landfill owner and operator during closure and post closure phases including future potential failures in liner or leachate collection.

The solid waste programs outlined in Subtitle D are largely managed by the states that must obtain authorization for solid waste permit provisions through the U.S. EPA. States must implement their permit program to at least meet the federal technical criteria, but may also elect to adopt requirements that exceed the federal regulations.

Hazardous waste may be said to have properties that are dangerous or harmful to human health or the environment. RCRA Subtitle C provides for a comprehensive cradle-to-grave regulatory system for hazardous waste that includes specific responsibilities and liabilities that apply to the waste generators, transporters and final treatment, storage or disposal facilities (TSDFs). Emphasis is given in this chapter to the key hazardous waste concepts and regulatory responsibilities associated with manufacturers as generators with brief reference to regulations for transporters and TSDFs.

The first step in a hazardous waste management system is the process of *hazardous waste determination*, which is the responsibility of all generators of wastes that may meet the formal definition. The EPA's *RCRA Orientation Manual* provides four key questions as the starting point for a hazardous waste determination:

1. Is the material a solid waste?
2. Is the material excluded from the definition of solid or hazardous waste?
3. Is the waste a listed or characteristic waste?
4. Is the waste delisted?

The regulatory definition of solid waste includes any waste (not limited to physical solids) that is abandoned, inherently waste-like, military munitions, or recycled material. Wastes that are not specifically excluded from the definition of solid or hazardous waste may be determined as hazardous waste, either on the basis of having hazardous waste *characteristics* or by being *listed* as hazardous waste. Some materials may be exempt if they have specifically been *delisted* from hazardous waste regulation.

There are five categories of hazardous waste exclusions. In the solid waste category, for example, there are 19 exclusions from the solid waste definition that have been specified. Examples include domestic sewage (including industrial wastes) and point source discharges to surface waters that are regulated by the provisions of the Clean Water Act. Radioactive wastes are exempt from RCRA, but regulated through the Nuclear Regulatory Commission or through the provisions of the Atomic Energy Act. Other examples include spent wood preservatives, processed scrap metals, and spent caustic solutions from petroleum refining. In the hazardous waste category, EPA exempts 17 types of solid waste from the definition of hazardous waste, including household hazardous wastes, certain agricultural wastes, fossil fuel combustion waste, cement kiln dust, trivalent chromium wastes, used oil filters, and others. It is the responsibility of the waste generator to determine if any of the five categories include exclusions pertaining to a specific type of waste.

Once it is determined that a solid waste is not exempt or excluded as a hazardous waste, then it must be determined if the waste is listed or has hazardous waste characteristics. Waste materials may be identified as listed hazardous waste if they are determined to contain toxic chemicals, have dangerous chemicals that are determined to be acutely hazardous, or exhibit one of the hazardous waste characteristics of ignitability, corrosivity, reactivity, and toxicity.

There are four lists located in 40 CFR Part 261. The F list is codified in 40 CFR §261.31 and contains waste from common manufacturing processes such as spent solvents (codes F001 through F005) and electroplating and metal finishing waste (codes F006 through F012 and F019), for example. The K list has wastes from 13 specific industries (Table 2) codified in 40 CFR §261.32. The P and U lists are codified in 40 CFR §261.33 and consist of chemicals of pure or commercial grade or as constituents of formulations. The P list chemicals are

Table 2 Industries that Generate K Listed Wastes

Coking process
Explosives manufacturing
Ink formulation
Inorganic chemicals manufacturing
Inorganic pigment manufacturing
Iron and steel manufacturing
Organic chemicals
Pesticides manufacturing
Petroleum manufacturing
Primary aluminum production
Secondary lead processing
Veterinary pharmaceuticals manufacturing
Wood preservation

acutely toxic, which may be fatal or seriously harmful in low doses. The U list chemicals are generally toxic but may also exhibit other hazardous characteristics such as ignitability, reactivity, or corrosivity.

After a manufacturer completes the assessment of the solid waste, determines if exclusions apply, and examines listed wastes, it must then examine whether the waste meets EPA's definition of characteristic waste by examining whether the waste meets any of the four hazardous waste characteristics established by EPA: ignitability, corrosivity, reactivity, and toxicity. *Ignitability* means that the waste may readily catch fire, and the wastes are coded D001. *Corrosivity* applies to wastes that are strong acids or bases and can cause damage on contact with skin, metals, or other materials. Corrosive wastes are coded D002. *Reactive* wastes, coded D003, may readily explode or violently react when exposed to certain conditions, such as contact with water or corrosive materials. *Toxicity* is examined from the perspective of the potential of toxic leachate forming if the waste is exposed to water passing through a land disposed waste. EPA has established a laboratory procedure called the *toxicity characteristic leaching procedure* (TCLP) that must be used to determine the potential for toxic leachate forming. EPA has identified 40 toxic chemicals coded D004 through D043 with thresholds for regulated leachate concentration based on the TCLP procedure (see Situation Example 4). The four hazardous characteristics are codified in 40 CFR §261.21-24.

Hazardous wastes may become mixed with other nonhazardous waste under various circumstances. EPA has established a *mixture rule*, which means that any amount of a listed hazardous waste mixed with any amount of nonhazardous waste will result in the mixed waste still being classified as a listed hazardous waste. A hazardous waste that is listed on the basis of characteristics or is a characteristic hazardous waste may become nonhazardous when mixed with a nonhazardous waste if the resulting mixture does not exhibit the hazardous characteristic. Exemptions under the *de minimis* rule may apply to very small amounts of listed hazardous waste in wastewater streams that facilities send to wastewater treatment plants.

Generators that may treat hazardous waste on site, and TSDFs are subject to the *derived from rule*, which regulates the status of residues from listed hazardous wastes. EPA has also established a *contained-in policy* used to determine the status of materials that have been contaminated by hazardous waste spills. For example, building materials, demolition debris, equipment, soil, and other materials that may have been contaminated by spills or contact with listed or characteristic hazardous wastes are subject to review by the contained-in policy to determine the appropriate management options.

A hazardous waste *generator* is any person who creates or produces a hazardous waste from an industrial process. Generators may be manufacturers, businesses, institutions, governments, or any facility that produces a hazardous waste. Once a generator completes the hazardous waste determination process, it is necessary to determine the generator status based on the amount of hazardous waste that is produced in a calendar month. *Large quantity generators* (LQGs) produce greater than 1,000 kg of hazardous waste per month *or* greater than 1 kg of acutely hazardous waste per month. *Small quantity generators* (SQGs) produce between 100 kg and 1,000 kg of hazardous waste per month *and* accumulate less than 6,000 kg of hazardous waste at any time. *Conditionally exempt small quantity generators* (CESQGs) produce less than 100 kg of hazardous waste per month *or* less than 1 kg of acutely hazardous waster per month.

All generators must complete identification and counting of the hazardous waste generation on a monthly basis. If a SQG exceeds the monthly limits for its status, it is subject to being reclassified by EPA as a LQG. All LQGs, SQGs, transporters of hazardous waste, and TSDFs must obtain an *EPA ID number* from EPA. LQGs and SQGs must comply with EPA requirements relating to on-site management of hazardous waste, emergency planning and personnel training, but the requirements are less stringent for SQGs. LQGs can accumulate hazardous waste for 90 days or less under most circumstances, while SQGs may accumulate waste for 180 days or less, or up to 270 days for hazardous wastes that are shipped to a TSDF more than 200 miles away. It is important to note that some states may set lower thresholds for qualification as a CESQG and may also place additional regulatory requirements on CESQGs such as a generator ID number, manifesting, record keeping, and reporting that may be similar to LQG and SQG requirements.

Hazardous wastes that are being transported are subject to the packaging and labeling requirements adopted by EPA from the Department of Transportation (DOT) regulations in 49 CFR Parts 172, 173, 178, and 179. In addition, all generators, transporters, and TSDFs must complete and maintain the information requirements of a *uniform hazardous waste manifest* that is to be used by all handlers of hazardous waste and the regulatory agencies as part of tracking hazardous wastes. Additional record keeping and reporting rules apply, including the requirement that LQGs must submit a *biennial report* in even-numbered years to the regional EPA administrator that provides the generator's EPA ID number and information about the generation and management of hazardous wastes.

CESQGs are not subject to most of the rules that apply to LQGs and SQGs unless otherwise indicated by more restrictive state rules, but must still comply with hazardous waste determination and monthly generation limits, storage limits, and proper disposal. Disposal in this case includes permitted TSDFs, but may also include other options such as state-permitted solid waste landfills, recycling facilities, and universal waste facilities.

Reuse, recycling, and reclamation of hazardous waste materials is preferable to disposal in a hazardous waste landfill under RCRA guidelines. However, these options for managing hazardous waste are still largely regulated by hazardous waste rules to assure that these alternatives to disposal do not threaten human health or the environment. Most hazardous wastes being recycled are still fully regulated by RCRA rules. Four exemptions are not RCRA regulated: industrial ethyl alcohol, scrap metal, waste derived fuel from refining processes, and unrefined waste-derived fuels and oils from refineries. Specialized standards have been established for another four categories of recycling, including use constituting disposal, precious metals reclamation, spent lead-acid battery reclamation, and burning hazardous waste for energy recovery. Use constituting disposal allows hazardous wastes that meet certain standards, including undergoing chemical reactions to prevent release of chemical hazards to the environment, to be placed on the land.

As a result of the Used Oil Recycling Act of 1980, followed by EPA rule making in 1985 and 1992, there is a comprehensive set of rules for management of used oil that are separate from hazardous waste rules. The used oil rules are found in 40 CFR 279, which encourages recycling with the presumption that this is the primary management strategy for used oil. Used oil is processed or redefined so that it can be burned for energy recovery or reclaimed as reusable oil. Use oil that is contaminated with more than 1,000 ppm of halogens is regarded as mixed with hazardous waste and is subject to hazardous waste rules. Used oil contaminated with PCBs is subject to regulation under the Toxic Substances Control Act (TSCA).

Following EPA's experience with the used oil recycling rules, similar rules were established for other categories of hazardous waste that are generally recycled in order to encourage recycling as a management strategy. Separate rules known as *universal waste rules* were established in 40 CFR Part 273 to manage hazardous waste batteries, pesticides that have been recalled or recovered from waste pesticide programs, hazardous waste lamps, and hazardous waste thermometers. Mercury is the common hazardous material in the lamps and thermometers.

After generators, the second major regulated sector in the cradle-to-grave hazardous waste management system is the transporter. *Transporters* engage in off-site transportation of hazardous waste in any circumstance that requires a hazardous waste manifest. This includes generators moving hazardous waste from one of their facilities to another facility, as well as transport to TSDFs. The rules for transporters were developed by EPA and DOT and are found in 49 CFR Parts 1171–179.

The third major regulated sector in the cradle-to-grave system is the *treatment, storage and disposal facility* (TSDF). TSDFs are permitted facilities that may engage in (1) treating the waste to alter the hazardous properties of the material, (2) storage of the hazardous waste for some temporary period, and (3) final disposal by some means on land or water where the material will remain after the TSDF discontinues operation. Generators should be diligent in inspecting and tracking the transportation and final disposition of hazardous wastes that they generate as their liability follows the waste to its final disposition. If a TSDF improperly manages or disposes of a hazardous waste resulting in regulatory and financial liabilities, the generator is subject to incumbent liability associated with their share of the waste that has been improperly managed by a TSDF.

TSDFs are subject to more comprehensive RCRA regulations than either generators or transporters. TSDFs that use combustion as a treatment must meet RCRA combustion standards that include the potential of hazardous waste byproducts in the bottom ash and fly ash. In addition, EPA issued a joint RCRA and CAA rule that requires more stringent emission standards for hazardous waste incinerators. These incinerators follow the maximum achievable control technology (MACT) with standards established through EPA research that are specific for the hazardous wastes. The full requirements of TSDFs are found in 40 CFR Parts 264 and 265 and a comprehensive summary can be found in EPA's *RCRA Orientation Manual*.

Land disposal restrictions (LDR) were established for generators, transporters, and TSDFs following the HSWA amendments to RCRA in 1984. The three major components of the LDR program include: (1) a *disposal prohibition*, which prohibits land disposal of inadequately treated hazardous waste, (2) a *dilution prohibition* to prevent circumvention of proper treatment of hazardous waste, and (3) a *storage prohibition* that prevents indefinite storage of hazardous waste in systems not intended for long-term storage. A major component of the LDR, which is articulated in the disposal prohibition, is the establishment of treatment standards that must be met before disposal is an option. The technology and cost efficiencies of treatment have changed over time, which has resulted in the concept of *best demonstrated available technology* (BDAT). Treatment must be performed in a manner that best reduces the hazardous properties of the waste. The technology that best meets the standards, or BDAT, is the result of ongoing research by EPA.

Another outcome of the HSWA amendments was the expansion of EPA's *corrective action* authority. The release of hazardous waste or materials into the air, water, or soil by a RCRA-regulated facility is subject to investigation and cleanup or remediation requirements through a corrective action process. Corrective actions may also be initiated by EPA when it is considering a RCRA facility permit application, or may be voluntarily initiated by the RCRA-regulated facility.

Enforcement of RCRA regulations has three major components: compliance monitoring, enforcement actions, and compliance assistance or incentives. Facility inspections generally include a review of operations and waste management practices, records review, visual inspection of the facility and sampling or testing requirements. Enforcement actions may take the form of *administrative actions* where the enforcement agency, the EPA, or authorized state agency issues an informal or formal action. In the case of formal actions, compliance orders and corrective action orders may include daily penalties for each day of noncompliance. Individuals responsible for RCRA-regulated facilities may be sued in court in a civil action for noncompliance. In civil actions, fines can be levied for each violation per day. In the most serious cases, criminal actions may be pursued in court where individuals could face both fines and prison terms. It should be noted that states that are authorized may have requirements that exceed the federal requirements and the enforcement procedures may differ from EPA.

Subtitle I addresses *underground storage tanks* (UST) that store petroleum or hazardous substances. USTs with petroleum can be found on farms, service stations, government facilities, and manufacturing facilities. This is the most common application, but manufacturers and others also use USTs (or USTs by definition) to store hazardous substances. The UST definition includes the tank and piping connected to the tank where at least 10 percent of the combined volume of the system is underground. That means that an above-ground tank with significant underground piping may meet the UST definition. Subtitle I addresses system design requirements, operation and monitoring, records, release investigation, corrective action, and closure. Financial responsibility rules apply to owners of petroleum USTs. Many states have been authorized to implement the UST regulations, and in those states the rules may exceed the federal rules.

3.3 Other Pollution Control Laws that Affect Industry

Comprehensive Environmental Response, Compensation, and Liability Act (CERCLA or Superfund) 42 U.S.C. s/s 9601 et seq. (1980)

The Comprehensive Environmental Response, Compensation, and Liability Act or CERCLA, was enacted on December 11, 1980. CERCLA created a tax on the chemical and petroleum industries for present and future clean-up of contaminated sites and properties. CERCLA provided broad federal authority to respond directly to releases or threatened releases of hazardous substances that may endanger public health or the environment. Over five years, $1.6 billion was collected. The tax went to a trust fund for cleaning up abandoned or uncontrolled hazardous waste sites. CERCLA also:

- Established prohibitions and requirements concerning closed and abandoned hazardous waste sites
- Provided for liability of persons responsible for releases of hazardous waste at these sites
- Established the trust fund to provide for clean-up when no responsible party could be identified

The law also authorized two kinds of response actions:

1. *Short-term removals:* Actions may be taken to address releases or threatened releases requiring prompt response.
2. *Long-term remedial response actions:* These permanently and significantly reduce the dangers associated with releases or threats of releases of hazardous substances that are serious, but not immediately life threatening.

These response actions can be conducted only at sites listed on EPA's National Priorities List (NPL). CERCLA also enabled the revision of the National Contingency Plan (NCP). The NCP provides the guidelines and procedures needed to respond to releases and threatened releases of hazardous substances, pollutants, or contaminants on the NPL.

Superfund Amendments and Reauthorization Act (SARA); 42 U.S.C.9601 et seq. (1986)

The Superfund Amendments and Reauthorization Act (SARA) amended CERCLA on October 17, 1986. SARA reflected EPA's experience in administering the complex Superfund program during its first six years and made several important changes and additions to the program:

- The importance of permanent remedies and innovative treatment technologies are stressed in cleaning up hazardous waste sites.
- Superfund actions must consider the standards and requirements found in other state and federal environmental laws and regulations.
- New enforcement authorities and settlement tools are provided.
- State involvement increases in every phase of the Superfund program.
- SARA increases the focus on human health problems posed by hazardous waste sites
- Greater citizen participation is encouraged in making decisions on how sites should be cleaned up.
- The size of the trust fund increased to $8.5 billion.

SARA also required EPA to revise the Hazard Ranking System (HRS) to ensure that it accurately assessed the relative degree of risk to human health and the environment posed by uncontrolled hazardous waste sites that may be placed on the NPL. Remediation and redevelopment of brownfields, or contaminated sites such as former manufacturing facility properties, is a desirable outcome, but companies or other investors that wish to develop such properties must practice due diligence. Failure to do so may result in acquisition of contaminated properties and the liabilities associated with the contamination.

Toxic Substances Control Act (TSCA); 15 U.S.C. s/s 2601 et seq. (1976)

The Toxic Substances Control Act (TSCA) of 1976 gave EPA the authority and ability to track industrial chemicals produced or imported into the United States. Each year, thousands of new chemicals are developed by industry, many with either unknown or dangerous characteristics. EPA repeatedly screens these chemicals and can require reporting or testing of those that may pose an environmental or human-health hazard. EPA also has the power to ban the manufacture and import of any chemical that poses an unreasonable risk. Probably the most familiar chemical example is polychlorinated biphenyls (PCB). PCBs were used in hundreds of industrial and commercial applications including electrical, heat transfer, and hydraulic equipment as plasticizers in paints, plastics and rubber products; in pigments, dyes, carbonless copy paper, and many other applications.

More than 1.5 billion pounds of PCBs were manufactured in the United States prior to cessation of production in 1977.

PCBs are steadily being removed and destroyed each year, but it is still likely to be a concern with many industrial facilities. PCBs, mercury, and many other chemicals of concern are also known as PBTs or *persistent bio-accumulative toxic* substances. They may be regulated under TSCA or RCRA in most cases. TSCA supplements other federal statutes, including CAA, CWA, RCRA, and EPCRA through the Toxic Release Inventory (see EPCRA section).

Oil Pollution Act of 1990 (OPA); 33 U.S.C. 2702 to 2761

The Oil Pollution Act (OPA) of 1990 streamlined and strengthened EPAs ability to prevent and respond to catastrophic oil spills. A trust fund financed by a tax on oil is available to clean up spills when the responsible party is incapable or unwilling to do so. The OPA requires oil storage facilities and vessels to submit to the federal government their plans detailing how they will respond to large discharges or releases. EPA has published regulations for above-ground storage facilities, and the U.S. Coast Guard has done so for oil tankers. The OPA also requires the development of area contingency plans to prepare and plan for oil spill response on a regional scale. Area contingency plans are created by a state emergency planning commission. Specific facility plans called a Spill Prevention Control and Countermeasures (SPCC) plan may be required for companies that store petroleum or other liquid chemicals in significant quantities.

Pollution Prevention Act (PPA); 42 U.S.C. 13101 and 13102, s/s et seq. (1990)

The Pollution Prevention Act focused industry, government, and public attention on reducing the amount of pollution through cost-effective changes in production, operation, and raw materials use. Pollution prevention or *source reduction* is much more desirable than waste management or pollution control because it tends to eliminate the problem, rather than deal with the problem at the end of pipe or emission point. Pollution prevention also includes other practices that increase efficiency in the use of energy, water, or other natural resources, and protects our resource base through conservation. Practices include recycling, source reduction, and sustainable agriculture.

In addition to many pollution prevention information resources available from the U.S. EPA, most states have an education and technical assistance unit in a university or other state agency that can help companies identify pollution prevention options. See the *Compliance Tools for Engineers* section for sample Web links to these resources.

Federal Insecticide, Fungicide and Rodenticide Act (FIFRA); 7 U.S.C. s/s 135 et seq. (1972)

FIFRA was enacted to provide federal control of pesticide distribution, sale, and use. EPA was given authority under FIFRA not only to study the consequences of pesticide usage but also to require users to register when purchasing pesticides. Later amendments to the law require users to take exams for certification as applicators of pesticides. All pesticides used in the United States must be registered (licensed) by EPA. Registration assures that pesticides will be properly labeled and that, in accordance with specifications, will not cause unreasonable harm to the environment.

Safe Drinking Water Act (SDWA); 42 U.S.C. s/s 300f et seq. (1974)
The Safe Drinking Water Act was established to protect the quality of drinking water in the U.S. This law focuses on all waters actually or potentially designed for drinking use, whether from above ground or underground sources. The act authorized EPA to establish safe standards of purity and required all owners or operators of public water systems to comply with primary (health-related) standards. State governments, which assume this power from EPA, also encourage attainment of secondary standards (nuisance-related). This law is important to industry because companies must make sure that their location or operations do not in any way impact the local sources of drinking water supply.

3.4 Information and Communication Laws

Emergency Planning and Community Right-to-Know Act (EPCRA); 42 U.S.C. 11011 et seq. (1986)
EPCRA was created under the provisions of Title III of the Superfund Amendments and Reauthorization Act (SARA) of 1986. SARA was a result of (CERCLA) the Comprehensive Environmental Response, Compensation, and Liability Act, which was enacted on December 11, 1980. The main purpose of EPCRA was to establish a system that increases public access and knowledge about toxic chemicals used by industry and released into the local environment, how they are being managed by the sources, and to provide for a state and local emergency response planning.

The EPA maintains a list of *extremely hazardous substances* (EHS) for which companies must report threshold levels of materials on site and requires that companies must notify the state emergency response commission if there is a *reportable quantity* release of a toxic substance to the environment. This includes air releases, as well as spills of liquid or releases into surface water or sewers above a reportable quantity.

Companies with 10 or more employees that manufacture or use toxic materials must complete an annual *Toxic Release Inventory* (TRI) report, commonly called Form R, that identifies which toxic chemicals and how much of those chemicals were release to the air, water, and land as a part of normal manufacturing operations. The TRI report also provides information on toxic chemicals sent to other facilities for waste management and describes pollution-prevention activities. The data from the TRI reports are accumulated by the EPA and made available to the public as the Toxic Release Inventory, which is accessible on the internet. EPCRA also requires another reporting form for chemicals on site that is linked with provisions of the Occupational Health and Safety Act (OSHA) with respect to the data accumulated from *material safety data sheets.*

Emergency planning for oil spills is also considered part of EPCRA and is related to the Oil Pollution Control Act. Most companies that store petroleum or other liquid chemicals in significant quantities are required to have a Spill Prevention Control and Countermeasures (SPCC) plan. EPA published a final rule on February 17, 2006, that extends the SPCC compliance dates for all facilities until October 31, 2007.

Freedom of Information Act (FOIA); U.S.C. s/s 552 (1966)

The Freedom of Information Act provides specifically that any person can make requests for government information. Citizens who make requests are not required to identify themselves or explain why they want the information they have requested. The intent of Congress in passing FOIA was that the workings of government are "for and by the people" and that the benefits of government information should be made available to everyone. Since the government holds information concerning manufacturing facilities, this makes that information available to the public under FOIA. The regulatory information collected and maintained by EPA or state governments are subject to FOIA, with certain protections for confidential business information that may be considered proprietary.

Occupational Safety and Health Act (OSHA); 29 U.S.C. 651 et seq. (1970)

Although not necessarily considered an environmental law, OSHA requirements are very closely related. The Occupational Safety and Health Act provided health and safety protections against physical and chemical hazards that may adversely affect employees. A critical part of OSHA is the *hazard communication standard* (HCS), which provides a focus on communicating the risks associated with chemicals in the workplace. The HCS requires information and education for employees on use and protections against hazard substances when working with regulated chemicals. This includes labeling standards for containers of hazardous materials and the ready availability to employees of *material safety data sheets* (MSDS). The manufacturers and distributors of hazardous materials must provide MSDSs that address the chemical identification of the material, its hazardous properties, safety precautions, exposure limits and procedures for first aid along with disposal information and chemical properties often used in complying with other environmental requirements.

3.5 Other Federal Environmental Laws Affecting Manufacturers

Endangered Species Act (ESA); 7 U.S.C. 136;16 U.S.C. 460 et seq. (1973)

The Endangered Species Act provides a program for the conservation of threatened and endangered plants and animals and the habitats in which they are found.

This is usually not an issue to a company unless relocating or rebuilding a facility that may have some adverse impact on an endangered species.

The U.S. Department of Interior Fish and Wildlife Service maintains the list of endangered and threatened species, which includes birds, insects, fish, reptiles, mammals, crustaceans, and flowering plants, grasses, and trees. The law prohibits any action, administrative or real, that results in a "taking" of a listed species, or adversely affects habitat. Likewise, import, export, interstate, and foreign commerce of listed species are all prohibited.

EPA's decision to register a pesticide under FIFRA is also based in part on the risk of adverse effects on endangered species, as well as environmental fate of the pesticide. Under FIFRA, EPA can issue emergency suspensions of certain pesticides to cancel or restrict their use if an endangered species will be adversely affected.

Coastal Zone Management Act (CZMA); 16 U.S.C. 1451 et seq

The Coastal Zone Management Act was first enacted in 1972 for the purposes of protecting coastal zone resources from continuing losses and degradation from development pressures and to restore coastal zone environments where possible. CZMA was reauthorized in 1990 and was designed to be implemented by each state that contains areas that meet the definition of coastal zone. The coastal zone includes the U.S. territorial seas and adjacent land defined as three nautical miles from the shore, although this inland distance may vary by state and locality.

The primary focus of the CZMA is to protect estuaries, wetlands, beaches, dunes, coastal floodplains, and other habitats for wildlife and coastal biological communities. Under the provisions of CZMA, states are to develop their own coastal zone management plans that include intergovernmental cooperation and citizen participation in the decision-making process. The state management plans must address protection and restoration of natural resources, as well as other interests including public access, cultural and historical sites, and other necessary uses of the coastal zone. Permits are employed by many states to regulate or manage coastal zone development or uses. Industry and local governments that plan projects or alterations in the designated coastal zone must determine the applicability of state requirements or permits under the guidelines of the federal CZMA.

4 STATE REGULATORY REQUIREMENTS

The EPA has the authority and is the responsible agency for promulgating most of the regulations resulting from federal environmental laws. However, in most states, the state environmental regulatory agency is responsible for at least some of the environmental regulatory programs if the state has been awarded statutory authority by EPA, or what is commonly referred to as *control authority* for those programs. States vary in what federal programs they are authorized to carry out, so it is important for companies to contact the regulatory agencies of the states in which they are located. In all instances, state regulations must be as restrictive, or more restrictive, than what is contained in 40 CFR in order for the state to be authorized as the control authority.

It is the responsibility of a business or company to contact the appropriate state regulatory agency to determine compliance with all regulations. Table 1 identifies the Web URLs that will provide access to appropriate state regulatory agencies. Additional help can be found at http://www.envcap.org/statetools/. In some states or for some specific media or programs, EPA is responsible for enforcing regulations at the state level. It takes considerable time and study to determine all regulations that apply directly to a company.

It may be necessary to do an audit of the company first to determine potential regulatory responsibilities. This can be done by obtaining consulting assistance, doing the research internally, or asking for the assistance from the regulatory or state technical assistance program. No matter how a company approaches the regulatory issue, the responsible engineer or other parties must use due diligence to ensure that the company has addressed all regulations, whether federal, state, or both.

The process of determining regulatory compliance by a company is frequently referred to as *due diligence* by the regulatory agencies. Due diligence is simply making the very best effort to identify and then comply with environmental regulations. In most instances, if due diligence is used and a mistake is made, the regulatory agency is much more likely to allow correction of the mistake, often without penalty, rather than taking an enforcement action. Although the company engineer or engineering staff may not be directly responsible for compliance with environmental regulation, it is important for staff to know where to find the regulations and how they apply to the company.

Where states are the control authority, they may vary in their assessment of the due diligence practiced by the company, even in cases when EPA discovers that errors are made. The potential for penalty reduction with due diligence makes a strong case for self-audits, or third-party environmental auditing, and is one major reason why an effective environmental management system (EMS) is a smart business management option. Auditing and EMS are discussed later in the chapter as approaches for assuring compliance.

In recent years many companies, certainly proactive companies, have invested a great deal of time and planning to make sure that they are in basic compliance with environmental regulations. There are many tools available to do that. Certainly with the advent of the Internet, at least accessing regulations has been made much easier. We suggest that there are at least four important approaches to compliance with environmental regulations:

1. Management or corporate commitment stated in written policy or philosophy
2. Resources in the form of staff time and funded activity at the company

3. Actions that are obvious within the company and to the regulatory agency
4. Continuous improvement in practices

With these four approaches in place, it is more likely that a company can do what is necessary to practice due diligence and ensure compliance at its facilities and that individuals can implement appropriate practices at least in the area, programs, projects, services, or products for which they may be responsible within the company.

It is always possible that an engineer may wind up in a company that does not practice environmentally conscious manufacturing or tends to avoid making decisions about environmental compliance. In those cases, the engineer may face a serious professional decision on whether to press the company for appropriate compliance practices, accepting the consequences of doing nothing, or leaving an unacceptable work environment.

5 COMPLIANCE TOOLS FOR THE ENGINEER

By now it is evident that environmental regulation may be a very difficult challenge. However, the good news is that there are many resources available today to help companies meet that challenge, most of which are available through the Internet. One thing to remember is that personal contact with state or federal regulators is always a good idea. Developing a relationship with the regulatory air permit engineer, for example, can be a time-saving step that may pay off when permit renewal time comes around. The following tools and information will help with that process. Other options include membership in trade or professional organizations and attending environmental conferences and meetings where federal, state, and industry professionals share knowledge and information.

Every state in the United States also has something called a *technical assistance program*. These programs are funded by state and federal sources to provide assistance for businesses, usually at or no cost. They are an excellent source of information on many environmental topics including, pollution prevention, waste reduction, recycling, and compliance assistance. A good place to find out how to reach them is through the applicable state regulatory agency small business assistance program and the National Pollution Prevention Roundtable yellow pages at http://www.p2.org/inforesources/nppr_yps.html.

5.1 National Environmental Compliance Assistance Centers

The compliance assistance centers are a very good source of regulatory information for a specific industry or entity. Currently there are 15 separate industries or entities with more planned (http://www.assistancecenters.net/). Many have direct links to state programs similar to what we have provided in Table 1. In addition to compliance information, these centers also contain many pollution prevention, waste reduction, and recycling tips and information.

5.2 Environmental Management Systems (EMS)

Today one of the mainstream concepts of environmental management is the Environmental Management Systems model. EMS essentially evolved from European models to an international standard, ISO-14001, in 1996. The 1996 standard has subsequently been refined with the release of ISO 14001:2004. An EMS, whether ISO or developed separately, combines all of the environmentally related elements of operating a business in one systematic program of continuous improvement. This includes environmental compliance, as well as pollution prevention and natural resource conservation, among many of the 17 elements involved with the EMS.

Many state and national voluntary cooperative programs include EMS as a core element of participation. U.S. EPA has recognized EMS as a desirable method of environmental management, which should be encouraged. Two recent examples include the Wisconsin Green Tier program, which was approved for use by U.S. EPA, and the "National Performance Track Partnership." The *National Performance Track* is a voluntary partnership that recognizes top environmental performance among participating U.S. facilities of all types, sizes, and complexity, public and private. Program partners are providing leadership in many areas, including preventing pollution at its source.

The *P2 Framework* is an approach to risk screening that incorporates pollution-prevention principles in the design and development of chemicals. The objective of the P2 Framework approach is to inform decision making at early stages of development and promote the selection and application of safer chemicals and processes. One very helpful tool that has been developed by EPA and the states is the *PBT Profiler*. This is a tool that uses a series of risk factors to screen chemicals that may be used for existing or more frequently, new products to eliminate or avoid using chemistry that contains persistent bio-accumulative toxic (PBT) substances.

Many of these tools are used by industry in working on a cooperative program called '*design for the environment.*' The DfE program is one of EPA's premier partnership programs, working with individual industry sectors to integrate cleaner, cheaper, and smarter solutions into everyday business practices. EPA also supports using "benign by design" principles in the design, manufacture, and use of chemicals and chemical processes—a concept known as *green chemistry*.

REFERENCES

References and materials for this chapter have come from the public domain including United States Code, the Code of Federal Regulations and largely from the U.S. Environmental Protection Agency. The authors' intent is to encourage readers to effectively use and access widely available resources and to access appropriate regulatory agencies for information and assistance.

References

U.S. Environmental Laws: http://www.epa.gov/epahome/laws.htm

The U.S. Code of Federal Regulations (CFR): http://www.epa.gov/epahome/cfr40.htm

The U.S. Environmental Protection Agency, 40 CFR Chapter 1 Sub-Chapter A-U, Parts 1–1068: http://www.epa.gov/epacfr40/chapt-I.info/chi-toc.htm

National Environmental Policy Act of 1969 (NEPA); 42 U.S.C. 4321–4347: http://www.epa.gov/region5/defs/html/nepa.htm

The Clean Air Act (CAA); 42 U.S.C. s/s 7401 et seq. (1970): http://www.epa.gov/region5/defs/html/caa.htm

The Plain English Guide to the Clean Air Act: http://www.epa.gov/air/oaqps/peg_caa/pegcaain.html

Clean Water Act (CWA); 33 U.S.C. ss/1251 et seq. (1977): http://www.epa.gov/region5/water/cwa.htm

Watershed Academy (2003). Introduction to the Clean Water Act: http://www.epa.gov/watertrain/cwa/rightindex.htm

Resource Conservation and Recovery Act (RCRA); 42 U.S.C. s/s 321 et seq.(1976): http://www.epa.gov/region5/defs/html/rcra.htm

US-EPA RCRA Orientation Manual: http://www.epa.gov/epaoswer/general/orientat

Comprehensive Environmental Response, Compensation, and Liability Act (CERCLA or Superfund) 42 U.S.C. s/s 9601 et seq. (1980): http://www.epa.gov/superfund/action/law/cercla.htm

U.S. EPA National Priorities List (Superfund Sites): http://www.epa.gov/superfund/sites/npl/index.htm

Superfund Amendments and Reauthorization Act (SARA); 42 U.S.C.9601 et seq. (1986): http://www.epa.gov/superfund/action/law/sara.htm

Hazard Ranking System (Used for Superfund Site Selections): http://www.epa.gov/superfund/programs/npl_hrs/hrsint.htm

Toxic Substances Control Act (TSCA); 15 U.S.C. s/s 2601 et seq. (1976): http://www.epa.gov/region5/defs/html/tsca.htm

U.S. EPA Polychlorinated Biphenyls (PCB): http://www.epa.gov/opptintr/pcb/

Oil Pollution Act of 1990 (OPA); 33 U.S.C. 2702 to 2761: http://www.epa.gov/region5/defs/html/opa.htm

U.S. EPA Oil Spills Web site and Links to SPCC: http://www.epa.gov/oilspill/index.htm

Pollution Prevention Act (PPA); 42 U.S.C. 13101 and 13102, s/s et seq. (1990): http://www.epa.gov/region5/defs/html/ppa.htm

Federal Insecticide, Fungicide and Rodenticide Act (FIFRA); 7 U.S.C. s/s 135 et seq. (1972): http://www.epa.gov/region5/defs/html/fifra.htm

Safe Drinking Water Act (SDWA); 42 U.S.C. s/s 300f et seq. (1974): http://www.epa.gov/region5/defs/html/sdwa.htm

Emergency Planning & Community Right-To-Know Act (EPCRA); 42 U.S.C. 11011 et seq. (1986): http://www.epa.gov/region5/defs/html/epcra.htm

Toxic Release Inventory (TRI) Web Site: http://www.epa.gov/tri/

Freedom of Information Act (FOIA); U.S.C. s/s 552 (1966): http://www.epa.gov/region5/defs/html/foia.htm

Occupational Safety and Health Act (OSHA); 29 U.S.C. 651 et seq. (1970): http://www.epa.gov/region5/defs/html/osha.htm

Endangered Species Act (ESA); 7 U.S.C. 136;16 U.S.C. 460 et seq. (1973): http://www.epa.gov/region5/defs/html/esa.htm

Coastal Zone Management Act (CZMA); 16 U.S.C. 1451 et seq.
US-EPA Home Page for Environmental Management Systems: http://www.epa.gov/ems/
US-EPA Industry Sectors Strategies Using EMS: http://www.epa.gov/sectors/ems.html
National Compliance Assistance Centers: http://www.assistancecenters.net/
Wisconsin "Green Tier" Program: http://www.dnr.state.wi.us/org/caer/cea/environmental/background/index.htm
National Performance Track Partnership: http://www.epa.gov/performancetrack/index.htm
The P2 Framework: http://www.epa.gov/oppt/p2framework/
The PBT Profiler: http://www.epa.gov/oppt/pbtprofiler/
US-EPA Design for the Environment Program: http://www.epa.gov/opptintr/dfe/index.htm

Situation Example 1. Notice of Noncompliance

The Jones Metal Finishing Company has just received a notice of an administrative action from the Environmental Protection Agency. The notice declares that the company has failed to properly store hazardous wastes that have resulted from its painting operations. It further indicates that the company has failed to properly dispose of its hazardous waste in a timely manner as specified by its SQG status.

The plant manager is asking the plant engineer what to do. Points to consider:
1. What are the potential legal and financial liabilities of noncompliance?
2. What must the plant engineer do to correct these issues?
3. What are the roles of the plant engineer and production personnel in the response?
4. How might this liability affect the competitive position of the company?
5. How could this situation be avoided in the first place?

Situation Example 2. Air Emissions

The Jones Metal Finishing Company wants to add a liquid spray-painting option to its other customer services. The newly hired engineer responsible for implementing the project is aware that this will generate some air emissions and waste but is not sure what to do for compliance. Here are a few, *not all*, basic steps to follow:
1. Determine the VOC and HAP content of paint(s) in pounds per gallon.
2. Calculate the MTE and PTE for the estimated total annual emissions.
3. Determine if the emissions exceed any threshold set by the state or EPA.
4. If not exempt, apply for a state construction permit for the paint booth and a Part 70 Operational Permit.

Situation Example 3. Wastewater

The Jones Metal Finishing Company engineer discovers that the parts that will be spray painted must also be washed, rinsed, and dried to remove oil prior to painting. He does not know what to do about the wastewater from the system. Here are a few, *not all*, basic steps that relate to the wastewater issue:

1. The wastewater will have to be tested to determine if it exceeds local, state, or federal limits for any specific pollutants.
2. If it does, the engineer must contact the local wastewater treatment plant to see if that will exceed local limits.
3. He then has to contact the state to obtain a NPDES wastewater discharge permit if needed.
4. Because his company is a metal-finishing company, he may also be required to meet categorical pretreatment standards.
5. In this case, the plating division of Jones Metal Finishing has an NPDES wastewater permit and operates a pretreatment system.
6. Our engineer can send the washwater to the plating division pretreatment system with no further action.

Situation Example 4. Hazardous Waste Determination

The Jones Metal Finishing Company engineer now has to decide whether or not waste paint, paint filters, clean-up solvent, and other waste materials from the paint operation are or are not hazardous waste. Here are a few, *not all*, basic steps:

1. Look at the paint MSDS for chemical information that would identify it as a hazardous waste on a 40CFR list or because of a characteristic.
2. Check with the paint and solvents manufacturer to see if either is known to be a hazardous waste.
3. Have all of the wastes tested using a TCLP test to be sure.

Index

A

Accelerated thermal cycling (ATC) test, 199
ACGIH. *See* American Conference of Governmental Industrial Hygienists
Action
  cycle. *See* Plan, do, check and act phase, 84
Active disassembly, 225–227
Active Disassembly using Smart Materials (ADSM), 225–227
Activity-based costing (ABC), 72
Activity-based management (ABM), 72
Actual emissions, 299
Actual Value Precedence (AVP), 215, 216
Acute poisoning, 154
Additional Disassembly Time, 250
Adhesives, usage, 189
Administrative actions, 311
Administrative costs, 272
ADSM. *See* Active Disassembly using Smart Materials
AEHA. *See* Association Electric Home Appliances
Aerosols
  composition, 152–153
  definitions, 149–150
  harm, 165
  sampling/analysis instrumentation, 153
Aging substrates, experimental results, 203
Agricola, 48
Airborne particles
  origin, 156–161
  standards, OSHA publication, 155–156
Airborne particulate, production, 146–147
Air compressors
  intake, 287
  usage, 287–288
Air emissions, 133
  inventory, 303
Air-jet systems, usage, 106
Air leaks, repair. *See* Compressed air leaks
AirMaster+, 268
Air permits, control authority, 299
Air quality. *See* Manufacturing
  health effects/regulations, 153–156
  introduction, 145–148
  references, 172–178
  regulations. *See* Workplace air quality regulations
  trends, 156
  standards, trends, 156
American Conference of Governmental Industrial Hygienists (ACGIH), 156
American National Standards Institute (ANSI), 149
  definitions. *See* Dust; Fumes; Mists
American Society of Mechanical Engineers, examination. *See* Products
Ammonia-based mixture, replacement, 188
AND/OR charts
  methodology, 248
  usage, 247
AND/OR graph
  basis. *See* Disassembly process layout model, 231
Annotation, 250
ANSI. *See* American National Standards Institute
Anti-degradation policies, 303
AP-42 handbook, 302
API. *See* Application programming interface
Application programming interface (API), 252
AR. *See* Assessment recommendation
ARC. *See* Assessment Recommendation Code
Arc-welding processes, 17–18
  environmental impacts, 18
Area array packages, manufacture, 194
Art of War, The (Tzu), 49
Asbestosis, 154
ASHRAE Handbook, usage, 287
Assemblies, 197–198
AssemblyDoc, 254
Assessment
  planning, 87–88

325

Assessment *(Continued)*
  synopsis, 91
  two-way communication, 92
  usage, 86–92
Assessment recommendation (AR), 92
Assessment Recommendation Code (ARC), 277–278
Associate (employee/worker synonym), 50
Association Electric Home Appliances (AEHA), 229
ATC. *See* Accelerated thermal cycling
Automatic disassembly, 227
Automatic flow controls, installation, 139
AVP. *See* Actual Value Precedence

B

Backlog, limitation, 71
Baldrige, Malcolm, 54
Ball grid array (BGA), 200. *See also* Plastic ball grid array
  assembly, 199
Barrel-plating operations, 125
BAT. *See* Best available technology economically feasible
BDAT. *See* Best demonstrated available technology
Bearing outer shell, disassembling
  operation method, 257–258
  tool (usage), 257–258
BEIs. *See* Biological exposure indices
Ben & Jerry's, principles, 65
Best available technology economically feasible (BAT), 304
Best demonstrated available technology (BDAT), 311
BestPractices, 267
  tools, 280
Best Value Precedence (BVP), 214, 215–216
BGA. *See* Ball grid array
Bill of materials (BOM), 254
  tree, 224
Bioaerosols, 170–171
Biochemical oxygen demand (BOD), 100. *See also* Metalworking fluids
  loadings, 109
  removal, 108
Biocides, 117
Biological exposure indices (BEIs), 156
Biological inhibition, 117–118
Biomimcry, 65
Bipartite directed graph, 223
BOD. *See* Biochemical oxygen demand

Boeing, specialization, 68
Boilers
  air-fuel ratio adjustment, 281–282
  bare surfaces, insulation, 280–281
  combustion air, preheating (hot flue gas usage), 281
  energy efficiency measures, 280–283
  waste heat recovery, 281
Branding, 31
Brazing, 20
Building-block design, 218
Build-to-order process, 68
BVP. *See* Best Value Precedence

C

CAA. *See* Clean Air Act
CAAA. *See* Clean Air Act
CAD. *See* Computer-assisted design
Cadmium plating, 128
CAM. *See* Computer-aided machining
Canadian runner modeling, 23
Capacity planning, usage, 82
Carbon arc welding (CAW), 157
Carbon black (CB)
  manufacturing, 160
  usage, 160–161
Carbon monoxide, pollutant, 302
Carbon nanotubes (CNTs), exposure, 168–170
  assessment, 169–170
Carburizing flame, production, 17
Case hardening, 126
Casting, 159–160. *See also* Die casting
  surface quality, 11
Casting Emissions Reduction Program (CERP), 160
Categorical pretreatment standards, 304
Cavity block, 22
CAW. *See* Carbon arc welding
CBD. *See* Carbon black; Chronic beryllium disease
Centrifugation, 111–113
Ceramic process, 12
CERCLA. *See* Comprehensive Environmental Response, Compensation, and Liability Act
CERP. *See* Casting Emissions Reduction Program
CESQGs. *See* Conditionally exempt small quantity generators
CFCs. *See* Chlorofluorocarbons
CFEST. *See* Cutting Fluid Evaluation Software Testbed
Change, stimulation, 54–57

Check
  cycle. *See* Plan, do, check and act phase, 83–84, 86–87
Chemical conversion, 125–126
Chemical recovery, 141
Chemicals manufacturing, energy efficiency case studies, 290–291
Chemical treatment, 163–164
Chillers, usage, 286
Chip scale packages (CSPs), 200
Chlorofluorocarbons (CFCs), 134
  elimination, 190
  material substitution, 167
CHP. *See* Combined heat and power
Chromating, 125
Chromium, usage, 187–188
Chromium plating, 129
Chronic beryllium disease (CBD), 154
Chronic obstructive pulmonary disease (COPD), 154
Chronic poisoning, 154
Cladding process, 126
Clean Air Act (CAA), 134, 298–303
  amendment (1990) (CAAA), 133–134
  regulation, 7
Cleaners, usage, 132
Cleaning chemicals, usage, 188
Cleaning processes, 190
Clean Water Act (CWA), 134–135, 303–304
  regulation, 7
Closed-loop process, 118
CNTs. *See* Carbon nanotubes
Coalescers, 110–111
Coastal Zone Management Act (CZMA), 304, 317
Coefficient of thermal expansion (CTE), 204
Cold-chamber die casting, 11
Collected MWF mist, returning, 162
Combined heat and power (CHP), 288
Commons, concept, 65
Communications
  facilitation, 57
  laws, 315–316
Company bottom-line, energy efficiency (importance), 268–269
Component companies, lead-free technology change (reaction), 191
Components
  production/supply, 40–41
  reclamation, 212
Comprehensive Environmental Response, Compensation, and Liability Act (CERCLA), 305, 312–313, 315
Compressed air leaks, repair, 287–288
Computer-aided machining (CAM), 23
Computer-assisted design (CAD), 231, 233
  set, 252
COM-supported programming language, 252
Conditionally exempt small quantity generators (CESQGs), 136, 309
Conductive adhesives, 204–205
Consciousness, superior levels (acquisition), 55
Conservation, 66
Consumer/customer charges, 272
Consumers, habits, 29
Control authority. *See* Air permits
Control systems, 83
Control technologies, 161–164
Conversion coatings, 129
COPD. *See* Chronic obstructive pulmonary disease
Core preparation/setting, 6
Corporate image, improvement, 85
Corrosivity, definition, 308
Cost-performance benefits, 63
Cost savings, IAC recommendation, 279
Cradle-to-grave regulatory system, 306
Creative destructions, 52
Creative process, usage, 59–61
Criteria pollutants, 302
Cross-disciplinary teams, 51
Cross-flow filtration, 114
Cross-flow membrane filtration, 115–117
Cross-training
  inhibition, 50
  needs, 71
Crown jewels, 67
CSPs. *See* Chip scale packages
CTE. *See* Coefficient of thermal expansion
Customer charges, 276
Customer delight, speed, 67
Customer demand, 27
Cutting Fluid Evaluation Software Testbed (CFEST), 104–105
Cutting fluids
  roles, 4
  use, 4
CWA. *See* Clean Water Act
Cyanide-based plating, 128
CZMA. *See* Coastal Zone Management Act

D

Das, tool bag (employment), 214
Data analysis, 91

Data collection, 87–90
DD. *See* Disassemblability degree; Disassembly direction
Dead-end filters, contrast, 115
Dead-end filtration, 114–115
Decision-making degrees of freedom, 166
Decisions
  macro level, 86
  micro level, 86
De-gating, 23
DEI. *See* Disassembly effort index
Dell, integration, 68
Dell, Michael, 68
Demand charges, 272–273
Deming, W. Edwards, 49
  ideas, 61–62
De Morbis Artificum Diatriba (Ramazzini), 153
De Re Metallica, 48
Derived from rule, 308
Designated use goal, 303
Design complexity, 247
  procedure simplification, 236
Design engineers, questions, 190
Design for Assembly (DfA) analysis, 249
Design for disassembly (DfD), 35
Design for ease of disassembly (DfD), 34, 39
Design for recycling (DfR), 32, 35
Design for reliability, 202
Design for the environment (DfE)
  application, 34–36
  applying, 37–40
  axioms, lists, 34
  checklist, modules, 36
  cooperative program, 320
  introduction, 29–36
  LCA, 27
  methods/instruments, 33
  necessity, 29–32
  perspective, 33
  proactive approach, 181
  procedures, LCA (inclusion), 34
  products, philosophy, 36
  reading, 44
  references, 42–44
  software, 42
  structure, 32–33
  summary, 42
  tools, selection, 37, 38
  web sites, 44
    usefulness, 40–42
Design for topics (DfX), 35
Design improvement suggestions, providing, 237

Desired Value Precedence (DVP), 214, 215
DfA. *See* Design for Assembly
DfD. *See* Design for disassembly; Design for ease of disassembly
DFD system, SolidWorks (application), 253–255
DfE. *See* Design for the environment
DfR. *See* Design for recycling
DfX. *See* Design for topics
Die casting, 10–12. *See also* Cold-chamber die casting; Hot-chamber die casting
  advantages/limitations, 11
Digital business (e-business), 67
Dilution prohibition, 311
Dip/galvanized coatings, application, 126
Directed graph model, 233–234
Directional graphs, basis. *See* Disassembly process
Direction of Disassembly, 249
Direct productivity enhancements, 278
Disassemblability degree (DD), 247–252
Disassembling
  models, 231–236
  system, composition, 236–237
  techniques, 229–231
Disassembly. *See* Active disassembly; Automatic disassembly; End-of-life electromechanical products
  analysis, 237
  assessment, 237
  design
    complexity, 247
    system, composition, 236–237
  direction, 245–247
  equipment, maintenance, 228–229
  estimation method, 247–252
  evaluation, 237–238
    chart, 248–250
    chart, categories, 248–250
    method proposal, 214–218
  expense, 239
  families, concept, 227
  objective aspects, relative index, 238–244
  objects, determination, 230
  optimization relationship, 220
  process
    energy consumption, 240–243
    environment, impact, 243–244
  programming, 224–225
  research, shortfalls, 229
  sequence, formation, 220
  standardization, degree (measurement). *See* Products

strategy, establishment, 230–231
subjective aspects, target correlations, 244–247
system control, 228–229
three-dimensional environment, development, 252–255
time, 239–240
  calculation, 240
tool, 228
  study, 255–260
Disassembly direction (DD) scope, 246
Disassembly effort index (DEI), 214
Disassembly Petri Net (DPN), 221
Disassembly Precedence Matrix (DPM), 221
Disassembly process, 220
  directional graphs, basis, 223
  graphs, basis, 224
  multiple graphs, basis, 223–224
  Touzanne proposal, 224
Disassembly process layout, 218–224
  AND/OR graph, basis, 222
  nondirected graph basis, 218–221
  Petri Net, basis, 221–222
Disassembly Process Plan (DPP), 221
Disassembly Tools, 249
Disk-stack centrifuge, operation, 112
Dismantling
  occurrence, 213
  product design, nonsuitability, 212
Disposal prohibition, 311
Distribution, 41
Do
  cycle. *See* Plan, do, check and act phase, 83
Document access, 88
Down-sizing, phrase, 62
DPM. *See* Disassembly Precedence Matrix
DPN. *See* Disassembly Petri Net
DPP. *See* Disassembly Process Plan
Drag-out, 131–132
  composition, 141
Drawing, 15–16
Dry machining, 105–106, 158
Due diligence, 318
Dust, ANSI definitions, 149
DVP. *See* Desired Value Precedence

E

Early manufacturing involvement (EMI), 53
E-business. *See* Digital business
ECD. *See* Environmentally conscious design

ECDM. *See* Environmentally conscious design and manufacturing
ECM. *See* Electrochemical machining; Environmental conscious manufacturing; Environmentally conscious manufacturing
ECO. *See* Energy Conservation Opportunities
Ecodesign, 32
  rules, 35
Edge-opposing punch, usage, 15
EDM. *See* Electrical discharge machining
Education programs, 70–71
EEE. *See* Electrical and electronic equipment
EEM. *See* Energy efficiency measures
EEO. *See* Energy efficiency opportunities
E&E products. *See* Electrical and electronic products
EERE. *See* Office of Energy Efficiency and Renewable Energy
Effluent Guidelines and Standards for Electroplating, 135
Effluent Guidelines and Standards for Metal Finishing, 135
EHS. *See* Extremely hazardous substances
EIS. *See* Environmental impact statement
Electrical and electronic (E&E) products, 24
Electrical and electronic equipment (EEE), 182–184
Electrical discharge machining (EDM), 4–5, 23
  cutting material suitability, 5
Electric furnaces, energy requirements, 8
Electric Household Appliance Recycling Law, 184
Electricity cost calculation, example, 274–275
Electricity energy bills, components, 271–272
Electrochemical conversion, 125–126
Electrochemical machining (ECM), 23
Electroless plating, 125
Electron-beam welding, 19–20
Electronic ballasts/reflectors, usage, 285
Electronic products, making (green approach), 211–212
Electroplating, 123–124
  process, 124–125
  references, 143
  shop, conductivity control system, 140
  standards, 135
Electropolishing, 127
Electroslag welding (ESW), 157

Elegance, subjective measure, 54
Emergency Planning and Community Right-to-Know Act (EPCRA), 315, 316
EMI. *See* Early manufacturing involvement
Emissions
 cap, 302
 reduction, 280
 reports, 89
Employee commitment, organization structure/development. *See* General Motors Saturn Project
Empowerment, 62. *See also* Workers
EMS. *See* Environmental management system; Environmental Management Systems
Endangered Species Act (ESA), 316–317
End-of-life, concept, 64
End-of-life cycle, 26
End-of-life electromechanical products, disassembly
 disassembly study, research/methodologies, 229–260
 introduction, 211–214
 product disassembly, 214–229
 references, 260–263
 research activities/overview, 214–229
End-of-pipe waste treatments, 179–180
End points, determination, 230–231
Energy
 benchmarking. *See* Industrial energy
 bills, 89, 90
 charges, 272
 consuming equipment, energy efficiency measures, 280–288
 consumption. *See* Disassembly module, 36
 cost, 81
  savings calculation, example, 275–276
 management, 278
 reduction, 80
 resources, conversion effectiveness, 270
 savings, 279
Energy Auditor's Handbook, 269
Energy Conservation Opportunities (ECO), 289
Energy efficiency. *See* Industrial energy efficiency
 DOE initiatives, 267–268
 economics, 268–269
 environmental benefits, 269
 importance. *See* Company bottom-line measures

 case studies, 288–291
 data analysis, 271–280
 term, 270
Energy efficiency measures (EEM), 267
Energy efficiency opportunities (EEO), 267
Energy expenditure per unit delivery, 3
Engineer, compliance tools, 319–320
Engineering Institute of Canada, climate change conference, 32
Enterprises, manufacturing systems, 69
Environment
 repercussion, 230
 term, 202
Environmental conscious manufacturing (ECM), 79
 decisions, best indicators, 86
 defects, indicators, 90
 influence, 85–86
 objectives, 86–87
 system effects, 84–85
Environmental consciousness, 64–67
Environmental design
 strategies/methodologies, categories, 32
Environmental hot spots, identification, 37. *See also* Product life cycle
Environmental image, concern, 31
Environmental impact, minimization, 4
Environmental impact statement (EIS), 298
Environmental laws, impact, 298–312
Environmentally benign manufacturing (EBM)
 introduction, 1
 issue, 2
 manufactured product, 26–27
 movement, 1–2
 principle, inclusion, 34
 processes, 3–26
 references, 27–28
 supply chain, relationship, 2–3
Environmentally conscious design and manufacturing (ECDM), 32
Environmentally conscious design (ECD), 32
 principles, 66–67
Environmentally conscious electronic manufacturing
 assemblies, 197–198
 boards, 191
 case studies, 187–204
 components, 191–194
 federal initiatives, 185
 future plans, 204–205
 history, 183–185

introduction, 179–180
LCA, usage, 185–186
legislation/regulation, 183–185
materials, usage, 187–189
processes, 189–191
products, 198–199
references, 206–209
reliability/performance, 202–203
solders, 195–197
state initiative, 185
summary, 205
terms, definitions, 180–182
testing/qualification, 199–202
tin whisker formation, 203–204
tools, 185–187
Environmentally conscious manufacturing (ECM), 3, 182–183
  components, 80–81
  incremental practices, 138–139
  practices, 137–140
  techniques, innovation, 140–142
  vendor roles, 142
Environmentally conscious process planning, model categories, 102–104
Environmentally responsible particulate mitigation/elimination, 164–167
Environmentally safe recycling, 185
Environmental management, stages, 38
Environmental management system (EMS), 318–319
Environmental Management Systems (EMS), 187, 320
Environmental pollutants, loadings, 109
Environmental regulations
  approaches, 318–319
  introduction, 295–296
EPA. See U.S. Environmental Protection Agency
EPCRA. See Emergency Planning and Community Right-to-Know Act
3EPlus, 268
EPS. See Expanded polystyrene
Equipment, requirement, 57
Equivalent to new (ETN) product, 66
ERM approach, 165–166
ESA. See Endangered Species Act
1E Series Electric Motor, disassembly
  mandrel design, 259–260
  sleeve design, 259
  tool design, 258–260
Estimation method, 247–252. See also Disassembly
ESW. See Electroslag welding
Etchants, 132
ETN. See Equivalent to new
European Union RoHS, 184–185

Exemplary plant practices, 91, 92
Expanded polystyrene (EPS), 160
Expert systems, 32
Exposure time. See Manufacturers
Extremely hazardous substances (EHS), 315
Extrusion, 15

F

FAHP. See Fuzzy Analytic Hierarchy Process
Fat, oil, and grease (FOG), 100
  loadings, 109
  removal, 108
FCA. See Fuel cost adjustment
Federal environmental laws, impact, 316–317
Federal Insecticide, Fungicide and Rodenticide Act (FIFRA), 314–315
Federal Mine Safety and Health Amendments Act of 1977, 155
Federal regulations, 296–297
  information/communication laws, 315–316
  overview, 297–317
Feedback loop, 181–182
Feedback sensors, 227
Feed system, 23
Fiber structure, production, 14
FIFRA. See Federal Insecticide, Fungicide and Rodenticide Act
File sharing, 253
Filler materials, absence, 20
Filtration, 114–117
  occurrence, 114
Final report, timeliness, 92
Finishing process. See Metal finishing
Fire/explosion, creation, 134
Fishbone diagrams, 60
Flotation, 111
Flow restrictors, installation, 139
Fluxes
  absence, 20
  usage, 189
FOG. See Fat, oil, and grease
Fog sprays, installation, 139
FOIA. See Freedom of Information Act
Force block, 22
Ford, Henry, 50
Forecasting, usage, 82
Forging, 14–15
  processes, 14–15
Foundry solid wastes, spent sand (percentage), 9
Freedom of Information Act (FOIA), 316

From the American System to Mass Production (Hounshell), 49
Front block, 22
Fuel cost adjustment (FCA), 273–275
Fuel-fired furnaces, energy requirements, 8
Full-mold casting, 12
Fumes, ANSI definitions, 149
Furnaces
 air-fuel ratio adjustment, 281–282
 bare surfaces, insulation, 280–281
 energy efficiency measures, 280–283
 preheating (hot flue gas usage), 281
 waste heat recovery, 281
Future of Life, The (Wilson), 65
Fuzzy Analytic Hierarchy Process (FAHP), 250–252
Fuzzy judgment matrix, 251

G

GaBi, web site, 42
Gas exchange, 147
Gas-flame processes, 17
Gas metal arc welding (GMAW), 157
Gas ovens, waste heat recovery, 281
Gas usage rate, 276
Gate reclamation, 7
Gaussian sphere, 246–247
GENAD. *See* Generic Assembly and Disassembly
General Electric, silicone development, 1889
General Motors Saturn Project, employee commitment (organization structure/development), 63
Generic Assembly and Disassembly (GENAD), 214
Geometric locking, 39
Glass manufacturing, energy efficiency case studies, 289
GMAW. *See* Gas metal arc welding
Goal-oriented concept, 180
Godkin, Edwin L., 58
Government Printing Office (GPO), 297
Grain structure, orientation, 14
Graphs, basis. *See* Disassembly process
Gravity settling, 110
Green design, 32
Greenhouse gases, production (decrease), 31
Greenpeace, 59
Grinding, 158–159
Groundwater monitoring, 306
Growth, life cycle, 51
Guilds, participation, 48
Gutenberg, printing operations, 48

H

Hand-operated lines, drain bar installation, 139
HAPs. *See* Hazardous air pollutants
Hazard communication standard (HCS), 316
Hazardous air pollutants (HAPs), 298–299, 302
Hazardous and Solid Waste Amendment (HSWA), 305, 311
Hazardous materials, definition, 81
Hazardous materials/processes, reduction, 80
Hazardous waste, 132–133
 determination, 306
 generator, 308–309
 properties, dangers, 306
HCS. *See* Hazard communication standard
Health effects. *See* Air quality
Heating, Ventilation & Air Conditioning (HVAC)
 bioaerosols, origination, 170
 equipment, usage, 286–287
 units, ventilation system configuration, 162
Heat pasteurization, 117–118
 limits, 117–118
HEPA. *See* High-efficiency particulate air
Hierarchal operations, 50
Hierarchies, handicap, 49–51
High-efficiency particulate air (HEPA), 162
Higher-level systems, failure, 84
High-level goals, 84
High pressure carbon monoxide (HiPCO), 169
High-pressure water cut, usage, 213
High-pressure waterjets, usage, 228
High-quality materials, usage, 39
HiPCO. *See* High pressure carbon monoxide
Holding pressure, 21
Homogeneous aerosol, definition, 149
Hot-chamber die casting, 10–11
Hot runner modeling, 23
Hot-working, 14
House of quality, 60
HSWA. *See* Hazardous and Solid Waste Amendment
Hughes Aircraft, change, 189
Human expense, 239

Human resource issues, 58
Human respiratory system, function, 147
HVAC. *See* Heating, Ventilation & Air Conditioning
Hydrocyclone, 113

I

IAC. *See* Industrial Assessment Center
IACs. *See* Industrial assessment centers
IBM, stability, 63
IDEALS. *See* Improved Design Life and Environmentally Aware Manufacturing of Electronics Assemblies by Lead-free Soldering
IE. *See* Industrial ecology
IEC. *See* International Electrotechnical Commission
Ignitability, definition, 308
Illinois Waste Management and Research Center, assessment, 129
Immersion plating, 125
Impression-die drop forging, 14
Improved Design Life and Environmentally Aware Manufacturing of Electronics Assemblies by Lead-free Soldering (IDEALS), 183
Improvement
  activity, design, 60
  impact, 59–61
Inception, life cycle, 51
Industrial Assessment Center (IAC)
  database, 271
    analysis, 277–280
    recommendations, 277–278. *See also* Cost savings
    implementation, 278–279
Industrial assessment centers (IACs), 267
Industrial ecology (IE), 32, 180
Industrial energy
  conservation, potential, 266–267
  consumption, benchmarking, 266
Industrial energy efficiency
  conclusion, 291
  introduction, 265–269
  literature review, 269–271
  references, 292–294
Industrial environmental compliance regulations, 295
Industrial Revolution, power density (increase), 48
Industrial Technologies Program (ITP), 267
Industries of Future (IOF) program, 267
  inclusion, 270
Industry consultants, reviews, 54
Information flow control, 228
Information laws, 315–316
Information technology (IT), growth, 67
Inhalable fraction, 152
In-house production, 41
Injection molding
  auxiliaries, 23
  equipment, 21–22
  involvement, 24
  materials, usage, 23
  principle, 21
  processes, environmental analysis, 24–25
  products, 23–24
  tooling, 22–23
  usage, 25
In-plant mists, reduction, 107
In-process recycling
  conclusions, 118–119
  usage. *See* Metalworking fluids
Insulated runner modeling, 23
Intellectual property, 67
International Electrotechnical Commission (IEC), 197
Inventory control system, implementation, 138
Inventory turnover reports, 89
Investment casting, 11–12
Investment costs, 272
Investment process, 12
IOF. *See* Industries of Future
Ishikawa diagrams, 60
ISO 14000, 187
ISO140001, 79–80
Isothermal rolling, 14
IT. *See* Information technology
ITP. *See* Industrial Technologies Program

J

Japan take-back policy, 184
JEDEC standard J-STD-020A Moisture Sensitivity Classification, 192
Joining
  elements, minimization, 39
  methods, 212
Jungle, The (Sinclair), 49
Just-in-time (JIT) manufacturing systems, 67

K

Kaizen, 68
Kan-bans, usage, 69
Kidder, Tracey, 52

Knowledge workers, 58
Kondratieff, ideas, 52
Kyoto Protocol, 31

L

Labor reduction, 62
Land disposal restrictions (LDRs), 311
Landfilling, 305
Large quantity generators (LQGs), 136, 309
Large-scale projects, usage, 47–48
Laser cut, usage, 213
Law on Promoting Green Purchasing, 184
LaySiD software, 224
LCA. See Life-cycle analysis
LCCCs. See Leadless ceramic chip carriers
LCD. See Life-cycle design
LCE. See Life-cycle engineering
LCI. See Life-cycle inventory
LCM. See Life-cycle management
LDRs. See Land disposal restrictions
Lead, pollutant, 302
Lead-based configurations, 201
Lead-based materials, alternatives, 195
Lead-based technologies, 200
Lead containment, addition, 201–202
Leadership
  evolution, 49–51
  issues, 58
Lead Exposure Reduction Act, 183
Lead-free alloys, reflow soldering temperature (increase), 191
Lead-free materials/processes, components, 193
Lead-free solder
  impact, 192
  lead, addition, 201
Lead-free technologies, switch, 195
Leadless ceramic chip carriers (LCCCs), 200
Lead plating, 129
Lead reporting level, adjustment, 184
Lead Tax Act (1983), 183
Lead termination finishes, 193
Lead-tin plating, 129
Leaks, prevention/containment, 138
Lean principles, 68
Legacy systems, 196
Legal jurisdictions, regulations, 31
LFC. See Lost foam casting
Liberty Brass, small company example, 69–70

Life-cycle analysis (LCA), 37. See also Streamlined life-cycle analysis
  inclusion. See Design for the environment
  origination, 186
  usage, 145
Life-cycle design (LCD), 32
Life-cycle engineering (LCE), 32
  efforts, 33
  focus, 32–33
Life-cycle inventory (LCI), generation, 186
Life-cycle management (LCM), systems/software (availability), 59
Life-cycle stages, 40–42
Lighting, energy efficiency, 285–286
Localization Requirement, 250
Logistic control, 228
Logistics, usage, 58
Long-range objectives, 70
Lost foam casting (LFC), 160
Lost-wax casting processes, 11–12
LQGs. See Large quantity generators
Lunar Society, discussions, 48

M

Machiavelli, 49
Machine tool, ECM, 95
Machining
  allowances, 10
  processes, 3–5
MACT. See Maximum achievable control technology
Magnetic ballasts, usage, 285
Magnetic separation, 113–114
Make to order, 85
Make to stock, 85
Management
  briefing, 90
  evolution, 49–51
  issues, 58
  styles, 49–50
Manorial system, 47
Manufactured product, 26–27
  alternative process plan, 164–166
  alternative process sequence, 166
  process change, 166–167
  processing conditions, 167
Manufacturers
  exposure time, 192
  federal environmental laws, impact, 316–317
  responsibility, extension, 66
Manufacturing
  air quality issues, 148

atomic level, 168
biotech applications, 170
embeddedness, 64–67
enterprise, responsibility, 148
environmental laws, impact, 298–312
environments, air quality, 153
facility, effectiveness, 86
industry, energy usage, 266
molecular level, 168
processes, environmental impact, 30
supra-molecular level, 168
trends, 167–171
Manufacturing systems, 81–84
components, 57–58
control systems, 83
levels, 82–83
metasystem, 82
strategic level, 82–83
tactical level, 83
Manufacturing systems, organization/management/improvement
change, 73–74
concepts, dichotomies, 72–73
future, 72–74
implementation, 67–72
introduction, 46–47
real-world examples, 68–70
references, 74–77
Manufacturing systems evaluation
introduction, 79–80
references, 93
Mass concentration, 152
Mass median aerodynamic diameter (MMAD), 158
Mass production
efficiencies, 49
norm, 30
Mass spectrometry, 152
Material flow analysis (MFA), 187
Materials
choice, 37
design, 39
module, 36
production/supply, 40–41
purchase orders, 89
recovery, 213
transformation, 57
variety, 212
Material safety data sheets (MSDSs), 89, 316
accompaniment. See Metalworking fluids
Matrixed cross-functional teams, 51
Maturation, life cycle, 51
Maximum achievable control technology (MACT), 303, 311

Maximum theoretical emissions (MTE), 299
Measurement, system, 54–57
Melting temperature delta, 197
Messaging software, 253
Metal, inclusions/impurity material (reorientation), 14
Metal casting, 3, 5–12
Metal-cutting operations, environmental considerations, 102–103
Metal-cutting process, 104
Metal-fabricated products manufacturing, pollution prevention, 96–97
Metal finishing, 123–124
environmental issues, 128–133
facilities, 133
federal regulations, impact, 133–137
operations
wastewater issues, 131–132
water issues, 129–131
post treatment, 128
processes, 124–126
alteration, 124
stages, 127–128
references, 143
regulatory issues, 128–133
standards, 135
surface preparation, 127
surface treatment, 127–128
Metal forming, 3
Metal-forming processes, 12–16
Metal fume fever, 154
Metal halide lamps, replacement, 285
Metal inert gas (MIG) arc welding, 157
Metal joining, 3
processes, 16–20
Metallic coatings, vapor deposition, 127
Metallic parts, separation, 39
Metal pouring, 7, 8
Metal preparation, 6, 8
Metal products, improvements, 14
Metal recovery, 141–142
site-specific savings, 141
Metals manufacturing, energy efficiency case studies, 289
Metal-to-metal contact, prevention, 15
Metalworking fluids (MWFs), 96–97
application, 147
bacterial byproducts, 101
BOD, 100
categories, 97–98
deterioration, microbial contamination, 101
environmental impact, 99–100
formulation considerations, 108

Metalworking fluids (MWFs), *(Continued)*
  hazardous chemical constituents, 99–100
  hazardous metal carry-off, 99
  health impact, 101–102
  incorporation, 157
  ingredients/contaminants, particle sizes (comparison), 116
  minimal application, 106–107
  minimum, research, 107
  mists
    production, 163
    reduction, 107–108
  MSDS, accompaniment, 99
  MWF-associated NTMs, 170
  nutrients, level elevation, 100
  oil/grease, impact, 100
  polymer additives, development, 108
  recycling, dead-end filtration (usage), 115
  state-of-the-art application, 98–99
  usage, 170
  uses/concerns, 97–99
Metalworking fluids (MWFs), pollution prevention
  conclusion, 119
  in-process recycling, usage, 109–119
  introduction, 95–96
  process modification, usage, 105–109
  process planning, usage, 102–105
  references, 120–122
Metalworking operations, straight oils/soluble oils (usage), 101
Metasystem, 82
  decisions, 83
Methanesulfonic acid (MSA), 188
MFA. *See* Material flow analysis
MIG. *See* Metal inert gas
Minimum energy consumption, design, 39–40
Minimum quantity lubrication (MQL), 167
Minimum Theoretical Number of Parts, 249
Mist reduction. *See* Metalworking fluids
Mists, ANSI definitions, 149
MMAD. *See* Mass median aerodynamic diameter
MMS. *See* Motor management system
Mobile phones, recycling, 227
ModelDoc, 254
Module checklists, 39
Mok, proposals, 214
Mold preparation, 6
Molten glass lubricants, usage, 15

Monodisperse aerosol, definition, 150
Motor management system (MMS), implementation, 283–284
MotorMaster, 283
Motor-related energy efficiency measures, 285
Motors, energy saving opportunities, 283–285
MQL. *See* Minimum quantity lubrication
MSA. *See* Methanesulfonic acid
MSDSs. *See* Material safety data sheets
MTE. *See* Maximum theoretical emissions
Multicavity die, 10
Multidisciplinary technology-intensive products, 213
Multiple graphs, basis. *See* Disassembly process
Multishift operation, 88
Mushroom effect, 52
MWFs. *See* Metalworking fluids
Mycobacterium tuberculosis, 170

N

NAAQS. *See* National Ambient Air Quality Standards
NAICS codes, usage, 277
Nanoparticles, 168–170
Nanotechnologies, 65
National Ambient Air Quality Standards (NAAQS), 148, 298
National Contingency Plan (NCP), 312–313
National Environmental Compliance Assistance Centers, 319
National Environmental Policy Act of 1969 (NEPA), 297–298, 304
National Institute for Occupational Safety and Health (NIOSH), 102
  establishment, 155
National Performance Track, 320
National Pollution Discharge Elimination System (NPDES), 134, 304
National Priorities List (NPL), 312–313
National Research Council of Canada (NRC), 40
Natural gas bills, components, 276–277
Natural gas ovens
  air-fuel ratio adjustment, 281–282
  preheating, hot flue gas (usage), 281
Nature, fair treatment, 30
NCP. *See* National Contingency Plan
Necessity factor, 66
Negative block, 22

NEPA. *See* National Environmental Policy Act of 1969
Neutral flame, production, 17
New products, innovation, 30
New Source Performance Standards, 304
Nickel-gold boards, 199–200
Nickel-gold printed circuit board finish, 200
Nickel-tungsten-silicon carbide, 188
Nitrogen dioxide, pollutant, 302
No-clean processes, 190–191
Nokia lead-free/lead-based mixed study, 200–201
Nokia Mobile Phones. *See* Plastic ball grid array
Nondirected graph, basis. *See* Disassembly process layout
Nonfatal workplace illnesses (U.S. Bureau of Labor Statistics), 154
Nonhazardous MWF, change, 99
Nonionic wetting agents, usage, 139
Nonpoint sources, 304
Nontuberculosis mycobacteria (NTM), 170. *See also* Metalworking fluids
NPDES. *See* National Pollution Discharge Elimination System
NPL. *See* National Priorities List
NRC. *See* National Research Council of Canada; Nuclear Regulatory Commission
NTM. *See* Nontuberculosis mycobacteria
Nuclear Regulatory Commission (NRC), 307
Number concentration, 152
Numbers, management, 49

O

Occupancy sensors, installation, 285–286
Occupational Safety and Health Act of 1970 (OSHA), 155, 316
  employer requirement, 164
OEM. *See* Original equipment manufacture
Office of Energy Efficiency and Renewable Energy (EERE), 267, 269
Off-shore foundries, pollution reduction, 9
Off-site evaluation, 88
Oil Pollution Act of 1990 (OPA), 314
OLE-supported programming language, 252
OPA. *See* Oil Pollution Act of 1990
Open-die hammer, 14
Operating agents, requirement, 58
Optical emission spectrometry, 152

Orchestral management, 53
Organic solderability preservative (OSP), 194, 196, 198
  boards, 199–200
Organizational behavior, 52–53
Organizational details, attention, 61
Organizational effectiveness, measurement, 54
Organizational maturity, 56
Organizations
  design, 59
  dyslexia, 51
  measurement intervals, 56–57
  structures, 56–57
    design, 59
  system, 54–57
  turbulence/internal rearrangement, 56
Organizing tools, 32
Original equipment manufacture (OEM), 166
  bidding wars, 69
OSHA. *See* Occupational Safety and Health Act of 1970; U.S. Occupational Safety & Health Administration
OSP. *See* Organic solderability preservative
Outside air, usage, 287
Outside facilitator, usage, 71
Ovens
  bare surfaces, insulation, 280–281
  energy efficiency measures, 280–283
  preheating (hot flue gas usage), 281
  waste heat recovery. *See* Gas ovens
Oxidizing flame, production, 17
Ozone, pollutant, 302

P

Packing pressure, 21
Palladium-nickel-finished components, 193–194, 196
Parent organizations, expectations, 52
Pareto plots, 60
PartDoc, 254
Part-fastener assembly relationship, 220
Particle composition, identification devices, 152
Particle size, 150–152
  ISO classifications, 152
Particulate
  characterization, 149–153
  classification, 149–150
  concerns, 167–171
  generation, 151

Particulate matter (PM), 147, 299
  pollutant, 302
Particulates not otherwise regulated (PNOR), 170
Part Number, assignation, 248–249
Parts
  cleaning, 7
  reclamation, 212
PAW. See Plasma arc welding
PBGA. See Plastic ball grid array
PBT. See Persistent bio-accumulative toxic
PBT Profiler, 320
PCBs. See Polychlorinated biphenyls
PDCA. See Plan, do, check and act
PELs. See Permissible exposure limits
PEO. See Polyethylene oxide
PEP, 268
Performance, definition, 202–203
Permissible exposure limits (PELs), 155–156
Persistent bio-accumulative toxic (PBT) substances, 320
Personal protection systems, 164
Petri Net, 223
  basis. See Disassembly process layout
  disassembly model, 234–236
PF. See Phenolics; Power factor
P2 Framework, 320
PGA. See Purchase gas adjustment
Phase separation, usage, 110–113
PHAST. See Process Heating Assessment Tool
Phenolics (PF), 23–24
Phosphate-based lubricants, usage, 15
Phosphating, preparation, 125
PIB. See Polymer polyisobutylene
PIXE. See Proton-induced X-ray emission spectrometry
Plan
  cycle. See Plan, do, check and act phase, 83
Plan, do, check and act (PDCA) cycle, 83–84
Plasma arc welding (PAW), 17, 157
Plastication stage, 21
Plastic ball grid array (PBGA)
  Nokia Mobile Phone assembly, 199–200
  packages, 194
Plastic product, life cycle, 25–26
Plastics injection molding, 3, 20–26
  environmental impact, 26
Plastics manufacturing, energy efficiency case studies, 289–290
Plating
  baths, 132
  chemicals, usage, 188
  processes, 189–190
Playing field, 70–71
PM. See Particulate matter
Pneumoconosis, 154
PNOR. See Particulates not otherwise regulated
Pollutants, inclusion, 8
Pollution control laws, impact, 312–315
Pollution Prevention Act of 1990 (PPA), 185, 314
Pollution prevention (PP), 38
  engagement, reasons, 180–181
  focus, 181
  goal, 180
Polychlorinated biphenyls (PCBs), 313
Polydisperse aerosol, definition, 150
Polyethylene oxide (PEO), usage (investigation), 164
Polymer polyisobutylene (PIB), addition, 163
Polymer resin, development, 25
Polypropylene, affinity, 110
Polytetrafluoroethylene-PTFE, 159
Popcorn impact, 197
Pore sizes, usage, 115
Positive block, 22
Post-solder assembly, 190
Post treatment. See Metal finishing
Potential, development, 55
Potential theoretical emissions (PTE), 299
POTWs. See Publicly owned treatment works
Powder manufacture, 160–161
Power factor (PF), 275
PP. See Pollution prevention
PPA. See Pollution Prevention Act of 1990
Precision-casting processes, 11–12
Press forging, 14
Prince, The (Machiavelli), 49
Principles of Scientific Management, The (Taylor), 49
Printed wiring board (PWB), 189, 191
  surface finish, 196, 198
Problem solving
  journey, 61
  usage, 59–61
Process bath, type, 141
Process data, feed-forward, 57
Process design, 212
Process-energy use targets, development, 270
Process equipment, requirement, 57

Process Heating Assessment Tool
 (PHAST), 268, 281
Process modification, usage. *See*
 Metalworking fluids
Process planning, 85–86
 levels, 86
 macro level, 86
 micro level, 86
Process plans, 165. *See also*
 Manufactured product
Process tank ingredients, concentration
 (reduction), 139
Production levels, 62
Productivity
 improvements, 63
 studies, 52
Product life cycle
 consideration, 146
 environmental hot spots, identification,
  37, 38
 importance, 24
Products
 American Society of Mechanical
  Engineers, examination, 32
 appropriateness, determination, 37
 cleaning, 9
 components, 30
 configuration graph, 223
 connection graph, 223
 delivery, 56
 design, 30, 39
  environmental impact, 30
  modules, 36
 disassembly, 36
  standardization, degree
   (measurement), 245
 environmental impact, minimization,
  29
 green manufacture, 198–199
 innovation. *See* New products
 lifespan, segments, 212
 lifetime, stages, 198–199
 recycling, shredding (usage), 36
 reutilization, 212
 usage, 41
 whole life cycle, 64
Program-focused teams, 51
Project-focused teams, 51
Proton-induced X-ray emission
 spectrometry (PIXE), 152
PTE. *See* Potential theoretical emissions
Publicly owned treatment works
 (POTWs), 134, 303, 304
 water discharge, 136
Purchase gas adjustment (PGA), 276
PWB. *See* Printed wiring board

Q

QFD. *See* Quality function deployment
Quality
 enhancement, 55
 experts, ideas, 61–62
Quality defect reports (QDRs), 89
Quality function deployment (QFD),
 60
Quantity generators. *See* Conditionally
 exempt small quantity generators;
 Large quantity generators; Small
 quantity generators
Quasi-island economy, 50

R

Rack-1 operations, 125
Radioactive wastes, exemptions, 307
Ramazzini, Bernardino, 153
Rating/ranking tools, 32
RCRA. *See* Resource Conservation and
 Recovery Act
Reachability, 249–250
Reach capability, 244–245
Real-time design information, distillation,
 254
Rear block, 22
Reclaiming self-cost, optimization, 218
Reclamation, waste treatment
 classification, 141
Recognition issues, 58
Recovering models, 231–236
Recovery/disposal, 41–42
Recyclable parts, 230
Recycled material, usage, 39
Recycling
 consideration, 109
 product design, nonsuitability, 212
Reengineering, phrase, 62
Reflow-processes components, 201
Reflow temperatures, 199
Regulatory compliance, 38
Relative index. *See* Disassembly
Reliability, definition, 202–203
Repetitive Operation Times, 249
Repetitive Times. *See* Summation of
 Difficulty Grades and Repetitive
 Times
Reporting/project formulation, 91–92
Resistance welding processes, 18–19
Resource Conservation and Recovery Act
 (RCRA), 134, 304–312
 explanation, 135–136
 Orientation Manual, 305, 306
 regulations, 311

Resources, effective use, 65
Respirable fraction, 152
Restoration, 66
Results, measurement, 71–72
Retrieval condition graph, 223–224
Reused components, usage, 39
Reward issues, 58
Right-sizing, phrase, 62
Rinse tanks, countercurrent configuration installation, 139
Rinse water, treatment, 132
Rinsewater purification/recovery techniques, 140–142
Rinsing baths, agitation, 139
Risk, reduction, 85
RoHS. *See* European Union RoHS
Roll forging, 15

S

Safe Drinking Water Act (SDWA), 315
Sand casting, 5–7
  environmental concerns, 7–9
  steps, 5–6
Sand reclamation, 7, 9
SARA. *See* Superfund Amendments and Reauthorization Act
SAW. *See* Submerged arc welding
Schumpeter, Joseph, 52
Screw connection, accomplishment, 241–242
SD. *See* Sustainable design
SDWA. *See* Safe Drinking Water Act
Self-managed teams, 52
SETAC. *See* Society of Environmental Toxicology and Chemistry
Sewage plant operations, interference, 134
Shakeout, 7
Shielded metal arc welding (SMAW), 157
Shielding gases, absence, 20
Shredding, usage. *See* Products
SIC codes, usage, 277
Simapro, web site, 42
Sinclair, Upton, 49
SIPs. *See* State implementation plans
Site visit
  inspection, 90
  observations, 91
Six Sigma, 70
Sketch Dimension, 254
Skimming, 110
SLCA. *See* Streamlined life-cycle analysis

SldWorks, 254
Sliding block, positioning, 257
Small blind bores (bearings), disassembly tool (usage), 256–257
Small companies, assistance, 87
Small quantity generators (SQGs), 136309
SMAW. *See* Shielded metal arc welding
Smith, Adam, 49
Smoke, definition, 149
Smoothness, 60
Snap connection, accomplishment, 242–243
Social engineering, 61–64
Society of Environmental Toxicology and Chemistry (SETAC), 186
Software systems, 32
Soldering, 20
Solders, 195–197
Solid waste, 132–133
  definition, 305
Solid Waste Disposal Act (1965), 304
SolidWorks, 252–253
  application. *See* DFD system
  capability, 253
  dimension-driven system, 254
Soul of a New Machine, The (Kidder), 52
Source reduction, 185, 305, 314
SPCC. *See* Spill Prevention Control and Countermeasures
Special Disassembly Problems, 250
Special projects, attention, 53
Spill Prevention Control and Countermeasures (SPCC), 314
Spills, prevention/containment, 138
Spray, definition, 149
Sprue reclamation, 7
SQGs. *See* Small quantity generators
SSAT. *See* Steam System Assessment Tool
SSST. *See* Steam System Scoping Tool
Stability, quest, 52–53
Standardization, degree (measurement). *See* Products
Standby charges, 276
Star systems, 53
State implementation plans (SIPs), 298–299
State of business session, usefulness, 71
State regulatory requirements, 317–319
Stationary platen, usage, 22
Statistics, benchmarking, 89
Steam engine, improvement, 48
Steam-related energy efficiency measures, 283

Steam System Assessment Tool (SSAT), 268
Steam systems
  bare surfaces, insulation, 280–281
  energy efficiency measures, 280–283
Steam System Scoping Tool (SSST), 268
Steam traps,
    inspection/repair/maintenance, 282–283
Storage prohibition, 311
Streamlined life-cycle analysis (SLCA), 36
  usage, 37
Stripping rate, 188
Subjective aspect, target correlations. *See* Disassembly
Submerged arc welding (SAW), 157
Suga, quantitative evaluation method, 216–217
Sulfur dioxide, pollutant, 302
Summation of Difficulty Grades, 250
Summation of Difficulty Grades and Repetitive Times, 250
Superfund Amendments and Reauthorization Act (SARA), 313, 315
Supply chain, relationship. *See* Environmentally benign manufacturing
Supply-chain model, 181–182
Surface finishing technologies, 126–127
Surface grinding, 159
Surface preparation. *See* Metal finishing
Surface treatment. *See* Metal finishing
Sustain, definition, 64
Sustainability, 64–66
Sustainable design (SD), 32
Sustainable development, 179
Swaging, 15
Synthetic minor permits, 302
Systems design, 59–61
System wastes, detection methods, 84

T

Taguchi design-of-experiments method, application, 197–198
Taguchi principles, 60
Taylor, Frederick, 49
TCLP. *See* Toxicity characteristic leaching procedure
Team decomposition, 54–55
Team interdependencies, management recognition, 53
Team members, pre-interview, 88
Technical criteria, EPA establishment, 305–306
Technology, change, 56
TEM. *See* Transmission electron microscopy
Thermal emission, 19–20
Thermoplastic materials, 192
Thermoplastic resins, requirement, 23
Thermosetting resin systems, 23–24
Thin small outline package (TSOP), 194
Thoracic fraction, 152
Three-dimensional design software, characteristics, 252
Three-dimensional environment, development. *See* Disassembly
Three-phase separation capability, 112–113
Threshold limit values (TLVs), 156
Tier I/II supplier
  environmental issues, 2
  manufacturing processes, 3
TIG. *See* Tungsten inert gas
Tin-lead eutectic alloy, 197
Tin-silver-bismuth solder joints, 199
Tin-silver-copper solder paste, 196
Tin whiskers, formation, 203–204
Title V Permit Program, 302–303
T12 lighting fixtures, replacement, 285
TLVs. *See* Threshold limit values
TMDLs. *See* Total maximum daily loads
TMY data, 287
Tooling, requirement, 57
Toshiba Chemical Corporation, board testing, 191
Total life-cycle basis, 179, 180
Total maximum daily loads (TMDLs), 303–304
Total quality management (TQM), 61–62
Total quality objectives, 56
Toxicity, definition, 308
Toxicity characteristic leaching procedure (TCLP), 308
Toxic reactions, promotions, 153
Toxics Release Inventory (TRI), 134, 315–316
  reporting, 137
Toxic Substances Control Act (TSCA), 313–314
Toxic Use Reduction Institute (TURI), 44
Toyota Production System (TPS), 68
TQM. *See* Total quality management

Tramp oils, 110
Transmission electron microscopy (TEM), 159
Transportation charges, 276
Transporters, 3
Treatment, storage or disposal facilities (TSDFs), 306, 310
TRI. See Toxics Release Inventory
TSCA. See Toxic Substances Control Act
TSDFs. See Treatment, storage or disposal facilities
TSOP. See Thin small outline package
T8/T5 lighting fixtures, usage, 285
Tungsten inert gas (TIG) arc welding, 157
TURI. See Toxic Use Reduction Institute
Type of Disassembly Tasks, 249
Tzu, Sun, 49

U

Ultraviolet (UV) radiation, 118
Underground storage tanks (UST), 312
Undirected graph model, 231, 233
Uniform hazardous waste manifest, 309
United States Code, impact, 296–297
Universal waste rules, 310
University of Washington, web site, 44
Unsaturated polyester (UP), 23–24
UP. See Unsaturated polyester
Upset forging, 14
U.S. Department of Energy (DOE) initiatives. See Energy efficiency
U.S. Environmental Protection Agency (EPA), 40
  bibliographic report, production, 96
  computer projection, 211
  pollution prevention hierarchy, publication, 181–182
U.S. foundries, pollution reduction, 9
U.S. Occupational Safety & Health Administration (OSHA), mission, 148
Used Oil Recycling Act of 1980, 310
Users, quality-of-life, 66
UST. See Underground storage tanks
Utility bills
  analysis, 271–277
  demand charges, 272–273
  energy charges, 272
  rate schedules, 273

V

Vacuum systems, usage, 106
Value-added output, 69
Value of Disassembly Force, 250
Vapor deposition. See Metallic coatings
Ventilation systems, 161–162
  maintenance program, implementation, 162
Vertical integration, 67–68
Virtual Assembly and Disassembly (VIRAD), 214
Visual C++, 254
Viswanathan, studies, 214–216
Volatile organic compounds (VOCs), 299, 302
  emission, 167
  expected total amount, 302
Vortex-encouraging conical chamber, 113

W

Walsh-Healy Public Contract Act (1936), 155
Waste
  expenditures, 80
  material, level, 26
  minimization/pollution prevention, 278
  reduction, 80
  streams, segregation, 138
  tracking, 138
Waste electrical and electronic equipment (WEEE), 184–185
Waste Management and Public Cleansing Law, 184
Wastewater
  flow, obstruction, 134
  sources, 132
Water-based MWFs, 163
Water meters (body/core disassembling), equipment usage (research), 255–256
Water-soluble MWFs, 101, 117
WEEE. See Waste electrical and electronic equipment
Welding, 16–17
  operations, 156–157
  processes, 17–20
West Virginia University, Industrial Assessment Center, 289
Wet machining, 157–158
Wild and Scenic Rivers Act, 304
WIP, 71

Workers
  empowerment, 50
  exposure, 161
Workforce
  considerations, 61–64
  reduction, considerations, 62–63
Workpieces
  melting, 156–157
  withdrawal, 138

Workplace air quality
  maintenance, 163
  regulations, 155–156

X

X-ray diffraction (XRD), 152
X-ray fluorescence spectrometry (XRF), 152